THE COAST

THE COAST

Editor

Sumantra Narang

scitus
academics

The Coast

Edited by **Sumantra Narang**

ISBN: 9781681178844
Library of Congress Control Number: 2017948911

SCITUS Academics LLC,
616, Corporate Way, Suite 2, 4766,
Valley Cottage, NY 10989

www.scitusacademics.com

Printed in United States of America.

Preface

The coast is one of the most dynamic parts of the earth surface. Since early civilization, coastal areas have been attractive settling grounds for human population as they provided abundant marine resources, fertile agricultural land and possibilities for trade and transport. This has led to high population densities and high levels of development in many coastal areas and this trend is continuing into the 21st century. About 23% of the world's population lives within 100 km of the coast and about 10 % of the population live in extremely low-lying areas (< 10 m above mean sea level). Coastal Hazards are physical phenomena that expose a coastal area to risk of property damage, loss of life and environmental degradation. Rapid-onset hazards last over periods of minutes to several days and examples include major cyclones accompanied by high winds, waves and surges or tsunamis created by submarine earthquakes and landslides. Slow-onset hazards develop incrementally over longer time periods and examples include erosion and gradual inundation. The predicted climate change is adding an extra risk factor to human settlement in coastal areas. Whereas the natural dynamics that shape our coastlines have been relatively stable and predictable over the last centuries, much more rapid change is now expected in processes as sea level rise, ocean temperature and acidity, tropical storm intensity and precipitation/runoff patterns. The world's coastlines will respond to these changes in different ways and at different pace depending on their bio-geophysical characteristics, but generally society will have to recognize that past coastal trends cannot be directly projected into the future.

This volume 'The Coast' emphasizes at coastal landforms, how they are created, and the hazards that threaten them including erosion from waves and storms, powerful tides, and sediment transport. It covers the range of coastal hazards that result from short-term low-frequency events like tsunamis and hurricanes, to longer-term processes like urban regeneration and changing agricultural practices and far-reaching impacts on coastal zones the world over. The intensity of each of these processes is likely to increase under changing scenarios of global climate. A number of methodologies have been proposed from time to time to assessment the vulnerability of coastal areas. However, most of these indices consider only the physical parameters. Vulnerability on the other hand includes a human component. It is proposed that any assessment of coastal vulnerability must consider both physical and human components of that particular coast.

Table of Contents

CHAPTER 1

Hydrography-Physical Description of the Bohai Sea

Changwei Bian [1], Wensheng Jiang [2], Thomas Pohlmann [3], and Jürgen Sündermann [3]

[1]*Key Laboratory of Physical Oceanography Ocean University of China Qingdao, Shandong 266100, P. R. China*

[2]*Laboratory of Marine Environment and Ecology Ocean University of China Qingdao, Shandong 266100, P. R. China*

[3]*Institute of Oceanography University of Hamburg Hamburg, Germany*

ABSTRACT

Bian, C. W.; Jiang, W. S.; Pohlmann, T., and Sündermann, J., 2016. Hydrography-physical description of the Bohai Sea.

The Bohai Sea used to be one of the most important fishing grounds in China. Recently, its ecological functions have declined rapidly due to the heavy anthropogenic activities which make ecological environment protection and restoration particularly topical in this region. Understanding the oceanographic conditions of the Bohai Sea is essential to evaluate the living environment (*e. g.*temperature and salinity) of planktons and transport fluxes of nutrients, sediments and contaminations. In this paper, we will analyze the published literature and data sets containing field observations, remote sensing reconstructions and numerical model results, and describe the principle features of the hydrodynamic system in the Bohai Sea, such as external forces, thermohaline patterns, tides, waves, currents and substance transport processes. The objective of this paper is to provide a general picture of the hydrography of the Bohai Sea; by this means the work may be beneficial for studies in other disciplines, too.

Keywords: Thermohaline, tide, circulation and transport processes.

INTRODUCTION

In the western Pacific, the marginal Bohai Sea (Figure 1) is a shallow semi-closed sea with a total area of 77,000 $km^2$and an average water depth of 18 m. The Bohai Sea used to be one of the most important fishing grounds in China and possesses a large amount of oil fields which provide natural resources for the economic development of the coastal regions. In recent decades, the massive human activities such as land reclamation, pollutant discharged from rivers and oil field exploitation caused a sharp deterioration of the marine ecological system in the Bohai Sea (Gao *et al.*, 2015; Pelling *et al.*, 2013; Wang, 2015). Today, as environmental problems attract more and more attention, a lot of research projects have been launched to study the physical, chemical, geological and biological characteristics of the Bohai Sea.

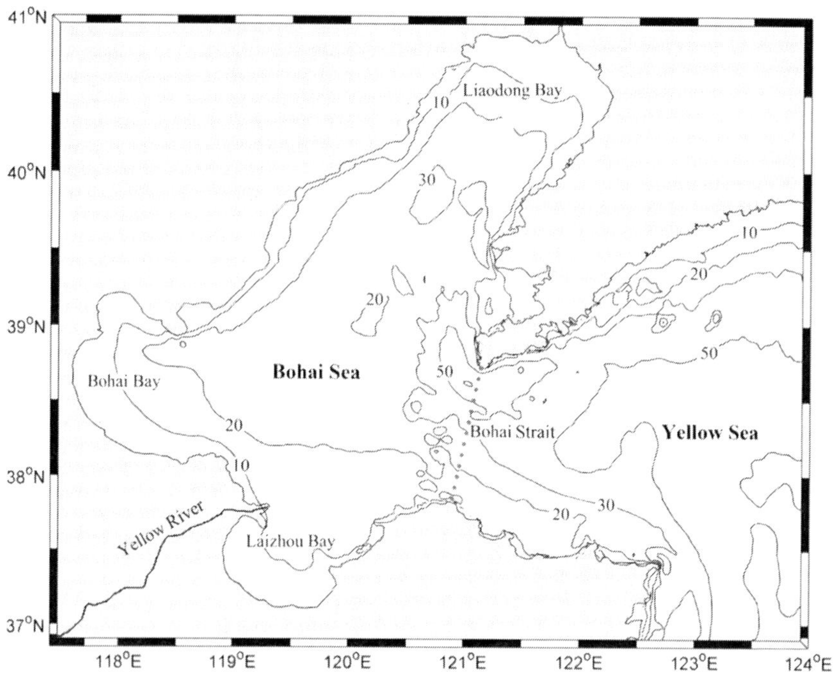

Figure 1. Bathymetry of the Bohai Sea (m). The dashed line indicates the boundary between the Bohai Sea and the Yellow Sea

In general, the Bohai Sea is divided into five regions: Liaodong Bay, Bohai Bay, Laizhou Bay, Central Bohai Sea and Bohai Strait. The Bohai Sea is characterized by a basin shape and deepens gradually from a depth of less than 20 m in the three coastal bays to the Central Bohai Sea. The northern

Bohai Strait with a maximum water depth of about 70 m plays an important role in steering the water exchange between Yellow Sea and Bohai Sea. As generally in the ocean, the physical system of the Bohai Sea is determined by the space-time distribution of seven macroscopic variables: temperature, salinity, density, pressure and the three components of the velocity vector. These variables help to distinguish one water mass from another, and to study and understand the motions of the sea water in response to the governing forces. Suspended and dissolved substances (*e.g.* suspended particle matter (SPM), contaminants and nutrients) are considered as physical properties as well, which is a reasonable assumption as long as chemical and biological processes can be neglected. These substances are of particular importance, because their spreading and deposition control the biological and morphological state of the Bohai Sea.

Bohai Sea dynamics are characterised by the regional interaction among the atmosphere, the hydrosphere and the lithosphere on the East Asian shelf exhibiting a broad spectrum of spatial and temporal scales (Figure 2 and Figure 3). Specific features of the physical system mainly include tides, surface waves, baroclinic eddies, internal waves, convection, turbulence, *etc.* Principally, the physical, chemical, biological and morphological compartments of the Bohai Sea system cannot be separated due to their strong interactions with each other. Certainly, the physical signal is essential for the transportation of chemical and biological substances within the whole ecosystem. But there is also a feedback depending on the scales considered, *e.g.* radiation flux, albedo, surface films, surface roughness and morphology.

Our knowledge of hydrology in the Bohai Sea is based on field observations, remote sensing reconstructions and model simulations. Since 1950s, comprehensive investigations have been conducted in the Bohai Sea to determine the temporal and spatial variability of physical, chemical and biological components. The State Scientific and Technological Commission of China organized the "Chinese National Comprehensive Oceanographic Survey (1958–1960)" which for the first time systematically expounded hydrographic and biogeochemical features of the Bohai Sea. After that, several comprehensive investigations were carried out, such as "Research on the Prediction of Water Quality and the Physical Self-Purification Capability of the Bohai Sea and the Ten Bays in China (1985–1986)", "Study of China Shelf Sea Circulation and Its Dynamics (1992–1995)", and "Analysis and Modelling of the Bohai Sea Ecosystem (1997–2000)" (Sündermann and Feng, 2004). These investigations provided fundamental knowledge of the temporal and spatial variability of the hydrography in the Bohai Sea. Since 1980s, remote sensing images have been used to study features of sea surface temperature (SST) and sea ice in the Bohai Sea (Jiang and Wen, 1987; Peng and Bernstein, 1985). In the recent two decades,

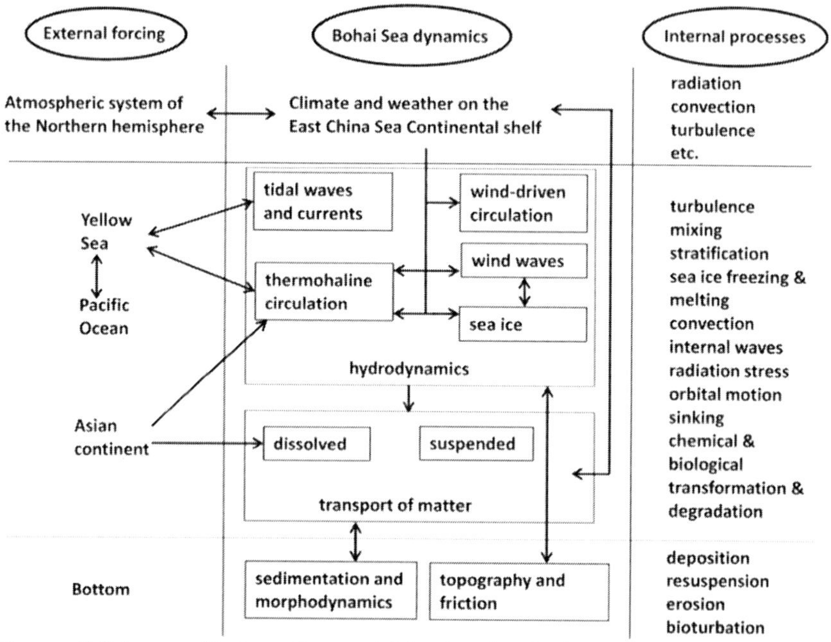

Figure 2. Scheme of interactions within the physical system

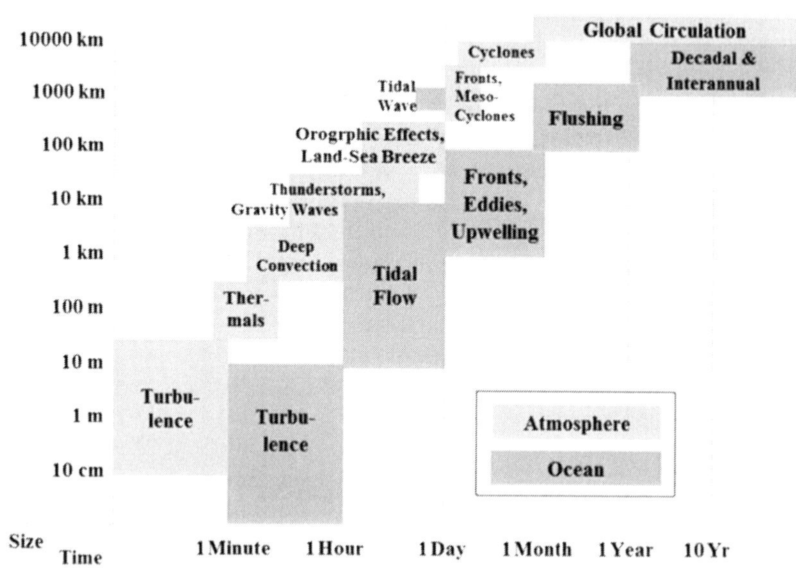

Figure 3. Scale spectrum of the physical system

remote sensing techniques were widely used to study sea level rise, SPM, wind and salinity distributions, because modern satellites provide continuous and high temporal and spatial resolution of these parameters (Liu *et al.*, 2010; Shi and Wang, 2012; Wang *et al.*, 2014). On the other hand, only the sea surface layer signals can be detected by satellite which limits the effect of remote sensing on hydrology studies. As a complete all-round tool, numerical modeling has become the main way to study the physical environment of the Bohai Sea since the 1980s. Although a number of previous numerical studies of the Bohai Sea have elaborated the state of one to several of these physical variables only a consensus has been reached concerning the basic characteristics of these variables. On the other hand, there are still unresolved issues regarding smaller scale features due to lack of sufficient field data which could help to verify the numerical results. Nevertheless, the best overall information to date is given by hydrodynamic models of the Bohai Sea, and hence, most physical results used in this paper are based on numerical models.

A lot of literature has been published concerning the hydrographic and hydrodynamic conditions of the Bohai Sea, most of which only focus on one or few variables, and in some cases may even have contradictory findings. Thus, a comprehensive literature synthesis is required to obtain a clear understanding of the physical features of the Bohai Sea and to notice the contradictions and confusions in existing studies. In the following, the external forces of the Bohai Sea are firstly introduced. Then the specific features of the physical system are described, namely thermohaline regime, wind-driven regime, tidal-driven regime and transport processes.

EXTERNAL FORCING

The Bohai Sea is a part of the East China Sea continental shelf, and its physical state is essentially determined by the adjacent Pacific Ocean, the Asian continent and the atmosphere of the northern hemisphere.

Pacific Ocean

The relatively narrow Bohai Strait allows the exchange of matter, heat and momentum between the Bohai Sea and the open ocean. The Yellow Sea forms the transition between the Bohai Sea and the Pacific Ocean. Planetary waves generated by astronomical and atmospheric forces in the ocean propagate over the shelf break through the East China Sea and the Yellow Sea into the Bohai Sea resulting in sea level changes, tides and water mass transports. In contrast, continental fresh water discharges from the Bohai Sea influence the water characteristics of the North Pacific.

The circulation system in the Bohai Sea (Figure 4) shows that the main inflow occurs along the northern Bohai Strait by the intrusion of the Yellow Sea Warm Current (YSWC) in winter. In the southern Bohai Strait, the Bohai Sea water flows out all the year(Bian *et al.*, 2013; Guan, 1994). Altogether, the net transport through the Bohai Strait is a superposition of continental outflow and Yellow Sea inflow. It has to be noted that the evaluation of the water exchange between the Bohai Sea and Yellow Sea is important for the import and export of substances. However, water exchange fluxes are rarely estimated by field investigations, and hence, most water exchange rates given for the study areas are based upon model results. Due to different models using different assumptions and forcings, the simulated results of water fluxes into the Bohai Sea differ in the range between 5×10^3 and 8×10^4 m^3/s (Bi *et al.*, 2015; Lin *et al.*, 2002; Wei *et al.*, 2003).

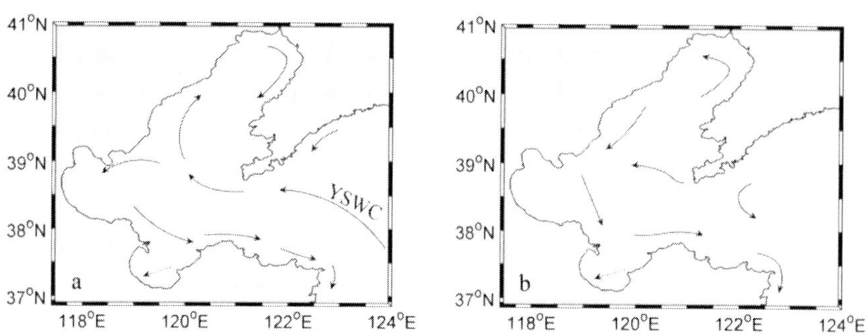

Figure 4. Schematic diagrams of the general circulation system in the Bohai Sea for (a) winter and (b) summer. Source: after Guan (1994)

Moreover, the interannual variability of the Pacific, mainly the ENSO (and other fluctuations), has also been observed in the Bohai Sea and affects the physical environment of this area. Studies based on satellite remote sensing data indicate that ENSO events dominate the interannual changes of sea level and are also correlated with the interannual SST anomaly in the Bohai Sea (Liu *et al.*, 2010; Wu *et al.*, 2005).

Terrestrial input

The continent affects the Bohai Sea by river discharged freshwater, dissolved and suspended matters. There are more than 40 rivers around the Bohai Sea, and about 7.2×10^{10} m^3 freshwater, 1.3×10^{10} t sediment,

3×10^{10} t waste water, 2×10^{4} t contaminants are directly discharged into the Bohai Sea annually (Zhang *et al.*, 2006). The freshwater input causes vertical and horizontal variations of salinity in estuarine areas, further forming stratification and fronts that act as barriers to water mixing. The Yellow River (also called Huanghe River), as the second largest river in the world in terms of sediment load, discharges 1.1×10^{9} t of sediment into the Bohai Sea annually (Milliman and Meade, 1983). Approximately 70 ~ 90% of the sediment transported to the sea is deposited at the mouth of the Yellow River (Bornhold *et al.*, 1986; Martin *et al.*, 1993), finally determining the morphological changes in the Bohai Sea. Sediment discharged from the Yellow River decreased sharply over the past 60 years with a reduction of about 75% due to the influence of human activities and climate changes (Zhang *et al.*, 2012), and the sediment load from the Yellow River will most likely remain small over the next 2 to 3 decades (Ren, 2006).

The Bohai Sea experienced rapid coastline changes due to natural developments of the Yellow River delta and large-scale anthropogenic land reclamation. Satellite remote sensing studies indicated that over the last three decades the newly formed land reclamation area in the Bohai Sea cover 2278 km^2 (Li *et al.*, 2013). These morphological changes induced by the coastal development have significant effects on the tidal regime in the Bohai Sea, leading to a rise of the tidal amplitude and the onshore sediment transport (Pelling *et al.*, 2013; Song *et al.*, 2013). And most importantly, the reclamation seriously damages the marine ecosystem, disturbs or takes away coastal wetland and reduces biodiversity.

Atmosphere

The atmosphere and the ocean form a coupled system, exchanging heat, momentum and water at the air-sea interface. Through the heat and momentum exchange, the atmosphere significantly affects the general circulation of the Bohai Sea as well as the vertical structure of the water column due to turbulence. The distribution of the wind direction and speed determines the major current patterns in the Bohai Sea. The wind further controls the spectrum of surface waves in the Bohai Sea, and cyclones and winter storms can lead to storm surges. Precipitation on the Northwest Pacific shelf influences the salinity of the Bohai Sea and its seasonal variability directly or via continental river discharge.

Table 1. Crucial factors of the atmosphere-sea interaction in the Bohai Sea (Sun, *2006*)

	Solar radiation (MJ/m^2)	Air temperature (℃)	Wind speed (m/s)	Precipitation (mm/month)
Winter	250 ~ 260	−4 ~ −2	5 ~ 7	2
Summer	600 ~ 650	24 ~ 25	4 ~ 5	140

Bottom sediment

The bottom represents a source or sink for sediment (by erosion, resuspension, deposition) and attached substances. As indicated, the Yellow River discharges large amounts of sediment into the Bohai Sea annually, and sediment transport processes in the Bohai Sea significantly influence the coastline and morphological features and as a consequence currents and sea surface waves. Related biogeochemical processes are important for the bottom sediment movement. For example, bioturbation in seabed sediments resulting from feeding, burrowing and habitat construction of benthic organisms can affect the sediment erosion threshold and transport rates.

THE THERMOHALINE REGIME

The thermodynamic state of the Bohai Sea is characterised by the 4D distributions of temperature, salinity and baroclinic circulation. The salinity and temperature characteristics of different areas are strongly influenced by heat, fresh water and momentum exchange with the atmosphere, and by the local freshwater supply via rivers.

Figure 5 shows the distribution of model simulated climatological SST in the Bohai Sea. In winter, under the control of Siberian high-pressure air masses, the SST of the coastal regions is low (−0.5 ~ 2°C), while the SST in the Central Bohai Sea is relatively high (2 ~ 4°C) (Figure 5a). These results are consistent with field observations and satellite reconstructed SST results (Bao *et al.*, 2002; Shi and Wang, 2012; Xu *et al.*, 2008). Due to the summer monsoon, the SST increases rapidly in summer and varies between 21°C and 25°C, showing higher SST values in the Central Bohai Sea compared to coastal areas. In contrast, both in-situ measured and satellite reconstructed SST distributions showed higher SST values in the coastal areas compared to the Central Bohai Sea in summer (Bao *et al.*, 2002; Shi and Wang, 2012; Xu *et al.*, 2008). Figure 5c presents the time series of the spatially averaged SST over the entire Bohai Sea, manifesting the strong seasonality. The SST variability ranges from the lowest values of 4°C to 5°C in February to the highest of 25°C in August which consists with results from field investigation (Liu *et al.*, 2003).

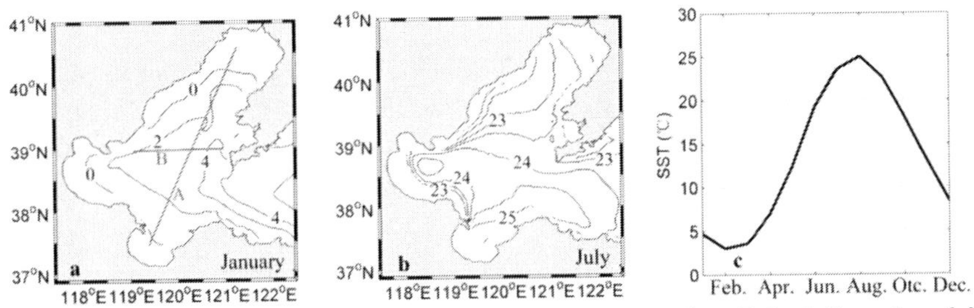

Figure 5. Climatological SST distribution (°C) in the Bohai Sea for (a) January (b) July and (c) time series of spatial mean. Source of data: HYbrid Coordinate Ocean Model (HYCOM dataset). The straight lines show the locations of SST vertical profiles in Figure 7

The sea surface salinity (SSS) in the Bohai Sea is subject to salt water intrusion from the YSWC, fresh water discharged by rivers, precipitation and evaporation. The spatial SSS distribution patterns are similar in winter and summer, showing high salinity in the Central Bohai Sea and Bohai Strait and low salinity in the coastal regions (Figure 6a, b). The tongue shape isohaline in the Bohai Strait indicates inflow of high salinity YSWC water while the salinity in the Yellow River estuary decreases dramatically in summer due to the river runoff. The time series of the spatial mean SSS over the entire region indicates that the salinity in the Bohai Sea is highest in February with a mean value of about 31.6 and decreases to 30.3 in August due to the strong precipitation and increased river runoff in summer (Figure 6c). The model results are able to reproduce the principle characteristics of salinity distribution in the Bohai Sea when compared with field observations(Bi *et al.*, 2015; Liu *et al.*, 2003; Lv *et al.*, 2009). The field investigation studies also show that the salinity of the Bohai Sea increased at a range of 1 to 5 in the recent decades (Bi *et al.*, 2015), and the salinity increased to around 10 from 1960s to 1990s in the Yellow River estuary due to the runoff reduction of the Yellow River (Wu *et al.*, 2004).

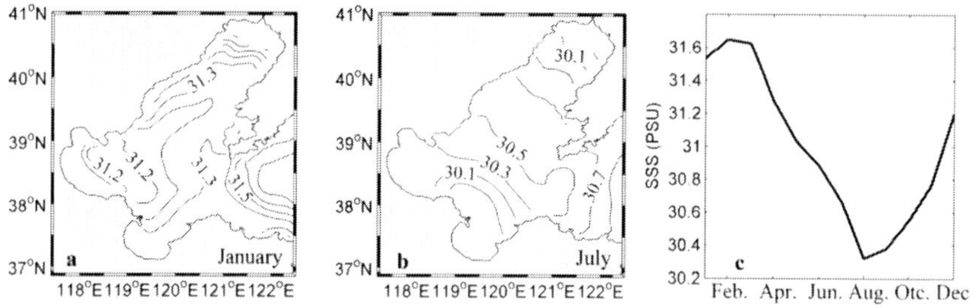

Figure 6. Climatological SSS distribution (without units) in the Bohai Sea for (a) January (b) July and (c) time series of spatial mean. Source of data: HYbrid Coordinate Ocean Model (HYCOM dataset)

The vertical temperature profiles at Transect A and B (Figure 5a) are shown in Figure 7. In January, the whole Bohai Sea is well mixed with vertically uniform temperatures. In April, as solar heat input increases, a thermocline emerges in the southern Bohai Sea (Figure 7a), and subsequently establishes over large areas of the western Bohai Sea (Figure 7b). The strength of thermocline reaches its maximum in July. After that, enhanced storm activity and seasonal cooling destroy the surface layer thermocline in October. The seasonal variation is mainly related to the heat flux input and wind intensity. In winter, heat is transported from the warm sea surface water to the colder atmosphere, leading to a higher surface water density and a resulting instability of water stratification, which in combination with strong winds entirely mixes the water column. In summer, warm surface water is lighter than the cold bottom water and the weaker winds are not strong enough to breakup the stable stratification, and thus the strength of thermocline reaches its maximum at the end of the summer. Compared to field observations in summer, the thermocline simulated by the models is weak especially in the upper layer (Ju *et al.*, 2013; Lin *et al.*, 2006), which could be attributed to insufficient surface mixing – a common problem of ocean models (Ezer, 2000).

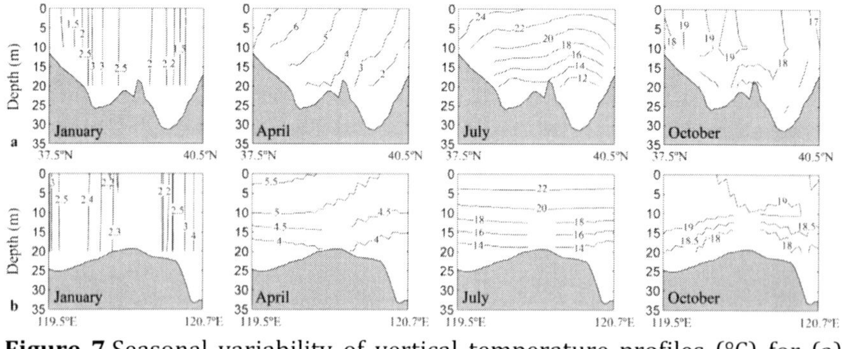

Figure 7. Seasonal variability of vertical temperature profiles (°C) for (a)

Transect A and (b) Transect B. The locations of Transect A and B a re shown in Figure 5a. Source of data: HYbrid Coordinate Ocean Model (HYCOM dataset)

The temperature and salinity distribution also influences the velocity fields by producing density gradients, which in turn cause baroclinic currents, representing the thermohaline-driven part of the total flow in the Bohai Sea. The latter is superimposed by the barotropic currents which are driven by wind and tides. Results of numerical circulation models show that the thermohaline-driven circulation in the Bohai Sea plays a more important role in the summer circulation system than the wind-driven circulation (Wan *et al.*, 2004; Wei *et al.*, 2003).

Seasonal sea ice is another feature that affects the SST, wave characteristics, ocean circulation, and stratification. The Bohai Sea coastal regions are continuously covered by sea ice during the winter season (Figure 8). In Liaodong Bay, the ice persists over 3 months with a coverage extending offshore over ~ 70 km. In contrast, sea ice occurs for about two months with a ~ 20 km offshore distance in Bohai Bay and Laizhou Bay. It has to be noted that the existence of sea ice in winter blocks the heat and momentum exchange between atmosphere and ocean and by this means affects all the above mentioned parameters.

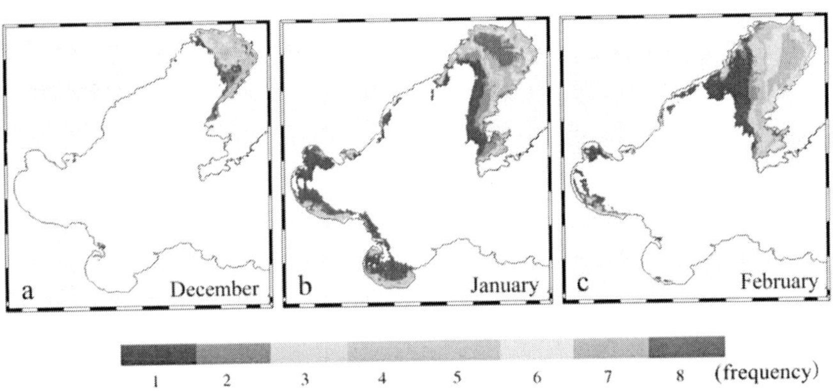

Figure 8. Sea ice occurrence (frequency) in the Bohai Sea from MODIS-Aqua observations between 2002 and 2010 for (a) December, (b) January, and (c) February. Blank areas mean no sea ice occurs during the month. Source: redrawn based on Shi and Wang (2012)

THE WIND-DRIVEN REGIME

The wind field in the Bohai Sea is dominated by the Asian monsoon. Strong northerly monsoon winds and relatively stormy conditions dominate the winter time with an average wind speed of 2.5 ~ 4 m/s, while weaker southerly monsoon winds generally prevail in summer with an average wind speed of 2 ~ 2.7 m/s (Figure 9a, b). It should be noted that the climatological winds may underestimate the actual wind speeds. The satellite wind data (QuikSCAT) show that the directly observed wind speeds in the Bohai Sea are ranging from 5.5 to 7 m/s and 5 to 5.5 m/s in winter and summer, respectively (Xu *et al.*, 2013).

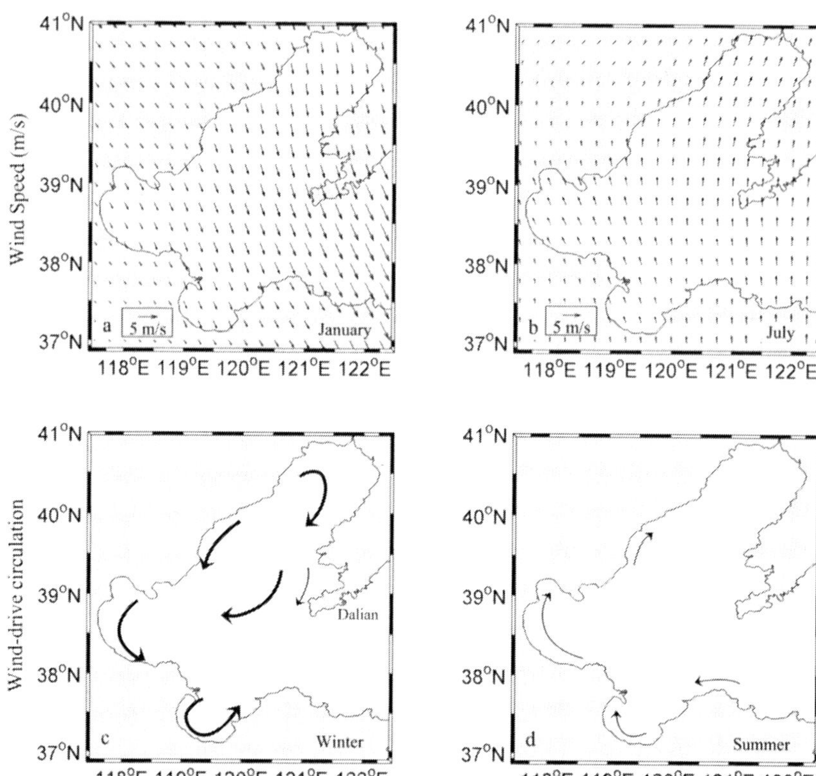

Figure 9. Climatological wind patterns in the Bohai Sea for (a) January and (b) July, and schematic wind-driven circulation for (c) Winter an d (d) Summer. Source of data: climatological wind data are from European Centre for Medium-Range Weather Forecasts (ECMWF). Wind-driven circulation results are from Huang *et al.* (1996)

Numerical studies have shown that the wind-driven circulation dominates the Bohai Sea circulation system in winter exhibiting a strong southward current along the western shore with current speeds of 8 ~ 20 cm/s and a south-westward anticyclonic flow in the Liaodong Bay with current speed of 4 ~ 10 cm/s (Figure 9c). Overall, these model results show a similar pattern that the observed circulation given in Figure 4a. In summer, in the Bohai Sea the wind-driven circulation is weaker, flowing westward along the southern shore and northward along the western shore with current speed of 4 ~ 10 cm/s (Figure 9d).

Storm surges resulting from wind and atmospheric pressure extremes constitute a serious hazard to the Bohai Sea by dramatically rising sea level within a short period of time. Storm surges in Chinese seas can be classified into two types: tropical storm surges and extratropical storm surges. Unlike other marginal seas of China, the Bohai Sea is less susceptible to tropical cyclones. But cold air out-breaks and extratropical cyclones in winter cause severe storm surges mainly affecting the southwestern part of the Bohai Sea (Zhao and Jiang, 2011). Tidal gauge data indicate that storm surges cause a sea level rise in the range of 0.7 ~ 0.8 m (1979 ~ 1995) at Dalian station (location in Figure 9c).

Surface waves are generated by the wind. Although their influence on the large-scale circulation and transport is relatively small, they are of significance for dynamical processes as they generate turbulence, shear stress by orbital motion and radiation stress. In turn these processes are important for the resuspension and vertical dispersion of bottom sediment, especially in the shallow Bohai Sea. The seasonal variation of the significant wave height in the Bohai Sea is dominated by the wind stress, with amplitudes ranging from 0.5 ~ 1.2 m in winter to 0.05 ~ 0.65 m in summer (Figure 10). Due to the bathymetry, both in winter and summer the largest wave heights are found in the Central Bohai Sea and the wave heights decrease towards the coast. These model simulated wave height patterns are consistent with observations, however, the latter show higher values, with observed maximum averaged wave height of 1.5 and 1.0 m in winter and summer, respectively (Chen, 1992). The dominant wave direction in the Bohai Sea is north and northwest in winter and south in summer, and the wave period is 2.6 ~ 3.5 s in winter and 2.5 ~ 2.7 s in summer (Chen, 1992).

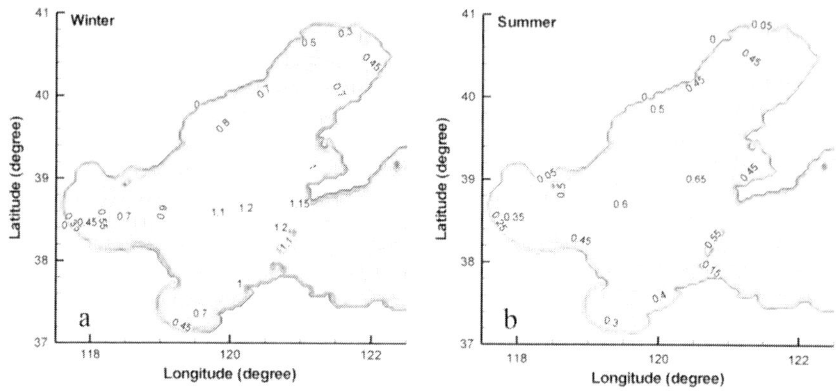

Figure 10. Characteristics of seasonally averaged significant wave height (m) for (a) winter and (b) summer. Source: Lv *et al.* (2014)

THE TIDAL-DRIVEN REGIME

The dynamics of the Bohai Sea is significantly influenced by astronomical tides. These are co-oscillations with the autonomous tidal waves of the Pacific Ocean. The specific geometry of the Bohai Sea basin implies eigen-periods and hence resonance in the semi-diurnal spectral range. The superposition of the semidiurnal principal lunar and solar tides M_2 and S_2 causes a significant spring-neap rhythm. Tides superimpose any other motion in the Bohai Sea and dissipate energy, mix water masses, prevent thermohaline stratification and strongly influence the sediment dynamics.

Tidal elevations are penetrating from the Pacific into the Bohai Sea moving anti-clockwise as a Kelvin wave through the entire basin. There are two amphidromic points of M_2(Figure 11a): one is near the mouth of the Yellow River at the south-western coast, and the other one is located offshore of Qinhuangdao at the north-western coast. The model simulated maximum amplitude of the M^2 tide is about 110 cm in Liaodong Bay and 70 cm in Bohai Bay (Hainbucher *et al.*, 2004) which are about 10% smaller than the M2 amplitude calculated from TOPEX/Poseidon altimetry data and other model studies (Bian *et al.*, 2013; Huang *et al.*, 1996; Pelling *et al.*, 2013).

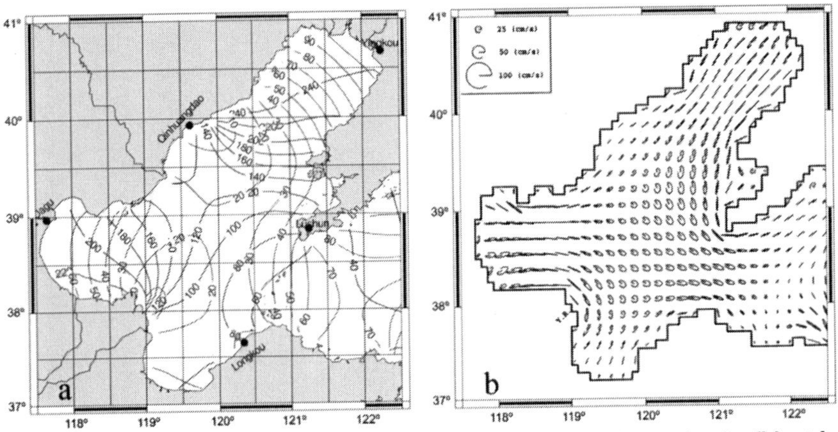

Figure 11.(a) Co-tidal lines (degrees) and co-range lines (cm), (b) tidal ellipses of M_2constituent in the Bohai Sea. Source: Co-tidal line s and co-range lines from Hainbucher *et al.* (2004), tidal ellipses adapted from Huang (1995)

The tidal currents are the most energetic feature and dominate any other flow in the Bohai Sea with current speed of 20 ~ 80 cm/s (Figure 11b). In the Central Bohai Sea and the Bohai Strait, the tidal current exhibits stretched ellipses of a counter-clockwise rotating flow, while tidal ellipses are distorted to a rectified flow in the three bays. There are three regions of strong tidal current in the Bohai Sea: the Bohai Bay, the Liaodong Bay and the northern Bohai Strait.

Nonlinear processes generate non-harmonic tidal motions and, as a consequence, non-vanishing residual currents and transports. The tide-induced residual circulation is an important component of the circulation in the Bohai Sea. Wei *et al.* (2004) calculated the tidal residual currents using the five major tidal constituents (M_2, S_2, N_2, K_1 and O_1) (Figure 12). Generally, the tidal residual current field is weaker than the wind-driven circulation. In the surface layer, the Lagrangian residual currents form a clockwise flow in the Bohai Bay and the Liaodong Bay. In contrast, it flows north-eastwards into the Laizhou Bay and the central Bohai Sea with current speeds of 1 ~ 5 cm/s. In the bottom layer strong residual currents flow north-eastwards into the Liaodong Bay showing current speeds between 1 ~ 5 cm/s.

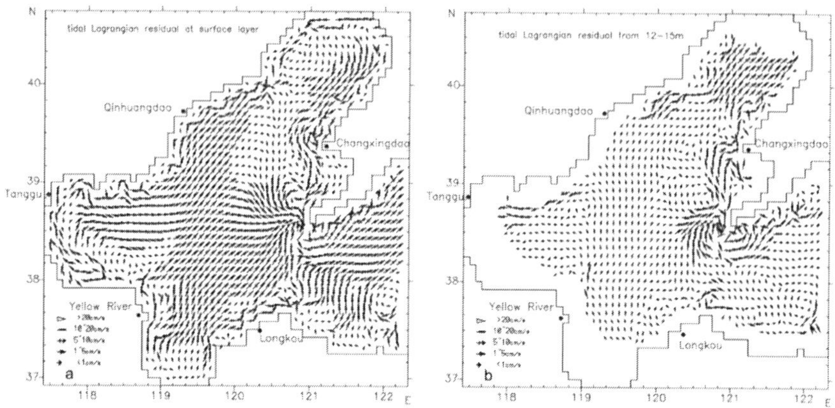

Figure 12. Lagrangian tidal residual current at (a) the surface layer and (b) between 12 and 15 m of the Bohai Sea. Source: from Wei *et al.* (2004)

It should be noticed that the Bohai Sea has experienced rapid coastline changes recently due to natural developments of the Yellow River delta and large-scale anthropogenic land reclamation which substantially alters the tidal dynamics of this area. A numerical tidal model showed that the M_2 tidal amplitudes have changed by up to 0.2 m over the last three decades due to the rapid coastal development (Pelling *et al.*, 2013), and another numerical model showed that besides tidal amplitudes, the tidal energy flux is also affected by the land reclamation in the Bohai Sea (Song *et al.*, 2013).

TRANSPORT PROCESSES

Dissolved and particulate, organic and inorganic substances in the Bohai Sea waters are transported within the current regimes described above. Their spreading and deposition control the biological and sedimentological dynamics, and hence, the ecological and morphological state of the Bohai Sea. To quantitatively understand the transport timescales of dissolved material discharged from large rivers into a semi-enclosed sea, the age of Yellow River water in the Bohai Sea was calculated using numerical models (Liu *et al.*, 2012). The results showed that the Yellow River water has a mean age of 3.0 years for the entire Bohai Sea, with a range of 1.2 ~ 3.9 years increasing from the Yellow River estuary towards the Liaodong Bay. Other model studies used different methods to evaluate the water exchange ability of the Bohai Sea, *i.e.* Wei *et al.* (2002) used a dispersion model to estimate the half-life time of the Bohai Sea (the time which is needed to reduce the initial concentration within the study area by 50%). The model results showed that Laizhou Bay has the highest exchange ability and the

Liaodong Bay has the lowest one with half-life times varying from 0.5 to 3.5 years. Turnover times are also a good measure to evaluate the water exchange ability of a marine system. They refer to the amount of time required to replace the entire water body of this system by inflowing water. Hainbucher *et al.* (2004)calculated turnover times of the Bohai Sea and found that the whole Bohai Sea needs almost half a year for a total renewal of water masses.

Besides advection by the flow field, the concentrations of substances are also determined by some other processes such as mixing, biogeochemical decay, deposition, resuspension, *etc.* Sinking, flocculation and bottom processes (*e.g.*bioturbation) are also important for the sediment dynamics. It is important to note, that the sediment transport budget finally determines the morphological changes in the Bohai Sea. Numerical studies of the sediment transport in the Bohai Sea (Figure 13) showed that the sediments discharged by the Yellow River are transported to the north-western Bohai Sea due to the residual currents (Bian *et al.*, 2013; Jiang *et al.*, 2004). After 12 months, the Yellow River discharged sediments almost cover the whole Bohai Sea. After 120 months, a large amount of sediment is deposited in the Yellow River estuary and the Central Bohai Sea area with a deposited sediment mass of more than 10 kg/m^2(equal to a sediment accumulation rate of 0.4 cm/yr).

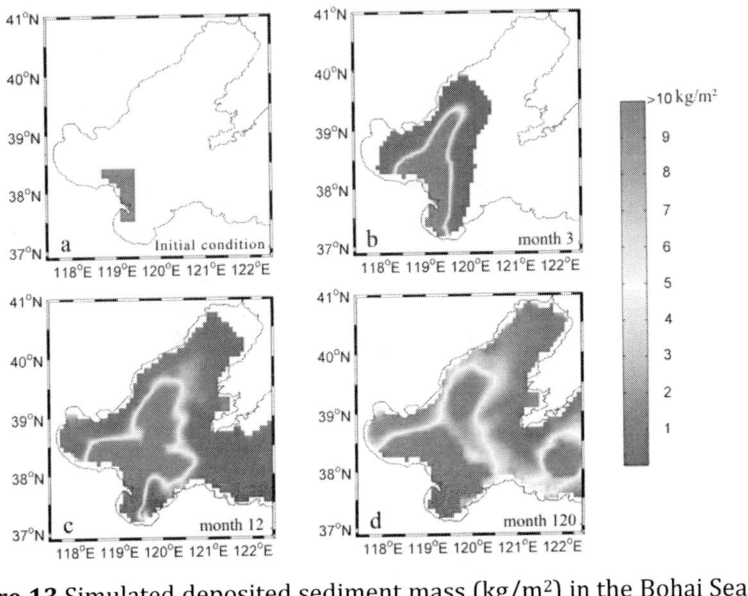

Figure 13. Simulated deposited sediment mass (kg/m^2) in the Bohai Sea for (a) Initial condition (b) after 3 months (c) after 12 months and (d) after 120 months. Source of data: model results after Bian *et al.* (2013)

CONCLUSIONS

Based on literature, remote sensing reconstructions and model data sets, this paper briefly describes the geographical features, external forces and principal dynamics of the Bohai Sea. To sum up, the astronomical tides are the dominant force in the Bohai Sea, they enhance all mixing processes and transport dissolved and suspended material through the residual currents. Beyond the tides, the main external force is the atmospheric flux of momentum resulting in a wind-driven circulation which on longer time scales plays a more important role in the substance transport than the tidal residual currents, and moreover, enhances mixing processes mainly in the upper ocean layer. The atmospheric heat fluxes dominate the seasonal variations of SST and in turn affect the mixing processes and the circulation system to a certain extent. The atmospheric fresh water flux and the Yellow River runoff determine the SSS pattern in the Bohai Sea. The intrusion water from the Yellow Sea also influences the thermohaline regime. Although there is general agreement on the basic dynamics of the Bohai Sea, uncertainties still exist regarding the water exchange at the Bohai Strait and the detailed circulation patterns in some specific regions.

It must be restated that due to the lack of high time-space resolution field data in the Bohai Sea, numerical results are used here to provide a qualitative understanding of the physical environment of the Bohai Sea. As indicated in the text, the model results show certain differences compared to observations probably due to underlying assumptions and simplifications in the model, especially with regard to the thermohaline regime. Recently, more and more projects have been launched to investigate the physical and biogeochemical characteristics of the Bohai Sea, and remote sensing is also widely used nowadays to support the reconstruction of these variables. These newly available data sets could be used to verify and improve existing oceanographic models, but furthermore, they could be employed for data assimilation which will be an additional method to provide a more accurate physical state of the Bohai Sea system.

ACKNOWLEDGMENTS

We are grateful to the HYCOM model group for providing useful dataset. WSJ was supported through the Natural Science Foundation project of Shandong province (Grant No. ZR2010DZ002). Thanks are also extended to anonymous reviewer for his constructive comments. We are grateful to Elsevier for permission to reuse the Figure 8 from Shi and Wang

(2012), Figure 10 from Lv *et al.* (2014), Figure 11a from Hainbucher *et al.* (2004), Figure 11b from Sündermann and Feng (2004) and Figure 12 from Wei *et al.* (2004).

LITERATURE CITED

1. Bao, X.; Wan, X.; Gao, G., and Wu, D., 2002. The characteristics of the seasonal variability of the sea surface temperature field in the Bohai Sea, the Huanghai Sea and the East China Sea from AVHRR data. *Acta Oceanologica Sinica*, 24(05), 125–133.
2. Bi, C.; Bao, X., and Wan, K., 2015. The effect of decadal salinity variations on circulation in the Bohai Sea. *Periodical of Ocean University of China*, 45(1), 1–8.
3. Bian, C. W.; Jiang, W. S., and Greatbatch, R. J., 2013. An exploratory model study of sediment transport sources and deposits in the Bohai Sea, Yellow Sea, and East China Sea. *Journal of Geophysical Research: Oceans*, 118, 1–16.
4. Bornhold, B.; Yang, Z.; Keller, G.; Prior, D.; Wiseman, W.; Wang, Q.; Wright, L.; Xu, W., and Zhuang, Z., 1986. Sedimentary framework of the modern Huanghe (Yellow River) delta. *Geo-Marine Letters*, 6(2), 77–83.
5. Chen, D. X., 1992. *Marine Atlas of Bohai Sea, Yellow Sea and East China Sea-Hydrology*, Beijing: China Ocean Press.
6. Ezer, T., 2000. On the seasonal mixed layer simulated by a basin-scale ocean model and the Mellor-Yamada turbulence scheme. *Journal of Geophysical Research: Oceans*, 105(C7), 16843–16855.
7. Gao, X.; Zhuang, W.; Chen, C. A., and Zhang, Y., 2015. Sediment quality of the SW coastal Laizhou Bay, Bohai Sea, China: a comprehensive assessment based on the analysis of heavy metals. *PloS One*, 10(3), e122190.
8. Guan, B. X., 1994. Patterns and structures of the currents in Bohai Huanghai and East China Sea, *In*: Zhou, D.; Liang Y. B., and Zeng, C. K. (eds.), *Oceanology of China Seas*. Dordrecht: Kluwer Academic Publishers, pp. 17–26.
9. Hainbucher, D.; Hao, W.; Pohlmann, T.; Sündermann, J., and Feng, S., 2004. Variability of the Bohai Sea circulation based on model calculations. *Journal of Marine Systems*, 44(3–4), 153–174.
10. Huang, D., 1995, Modelling studies of barotropic and baroclinic dynamics in the Bohai Sea. Hamburg: University of Hamburg, Ph. D. dissertation, 126 p.
11. Huang, D.; Chen, Z., and Su, J., 1996. Application of 3D shelf model on the Bohai Sea: interactions of tidal and wind-drive circulation. *Acta Oceanologica Sinica*, 18(05), 1–13.
12. Jiang, W.; Pohlmann, T.; Sun, J., and Starke, A., 2004. SPM transport in the Bohai Sea: field experiments and numerical modelling. *Journal of Marine Systems*, 44(3–4), 175–188.
13. Jiang, Z. and Wen, L., 1987. Application of NOAA satellite images on estimation of sec ice parameters in the Bohai Sea. *Journal of Waterway and Harbor*, (02), 1–7.
14. Ju, X. and Xiong, X., 2013. Distributions and seasonal changes of water temperature in the Bohai Sea, Yellow Sea and East China Sea. *Advances in Marine Science*, 31(01), 55–68.
15. Li, S.; Wang, N.; Zhang, Y., and Song, N., 2013. Ecological impacts of marine reclamation in Bohai sea from 1981 to 2011. *Marine Environmental Science*, 32(6), 926–929.
16. Lin, X.; Xie, S.; Chen, X., and Xu, L., 2006. A well-mixed warm water column in the central Bohai Sea in summer: Effects of tidal and surface wave mixing. *Journal of Geophysical Research-Oceans*, 111(C11017C11), 1–8.

17. Lin, X.; Wu, D.; Bao, X., and Jiang, W., 2002. Study on seasonal temperature and flux variation of the Bohai Strait. *Journal of Ocean University of Qingdao*, 32(03), 355–360.
18. Liu, X.; Liu, Y.; Guo, L.; Rong, Z.; Gu, Y., and Liu, Y., 2010. Interannual changes of sea level in the two regions of East China Sea and different responses to ENSO. *Global And Planetary Change*, 72(3), 215–226.
19. Liu, Z.; Wang, H.; Guo, X.; Wang, Q., and Gao, H., 2012. The age of Yellow River water in the Bohai Sea. *Journal of Geophysical Research: Oceans*, 117(C11), 1–19.
20. Liu, Z.; Wei, H., and Jiang, S., 2003. Characteristics of seasonal variation of monthly mean temperature and salinity fields in the Bohai Sea and analysis of the related dynamics. *Journal of Ocean University of Qingdao*, 33(01), 7–14.
21. Lv, C.; Bao, X.; Wu, D.; Chen, X., and Yu, H., 2009. Analysis of decadal variability and interannual variability of salinity field in Bohai and northern Yellow Sea in winter. *Periodical of Ocean University of China*, 39(S1), 267–273.
22. Lv, X.; Yuan, D.; Ma, X., and Tao, J., 2014. Wave characteristics analysis in Bohai Sea based on ECMWF wind field. *Ocean Engineering*, 91(0), 159–171.
23. Martin, J. M.; Zhang, J.; Shi, M. C., and Zhou, Q., 1993. Actual flux of the Huanghe (yellow river) sediment to the Western Pacific ocean. *Netherlands Journal of Sea Research*, 31 (3), 243–254.
24. Milliman, J. D. and Meade, R. H., 1983. World–wide delivery of river sediment to the oceans. *The Journal of Geology*, 91 (1), 1–21.
25. Pelling, H. E.; Uehara, K., and Green, J. A. M., 2013. The impact of rapid coastline changes and sea level rise on the tides in the Bohai Sea, China. *Journal of Geophysical Research: Oceans*, 118(7), 3462–3472.
26. Peng, C. and Bernstein, R. L., 1985. A study on comparing sea surface temperature patterns from satellite with that from ship data in the Huanghai and Bohai Seas. *Oceanologia et Limnologia Sinica*, 16(05), 347–353.
27. Ren, M., 2006. Sediment discharge of the Yellow River, China: past, present and future – A synthesis. *Advances in Earth Science*, 21(06), 551–563.
28. Shi, W. and Wang, M., 2012. Sea ice properties in the Bohai Sea measured by MODIS-Aqua: 2. Study of sea ice seasonal and interannual variability. *Journal of Marine Systems*, 95 (0), 41–49.
29. Song, D.; Wang, X. H.; Zhu, X., and Bao, X., 2013. Modeling studies of the far-field effects of tidal flat reclamation on tidal dynamics in the East China Seas. *Estuarine, Coastal and Shelf Science*, 133(0), 147–160.
30. Sun, X. P., 2006. *Regional Oceanography of China Seas*. Beijing: China Ocean Press, 376 p.
31. Sündermann, J. and Feng, S., 2004. Analysis and modelling of the Bohai sea ecosystem–a joint German-Chinese study. *Journal of Marine Systems*, 44(3–4), 127–140.
32. Wan, X.; Bao, X.; Wu, D., and Jiang, H., 2004. Numerical diagnostic simulation of summertime tide-induced, wind-driven and thermohaline currents in the Bohai Sea. *Oceanologia et Limuologia Sinica*, 35(01), 41–47.
33. Wang, D., 2015. Remote sensing-based study on the temporal variations in Chlorophyll-a concentration after Confo oil spill in Bohai Sea. *Journal of the Indian Society of Remote Sensing*, 43(1), 133–142.
34. Wang, H.; Wang, A.; Bi, N.; Zeng, X., and Xiao, H., 2014. Seasonal distribution of suspended sediment in the Bohai Sea, China. *Continental Shelf Research*, 90(1), 17–32.
35. Wei, H.; Hainbucher, D.; Pohlmann, T.; Feng, S., and Suendermann, J., 2004. Tidal-induced Lagrangian and Eulerian mean circulation in the Bohai Sea. *Journal of Marine Systems*, 44(3–4), 141–151.
36. Wei, H.; Tian, T.; Zhou, F., and Zhao, L., 2002. Numerical study on the water exchange of the Bohai Sea: simulation of the half-time by dispersion model. *Journal of Ocean University of Qingdao*, 32(04), 519–525.

37. Wei, Z.; Li, C.; Fang, G., and Wang, X., 2003. Numerical diagnostic study of the summertime circulation in the Bohai Sea and the water transport in the Bohai Strait. *Advances in Marine Science*, 21(04), 454–464.
38. Wu, D.; Li, Q.; Lin, X., and Bao, X., 2005. The characteristics of the Bohai Sea SST anomaly interannual variability during 1990 ~ 1999. *Periodical of Ocean University of China*, 35(02), 173–176.
39. Wu, D.; Mou, L.; Li, Q.; Bao, X., and Wan, X., 2004. The characteristics of long term salinity variations and its potential influencing factor in the Bohai Sea. *Progress in Natural Science*, 14(02), 73–77.
40. Xu, J.; Zhang, X.; Luo, Y., and Xu, M., 2013. The validation analysis of QuikSCAT wind speed and the wind distribution in China's offshore area. *Acta Oceanologica Sinica*, 35(05), 76–86.
41. Xu, L.; Lin, X., and Wu, D., 2008. The impacts of wind wave and topography on the summer temperature structure in the Bohai and Yellow Seas. *Periodical of Ocean University of China*, 38(02), 183–188.
42. Zhang, J.; Wang, H.; Zhang, Y., and Bi, S., 2012. Variation of sediment load at the major tributaries in the middle reaches of Yellow River and its impacts on the sediment flux to the sea. *Marine Geology & Quaternary Geology*, 32(03), 21–30.
43. Zhang, Z.; Zhu, M.; Wang, Z., and Wang, J., 2006. Monitoring and managing pollution load in Bohai Sea, PR China. *Ocean & Coastal Management*, 49 (9–10), 706–716.
44. Zhao, P. and Jiang, W., 2011. A numerical study of storm surges caused by cold-air outbreaks in the Bohai Sea. *Natural Hazards*, 59(1), 1–15.

CHAPTER 2

The Impacts of Wind and Air Pressure on Fluctuations of the Mean Bohai Sea Level

Zhigang Gao[1], Qinghe Zhang[2], Hui Wang[3], Tao Zou[4], Kexiu Liu[5], Shuangquan Wu[6]

[1]*State Key Laboratory of Hydraulic Engineering Simulation and Safety, Tianjin University, Tianjin 300072, P. R. China National Marine Data & Information Service, Tianjin, 300171, P. R. China*

[2]*State Key Laboratory of Hydraulic Engineering Simulation and Safety, Tianjin University, Tianjin 300072, P. R. China*

[3]*National Marine Data & Information Service, Tianjin, 300171, P. R. China*

[4]*Yantai Institute of Coastal Zone Research, Chinese Academy of Sciences, Yantai, Shandong 264003, P. R. China*

[5]*National Marine Data & Information Service, Tianjin, 300171, P. R. China*

[6]*National Marine Data & Information Service, Tianjin, 300171, P. R. China*

ABSTRACT

Gao, Z. G.; Zhang, Q. H.; Wang, H.; Zou, T.; Liu, K. X., and Wu, S. Q., 2016. The impacts of wind and air pressure on fluctuations of the mean Bohai Sea level

A numerical ocean model is employed to simulate mean sea level along coast of the Bohai sea for quantitative analysis on the contribution of wind and pressure to coastal mean sea level variations. It's shown that numerical results represent well the temporal and spatial character of mean sea level along coast of the Bohai Sea, and can be used to analyze mean sea level abnormity events. It's found that seasonal mean sea level variation caused by wind and air pressure accounts for about 70% of the variation of mean sea level. The three abnormal events of mean sea level occurred in Nov., 1988, Oct., 2006, and Feb., 2010 respectively indicate that sea surface air pressure and wind variation caused a wind-driven current. Consequently, water transportation is of great importance for the seasonal variations of mean sea level.

Keywords: Bohai Sea, mean sea level fluctuations, numerical modelling, wind and air pressure.

INTRODUCTION

Sea water expansion, melting of polar ice shelves and continental glacier caused by global warming are the main factors for global sea level rising. The distribution of global sea level rise indicates that there are obvious regional characteristics. The sea level rises in the western Pacific, eastern Indian Ocean and Atlantic, while it goes down in the eastern Pacific and western Indian Ocean (IPCC *et al.*, 2007). Fukumori *et al.* (1998) and Vivier *et al.* (1999) show that the large-scale and high-frequency sea level variability is mainly wind-driven and that the ocean adjusts via propagating barotropic Rossby waves. Other studies reported evidence of large-scale response of the ocean to wind forcing as barotropic Rossby waves (Brink, 1989; Luther *et al.*, 1990; Niiler *et al.*, 1993; Samelson, 1990).

China Sea locates in the region where the rate of sea level rise is higher. In addition of the variation of global sea level, the variation of regional sea level is also affected by the hydrological and meteorological factors such as air temperature, sea surface temperature (SST), currents, wind, atmospheric pressure and precipitation etc. (Yan *et al.*, 2008; Zhang *et al.*, 2009; Zheng *et al.*, 1999). Generally speaking, if air temperature and SST rise and atmospheric pressure reduces then the sea level would rise. The anomaly of a middle or shorter term of mean sea level is caused by non-astronomical factors such as atmospheric pressure, wind, precipitation, runoff and so on (Gu *et al.*, 2009; Wang *et al.*, 2012a, 2012b).

The Bohai Sea is a semi-closed shallow sea with a mean depth of around 18 m, composed of the Liaodong Bay, the Bohai Bay, the Laizhou Bay and the Bohai Basin, and connected to the North Yellow Sea through a latitudinal strait.

The consistent viewpoint is that the sea level variation of China Sea has a close relationship with atmospheric pressure and wind, whereas, the results of predecessors were focused on the qualitative analysis without quantitative model. In this paper sea level (driven by wind and atmospheric pressure) in the Bohai Sea was simulated using wind and atmospheric pressure data which was offered by ERA-Interim (Dee *et al.*, 2011) of European Centre for Medium-Range Weather Forecasts (ECMWF). Data of the ocean observing stations of China was also applied to do quantitative analysis about the effect of sea surface atmospheric pressure and wind on sea level variation in the coastal region of the Bohai Sea.

MODEL AND DATA

Estuarine Coastal Ocean Model (ECOM) (Blumberg *et al.*, 1983) was adopted to build the sea level numerical model in the Bohai Sea. The model applied rectangular grid, and the model domain is 117°E–123°E, 36.5°N–41.5°N, the horizontal grid resolution of the model is 5′×5′. The water depths of the model were interpolated data from ETOPO5 (Figure 1). The model was forced by 1/8° horizontal resolution and 3-hourly interval atmospheric pressure and wind which are the product of ERA-Interim (*http://apps.ecmwf.int/datasets/data/interim-full-daily*). The boundary condition is derived from water level of the bigger zone model covers northwest Pacific Ocean which is also only forced by wind and air pressure of ERA-Interim. The Bohai Sea is shallow which makes the barotropic effect more important and a 2-D numerical simulation was applied to study the sea level variation. The model is purely two-dimensional adopted a zero initial condition and ran for the time period of 1st. Jan., 1979 to 31th Dec., 2012. Then hourly water level and vertical mean currents were produced and monthly mean sea level and current were calculated basing on the hourly result. The monthly mean sea level data of six observation stations along the coast of the Bohai sea were used to do the verification. Observation data covers from 1980 to 2012 with quality controlled and calibration. All the sea level value is based on the datum of mean sea level of 1980 to 1998.

The quality of wind and air pressure of ERA-Interim was checked by coastal station observation. Because of different time interval, which the ERA-Interim is of 3hr interval but the observed data is of 6hr interval, monthly mean result of ERA-Interim and observed data are compared to test the quality. Root-Mean-Square Error (RMSE) is calculated as shown in Table 1, which shows that the average RMSE of monthly mean air pressure is 0.46 hPa ranging from 0.30 hPa to 0.74 hPa, the RMSE of monthly mean of wind velocity is 0.67 ms^{-1} ranging from 0.60 ms^{-1} to 0.86 ms^{-1} with a fractional error ranging from 8% to 25%, the RMSE of monthly mean of wind U-component is 0.67 ms^{-1} ranging from 0.61 ms^{-1} to 1.02 ms^{-1} with a fractional error ranging from 9% to 28%, the RMSE of monthly mean of wind V-component is 0.68 ms^{-1} ranging from 0.49 ms^{-1} to 0.84 ms^{-1} with a fractional error ranging from 1% to 24%. The result illustrates high agreement between the reanalysis data and the observation, which means the ERA-Interim data is reliable for the sea level numerical simulation.

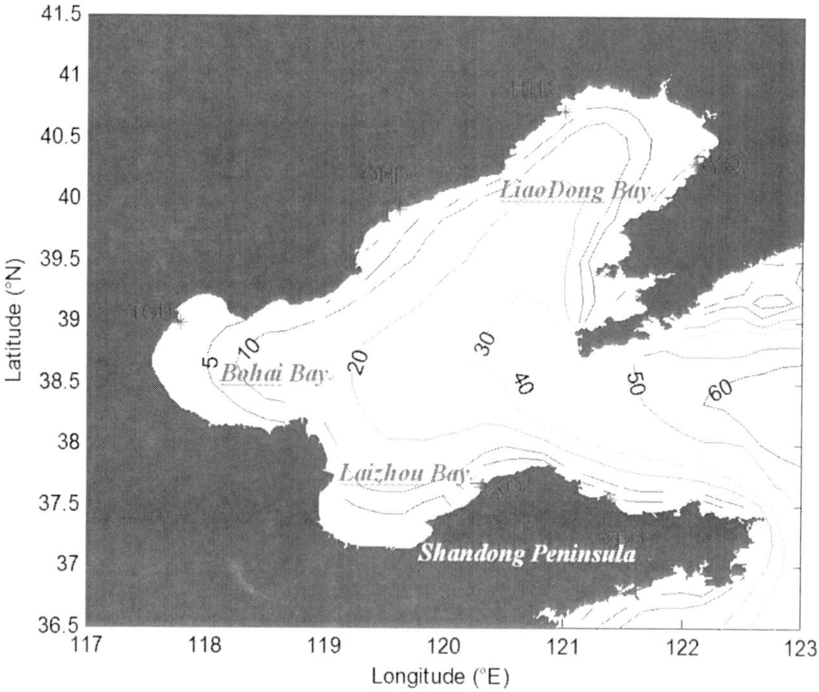

Figure 1. Study area and depth distribution

Table 1. Comparison of ERA-Interim and observed data of station along coast of the Bohai Sea

station	RMSE of monthly mean of air pressure /hPa	RMSE of monthly mean of 10m wind					
		Velocity /ms^{-1}	Fractional Error	U /ms^{-1}	Fractional Error	V /ms^{-1}	Fractional Error
BYQ (BaYuQuan)	0.30	0.65	8%	0.63	11%	0.84	18%
HLD (HuLuDao)	0.36	0.62	18%	0.43	12%	0.65	21%
QHD (QinHuangDao)	0.53	0.66	25%	0.64	28%	0.72	24%
TGU (TangGu)	0.74	0.61	18%	0.61	9%	0.49	23%
LKO (LongKou)	0.35	0.86	23%	1.02	24%	0.81	20%
ZFD (ZhiFuDao)	0.46	0.60	10%	0.71	25%	0.58	1%
average	0.46	0.67	17%	0.67	18%	0.68	18%

RESULT EXAMINATION OF THE MODEL

Arithmetic method was applied to get monthly mean sea level data which is from 1980 to 2012 at each node using hourly sea level calculated by the numerical model, and then climate monthly mean sea level could be calculated. The result of the verification between simulation and observed data is shown in Table 2. The contribution ratio referred in this essay is annual range ratio between simulation and observed data. It can reflect the contribution of seasonal sea level variation caused by atmospheric pressure and wind to the real variation if the precision of the simulation is good. The comparison of monthly mean sea level in each station can be seen in Figure 2, and seasonal variation in Figure 3.

(1) Table 2 and Figure 2 show that the simulation results of monthly mean sea level are consistent with the observations. The error statistics show the average root-mean-square error (RMSE) is 100 mm ranging from 93mm to 104mm and all the correlation coefficients are about 0.90.

(2) Table 2 and Figure 3 show that the simulation could reflect the characteristics of seasonal sea level variation. The amplitudes and phases have a good agreement with the observations. The annual range of simulated sea level is 391 mm and that of observed is 531 mm, which means the atmospheric pressure and wind are the main factors of sea level variation along coast of the Bohai Sea with a contribution ratio of 73%. BYQ and HLD are two stations with the largest contribution rate of more than 80%, followed by the TGU Station of 77%. As the three stations are located in the top of bay where water depth is relatively small, it seems comprehensible that wind and air pressure play an important role. Although the correlation coefficients are very high, the amplitude of simulated result is less than that of observed meanwhile it seems that the simulated result has a month ahead of observed data (Figure 3). Maybe that's because of the assumption of the numerical model, the atmospheric condition is taken into consideration only, which means some other factors affecting sea level such as sea temperature or salinity are not considered in the model. Thus the modeled amplitude is less than observed is comprehensible.

Table 2. Comparison of simulated result and observed data of station along coast of the Bohai Sea

station	monthly mean sea level comparison		sea level annual range comparison		
	RMSE /mm	correlation coefficient	simulation /mm	observation /mm	ratio
BYQ (BaYuQuan)	102	0.88	450	528	85%
HLD (HuLuDao)	97	0.91	466	578	81%
QHD (QinHuangDao)	93	0.92	404	552	73%
TGU (TangGu)	104	0.91	447	578	77%
LKO (LongKou)	104	0.89	297	487	61%
ZFD (ZhiFuDao)	98	0.89	283	460	62%
average	100	0.90	391	531	73%

All these above suggests that the simulation results could basically reflect the characteristics of monthly mean sea level variation along coast of the Bohai Sea taking the model assumption of only air condition forcing into consideration. The amplitudes and phases have a good agreement with the observations and the correlation coefficients of six stations are all over 0.85. The air pressure and wind are the main factors for the seasonal sea level variation of coastal area, and the contribution ratios of six stations are more than 60% viewed from annual range.

ANALYSIS OF SEA LEVEL ANOMALY

Extreme weather and climate events happened frequently along with the global warming, and also result in big variation of regional sea level. The big variation happens to China coastal area every year. *Chinese Sea Level Bulletin* 2012 pointed that the sea level along China coastal area in 2012 reached a maximum since 1980. And the sea level was higher than ordinary year in May, June, August and October. Sea temperature, air temperature, atmospheric pressure and wind were the main factors which caused the anomaly of the sea level.

The mean sea level abnormal events (short as MSLAE) defined in this essay are that when the monthly mean sea level is 100 mm higher than ordinary year. Three typical MSLAE occurred in the Bohai Sea are discussed using the simulation results and observed data as following.

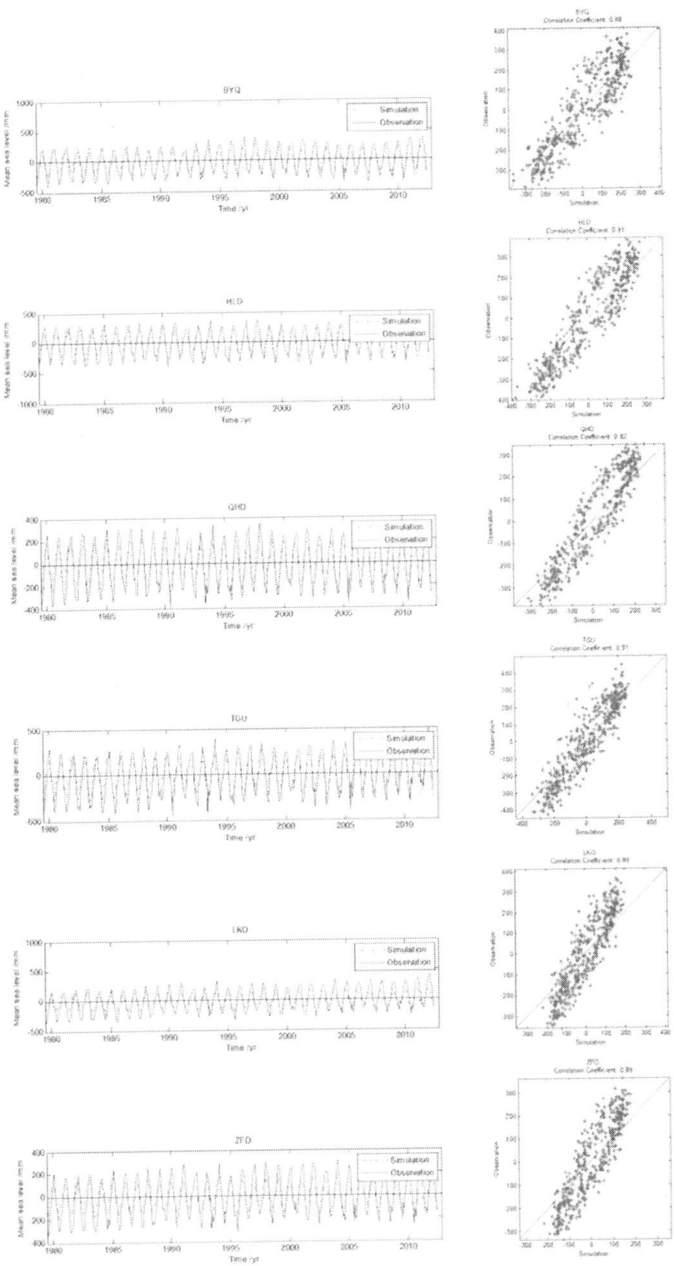

Figure 2. Comparison of monthly mean sea level between simulation and observation of station along coast of the Bohai Sea

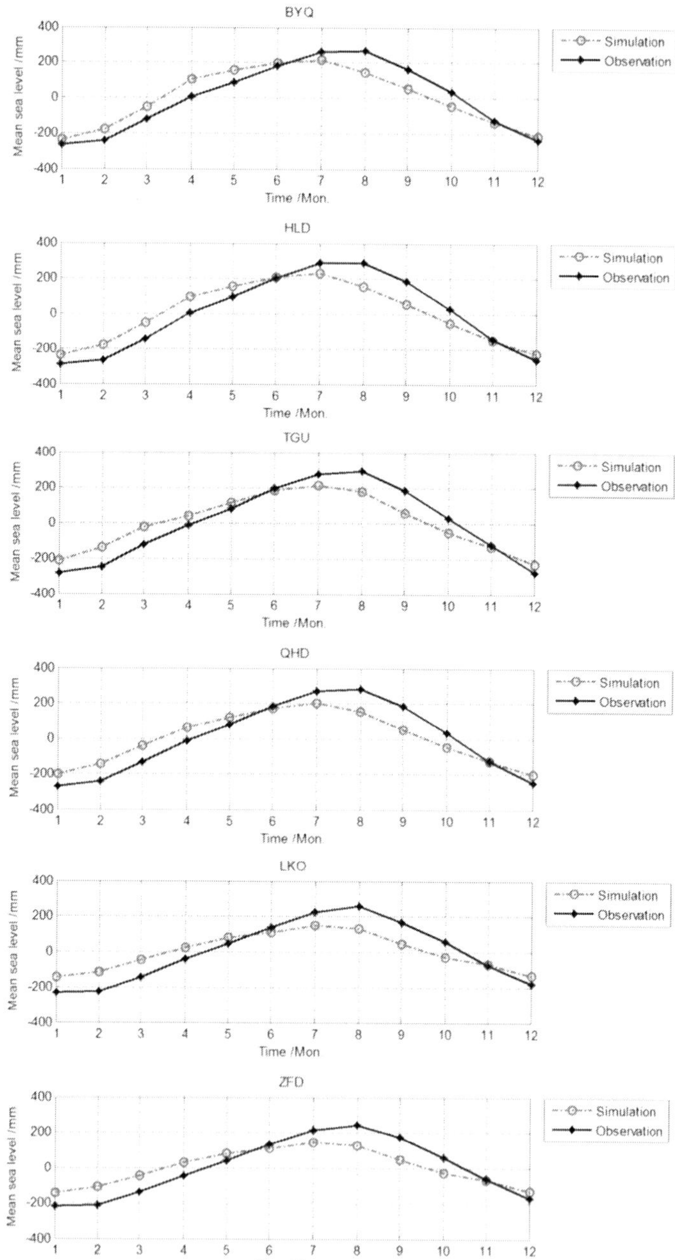

Figure 3. Comparison of seasonal mean sea level variation between simulation and observation of station along coast of the Bohai Sea

The sea level anomaly low along coast of the Bohai Sea in Nov., 1988

The sea level was abnormally lower along coast of the Bohai Sea in Nov., 1988. Monthly mean sea level observed from stations are of 100 mm lower than that of ordinary year while the lowest station is TGU which is more than 200 mm lower. The comparison between simulation and observation is in Figure 4. It shows that the simulation can reflect the abnormal variation of sea level in each station. Although the absolute value of simulated is less than that of observed, the spatial distribution trend is basically consistent. Both simulated and observed result indicates that the Bohai Bay has the lowest sea level variation.

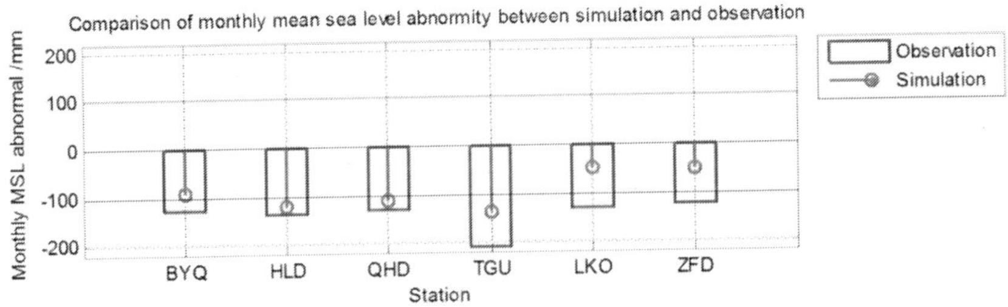

Figure 4. Comparison of monthly mean sea level abnormity between simulation and observation of Nov., 1988

The analysis of sea level along coast of the Bohai sea, the simulation of wind-driven flow and anomaly of wind and atmospheric pressure (Figure 5a) show that the direction of wind field anomaly is northward, and value is about 2 m/s. North-westward to northward wind was the prevail wind which means the north-westward wind was strengthen in this month. From the results of the simulation of wind-driven current (Figure 5c), we can see that there is a strong coastal current along the coast of eastern Liaodong Bay and northern Shandong Peninsula, the direction is from the Bohai sea to the Yellow Sea which means water transport out of Bohai leading to the mean sea level significantly lower; where in the Bohai Bay there are two circulation with opposite direction, the south wing of counterclockwise, the north wing of clockwise, the combined effect of the two circulation leads to outward transport of the top area of Bohai Bay, resulting in mean sea level anomaly low at the station of TGU.

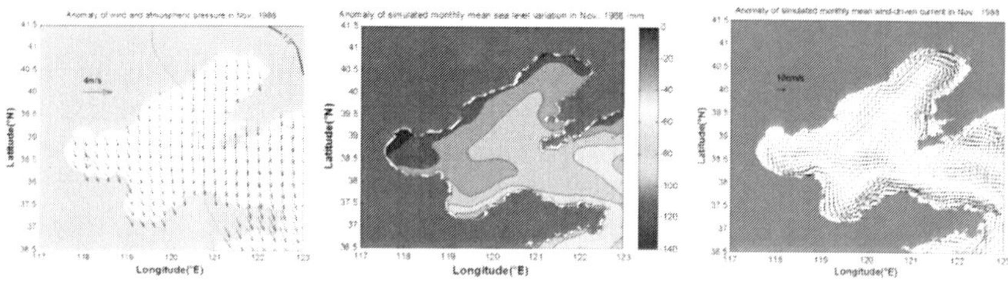

Figure 5. Anomaly of simulated monthly mean sea level variation and wind-driven current, anomaly of wind and pressure in the study area in Nov., 1988

From the pressure field anomaly distribution (Figure 5a), air pressure of the Bohai Sea is of 0.5 hPa ~ 1.5 hPa lower in Nov., 1988 than that of the same period of prevail year. Considering from the static effect positive pressure anomaly would lead to lower sea level while negative pressure anomaly would lead to higher sea level. In this case, air pressure over Bohai sea is negative anomaly, would cause the sea level rise, while the actual sea level is significantly negative anomaly, thus we could draw a conclusion that water transport effect caused by wind-driven current plays a dominant role for this MSLAE.

The sea level anomaly high along coast of the Bohai Sea in Oct., 2006

The sea level was abnormally higher along coast of the Bohai Sea in Oct., 2006. Monthly mean sea level of TGU and LKO stations are of 100 mm higher than that of ordinary year. The comparison between simulation and observation is in Figure 6. It shows that the simulation can reflect the abnormal variation of sea level in each station. Although the absolute value of simulated is less than that of observed, the spatial distribution trend is basically consistent. Both simulated and observed result indicates that the Bohai Bay has the highest sea level variation and the Liaodong Bay has the lowest sea level variation.

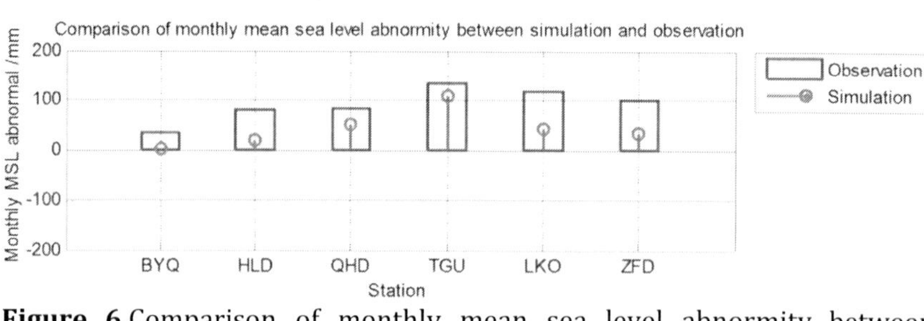

Figure 6. Comparison of monthly mean sea level abnormity between simulation and observation in Oct., 2006

The analysis of sea level along coast of the Bohai sea, the simulation of wind-driven current and anomaly of wind and atmospheric pressure (Figure 7a) shows that direction of wind anomaly is east-northeast with quantity value of about 2 m/s, and the west-northwest wind was the prevail wind in October of ordinary year, which indicates that compared to the same period of average year the west-northwest wind was significantly weaker resulting in water transported from the Bohai sea to the Yellow Sea transport decreasing, further leading to mean sea level higher than the same period of average year. There is a strong counter clockwise wind-driven circulation in the Bohai Bay, resulting in monthly sea level of TGU station significantly higher than average state; meanwhile, atmosphere pressure over the whole Bohai Sea is of 0.5 hP ~ 1.5 hPa lower than that of average year, which would result in a positive anomaly effect to the mean sea level. The above analysis shows that the combined action of wind and pressure is an important cause of this MSLAE.

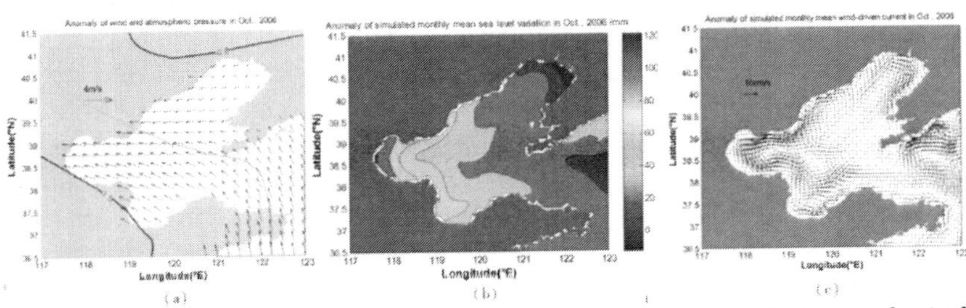

Figure 7. Anomaly of simulated monthly mean sea level variation and wind-driven current, anomaly of wind and pressure in the study area in Oct., 2006

The sea level anomaly high along coast of the Bohai Sea in Feb., 2010

The sea level was abnormally higher along coast of the Bohai Sea in Feb., 2010. Monthly mean sea level of TGU stations are of more than 200 mm higher than that of ordinary year. The comparison between simulation and observation is in Figure 8. It shows that the simulation could reflect the spatial character of sea level abnormal along coast of the Bohai Sea. Although the absolute value of simulated is less than that of observed, the spatial distribution trend is basically consistent. Both simulated and observed result indicates the Bohai Bay has the highest sea level variation and the Liaodong Bay has the lowest sea level variation.

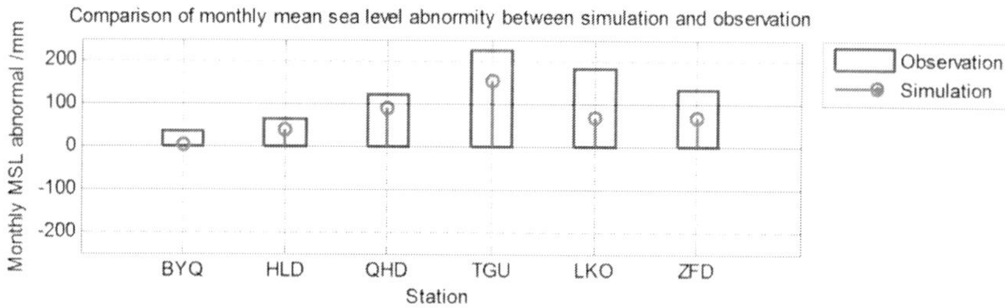

Figure 8. Comparison of monthly mean sea level abnormity between simulation and observation in Feb., 2010

The analysis of sea level along coast of the Bohai sea (Figure 9b), the simulation of wind-driven current (Figure 9c) and anomaly of wind and atmospheric pressure (Figure 9a) shows that direction of wind anomaly is east with value of about 4 m/s, and the northwest wind was the prevail wind in Feb., of average year. which indicates that the west wind was significantly weaker than that of average state, causing offshore water transport reducing in Bohai Bay. From the result of simulated wind-driven current (Figure 9c), there exists strong coastal current along the coast of northern Shandong Peninsula, the direction is from the Yellow Sea to the Bohai Sea, that means water transport to the Bohai is positive, resulting in mean sea level overall higher; wherein the Bohai Bay there are two circulation with opposite direction, one is at the north wing, with strong anti-clockwise circulation, the other is at the south wing with weak clockwise circulation, the combined effect of the two circulation leading to the water shore-ward accumulation, resulting in monthly mean sea level significantly higher in the Bohai Bay. Meanwhile, atmosphere pressure over the whole Bohai Sea is of 1.0 hP ~ 2.5 hPa lower than that of average year, which would result in a positive anomaly effect to the mean sea level. The above results show that this MSLAE was mainly caused by an abnormality of air pressure and wind.

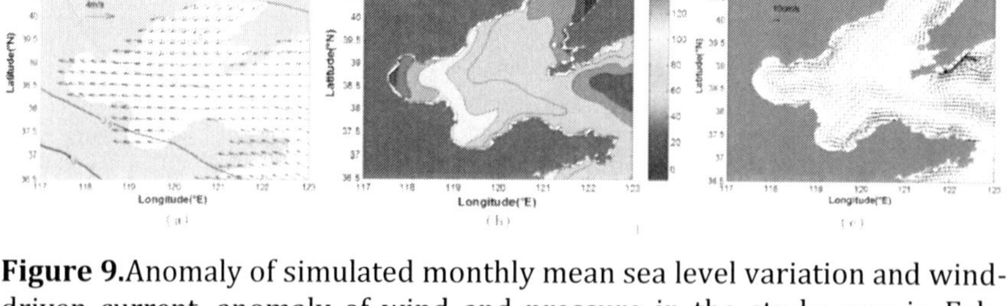

Figure 9. Anomaly of simulated monthly mean sea level variation and wind-driven current, anomaly of wind and pressure in the study area in Feb., 2010

In summary, model result could basically characterize the spatial features in coastal Bohai area during the three MSLAEs, although the absolute value of simulated is less than that of observed. Absolute bias between model and observation might due to the neglect of other factors such as sea temperature and salinity.

CONCLUSIONS

Numerical simulation was applied to quantitatively analyze the impact of meteorological factors (atmospheric pressure and wind) on the mean sea level variation in the coastal area of the Bohai Sea. The main conclusions are as follows.

(1) The comparison between simulation and observation suggests that the numerical model driven by air pressure and wind could characterize the spatial and temporal features of monthly mean sea level variation along coast of the Bohai Sea. The average RMSE of monthly mean sea level is 100 mm, and the average correlation coefficient is 0.90.

(2) The trend of seasonal sea level variation is basically consistent between the simulation and observation. The contribution ratio of air pressure and wind to the sea level variation in the coastal area of the Bohai Sea is about 73%, and they are the main factors for the seasonal variation.

(3) According to the analysis of three typical MSLAE occurred in Nov., 1988, Oct., 2006 and Feb., 2010, the simulation results can reflect the spatial characteristics of the sea level anomaly. It suggests that air pressure and wind variation leading wind-driven current and consequently water transportation are of great importance to the abnormal variations of mean sea level along coast of the Bohai Sea.

LITERATURE CITED

1. Blumberg, A. F. and Mellor, G. L., 1983. A Description of A Three-Dimensional Coastal Ocean Circulation Model in Three-dimensional Coastal Ocean Models. *American Geophysical Union*, 2–16.
2. Brink, K. H., 1989. Evidence for wind-driven current fluctuations in the western North Atlantic. *Journal of Geophysics Research*, 94, 2029–2044.
3. Dee, D. P.; Uppala, S. M.; Simmons, A. J.; Berrisford, P.; Poli, P.; Kobayashi, S.; Andrae, U.; Balmaseda, M. A.; Balsamo, G.; Bauer, P.; Bechtold, P.; Beljaars, A. C. M.; van de Berg, L.; Bidlot, J.; Bormann, N.; Delsol, C.; Dragani, R.; Fuentes, M.; Geer, A. J.; Haimberger, L.; Healy, S. B.; Hersbach, H.; Holm, E. V.; Isaksen, L.; Kallberg, P.; Kohler, M.; Matricardi, M.; McNally, A. P.; Monge-Sanz, B. M.; Morcrette, J-J.; Park, B-K.; Peubey, C.; de Rosnay, P.; Tavolato, C.; Thepaut, J-N., and Vitart, F., 2011. The ERA-Interim reanalysis: configuration

and performance of the data assimilationsystem. *Quarterly Journal of the Royal Meteorological Society*, 137, 553–597.

4. Fukumori, I.; Raghunath, R., and Fu, L., 1998. Nature of global large-scale sea level variability in relation to atmospheric forcing: A modeling study. *Journal of Geophysics Research*, 103, 5493–5512.
5. Gu, X. L. and Li, P. L., 2009. Pacific sea level variations and its factors. *Acta Oceanologica Sinica*, 31, 28–36.
6. IPCC, 2007. Climate Change 2007: The Physical Science Basis. In: Solomon, S.; Qin, D.; Manning, M.; Chen, Z.; Marquis, M.; Averyt, K. B.; Tignor, M., and Miller, H. L. (eds.), Contribution of Working Group I to the Fourth Assessment Report of the Intergovernmental Panel on Climate Change. Cambridge, UK: Cambridge University Press, 996 p.
7. Luther, D. S.; Chave, A. D.; Filloux, J. H., and Spain, P. F., 1990. Evidence for local and nonlocal barotropic responses to atmospheric forcing during BEMPEX. *Geophysics of Research Letter*, 17, 949–952.
8. Niiler, P. P.; Filloux, J.; Liu, W. T.; Samelson, R. M.; Paduan, J. D., and Paulson, C. A., 1993. Wind-forced variability of the deep eastern North Pacific: Observations of seafloor pressure and abyssal currents. *Journal of Geophysics Research*, 98, 22589–22602.
9. Samelson, R. M., 1990. Evidence for wind-driven current fluctuations in the eastern North Atlantic. *Journal of Geophysics Research*, 95, 11359–11368.
10. State Oceanic Administration People's Republic of China.2012, *Chinese Sea Level Bulletin*, 36–38.
11. Vivier, F.; Kelly, K. A., and Thompson, K., 1999. Contributions of wind forcing, waves, and surface heating to sea-surface height observations in the Pacific Ocean. *Journal of Geophysics Research*, 104, 20767–20788.
12. Wang, H.; Fan, W. J.; Li, Y.; He, Q.; Gao, Z. G.; Chi, Y. X., and Mu, L., 2012a. Analysis on sea level anomaly in the coastal area of Bohai Sea and Yellow Sea in February. *Marine Science Bulletin*, 31(3), 255–261.
13. Wang, H.; Fan, W. J., and Gao, Z. G., 2012b. Analysis on the sea level anomaly high in the coastal area of Bohai Sea and Yellow Sea. *Marine Science Bulletin*, 31(6), 613–620.
14. Yan, M.; Zuo, J. C., and Fu, S. B., 2008. Advances on sea level variation research in global and China sea. *Marine Environmental Science*, 27(2), 197–200.
15. Zhang, J. L.; Chen, M. C.; Chen, M. X.; Wang, H., and Zuo, J. C., 2009. Characteristics of SST variation/change in the coastal region of the East China Sea. International Offshore and Polar Engineering Conference.
16. Zheng, W. Z. and Chen, Z. Y., 1999. Distribution of Annual Rates of Sea Level and Variation of Long-Period Constituents in China. *Marine Science Bulletin*, 18(4), 1–10.

CHAPTER 3

Climate Change, Coastal Vulnerability and the Need for Adaptation Alternatives: Planning and Design Examples from Egypt and the USA

S Jeffress Williams [1,*] and Nabil Ismail [2]

[1]*U.S. Geological Survey, Woods Hole, Massachusetts and University of Hawaii, Honolulu, HI 96822, USA*

[2]*Coastal Engineering, Maritime Academy, Alexandria, Egypt and Director of Costamarine Technologies, Davis, CA 95616, USA*

ABSTRACT

Planning and design of coastal protection for high-risk events with low to moderate or uncertain probabilities are a challenging balance of short- and long-term cost *vs.* protection of lives and infrastructure. The pervasive, complex, and accelerating impacts of climate change on coastal areas, including sea-level rise, storm surge and tidal flooding, require full integration of the latest science into strategic plans and engineering designs. While the impacts of changes occurring are global, local effects are highly variable and often greatly exacerbated by geophysical (land subsidence, faulting), oceanographic (ocean circulation, wind patterns) and anthropogenic factors. Reducing carbon emissions is needed to mitigate global warming, but adaptation can accommodate at least near future change impacts. Adaptation should include alternatives that best match region-specific risk, time frame, environmental conditions, and the desired protection. Optimal alternatives are ones that provide protection, accommodate or mimic natural coastal processes, and include landforms such as barrier islands and wetlands. Plans are often for 50 years, but longer-term planning is recommended since risk from climate change will persist for centuries. This paper presents an assessment of impacts of accelerating climate change on the adequacy of coastal protection strategies and explores design measures needed for an optimum degree of protection and risk reduction. Three coastal areas facing similar challenges are discussed: Abu-Qir Bay, Nile River delta plain, Egypt; Lake Borgne, New Orleans, Louisiana delta plain; and the New York City region.

Keywords: climate change; sea-level rise; Egypt; New York City; New Orleans; Louisiana; coastal vulnerability; deltas; coastal protection; coastal management; adaptation

1. INTRODUCTION

For over one hundred years, coastal protection structures were designed and constructed based mostly on assumptions that parameters were fairly constant and predictable and conditions driving forces of the past would persist pretty much unchanged into the future. Results of climate science studies over the past three decades, especially the past decade, show that climate change is unequivocal, due largely to carbon dioxide increase and other greenhouse gas (GHG) concentrations in the atmosphere. Observations show the effects are global, but vary greatly on regional scales. The impacts of climate change most germane to coastal protection are: global mean sea-level rise (GSLR), increase in storm surge and flooding levels, increase in high tide nuisance flooding, change in wave characteristics due to more intense storm events, and increase in extreme weather events. Natural disasters are becoming more frequent and more severe on highly vulnerable deltas and lowland coastal areas. This is evident for barrier islands and river delta coasts, which are experiencing high rates of erosion, flooding, and marine transgression. Projected future GSLR poses further significant threat, especially to coastal landforms, which are already undergoing deterioration due to anthropogenic impacts. The impacts of climate change on coasts vary temporally and regionally but all coastal regions are facing similar issues.

Adaptation measures responding to climate change impacts should be implemented in a phased and progressive manner over time and all vulnerable countries can benefit by international cooperation and sharing of experiences, expertise, and resources.

2. OBJECTIVE

This paper presents an assessment of the impacts of accelerating climate change (GSLR, storm intensity, tidal nuisance flooding) on the adequacy of coastal protection strategies and explores additional measures needed for an optimum degree of protection and risk reduction. Three coastal regions with similar circumstances are considered: the Nile Delta Coast at Abu-Qir Bay, Egypt; the Louisiana coast including the Lake Borgne Surge Barrier; and the New York City region. Current plans are discussed and recommendations are made for employing adaptation to cope with projected climate change impacts and reduce risk to humans living in low coastal regions.

3. CLIMATE CHANGE—GLOBAL SEA-LEVEL RISE

Climate science is expanding in information and its understanding rapidly. Applying 50 to 100 year conditions into the future and beyond will be different than the historic past or the present. This requires a significant change in thinking about sustainable coastal management and protection of delta and lowland coastlines (Figure 1). Impacts of climate change should be fully integrated into coastal planning and engineering and uncertainties factored in. Risk and vulnerability need to be assessed to decide on the most cost-effective approach for projects designed for the next 50 years, to 2100 and well beyond.

The changes in climate already have significant consequences with regard to precipitation (extreme rain events, flooding, droughts), temperature extremes, and ocean levels (storm surge, sea-level rise, tidal flooding). Sea-level rise, increasingly recognized as important on local and regional scales, is a dominant driver of coastal change, along with storms.

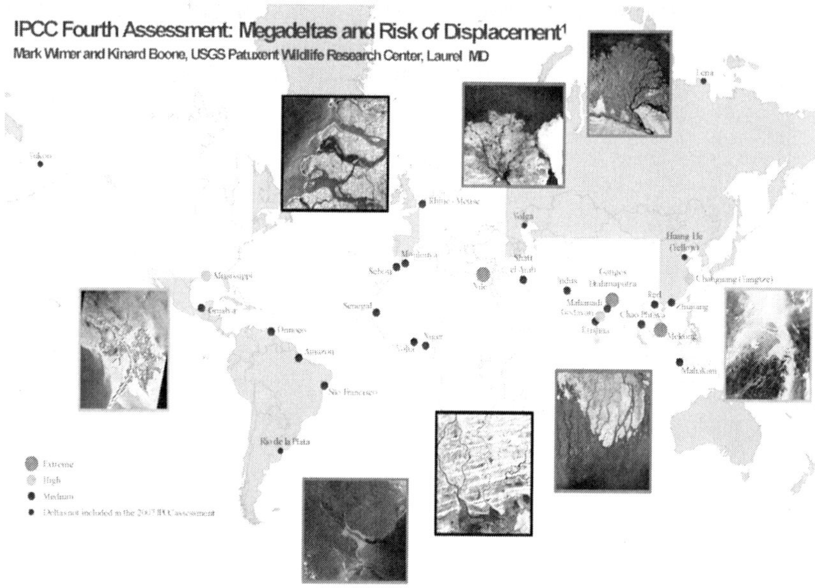

Figure 1. World Mega Deltas threatened by climate and human changes (open source poster modified by USGS from [1]).

Global mean sea level was relatively stable for the past several thousand years under a mild climate until the mid-19th century, but during the 20th century sea level began rising due to global warming resulting from human activities at a global average rate of 1.2 mm/year. The current average global rise rate is 3.2 mm/year, 2.5 times faster [2,3]. Many coastal regions, particularly deltas, however, are experiencing much greater local or relative sea-level rise (LSLR), defined as global mean sea-level rise plus subsidence, plus sediment compaction, and in minor cases land emergence. Example rates are: Nile delta region 2–5 mm/year, Louisiana delta region ~4–10+ mm/year, and New York City area 4 mm/year. These higher than global rates have significant areal variations and are due to local geophysical (subsidence, faulting), oceanographic, and human factors (oil, gas, and ground water extraction, as well as wetland reclamation). Combined with storms, this rise is resulting in greater surge elevations, more frequent tidal nuisance flooding, and record increases in damage to coastal infrastructure and loss of life [2]. Projected GSLR by 2100 [4,5,6,7] is 0.2 m to 2 m and would be in addition to local rise factors such as subsidence and oceanographic processes (Figure 2). The purple and red color bars simply denote the most recent published projections. The range of projections is large due to the uncertainty of continued global warming, which in turn depends on the extent of continued man-made emissions of greenhouse gases into the atmosphere. Even greater global sea-level rise is possible (3 m+) over the next several hundred years [8,9], unless carbon emissions, global warming, and ice-sheet melting are greatly slowed or ideally reversed.

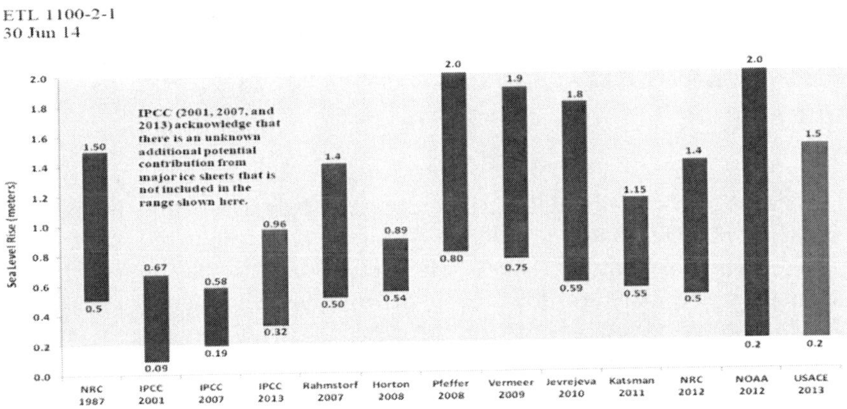

Figure 2. Comparison of maximum and minimum estimates of global sea level rise by 2100.

4. RECENT WORLD CATASTROPHIC STORM EVENTS

Hurricane Katrina, with a storm surge of up to 8.5 m and wave heights of 5.5 m (total 14 m), made landfall in eastern Louisiana and the Mississippi coast on 29 August 2005, resulting in major loss of lives from massive levee failures and flooding of New Orleans, loss of wetlands, and extensive erosion of barrier islands along the Gulf coast.

The Nile Delta Super Storm on 12 December 2010 is a striking example of the severity of more energetic storm events since 2003. The storm caused extensive flooding of the delta and coastal cities in Egypt. A surge of over 1.2 m and 7 m waves forced the closure of Alexandria's main harbor. The main damage resulted from wave and surge overtopping of coastal structures designed for smaller events.

Hurricane "Super storm" Sandy made landfall along the New Jersey–New York coast on 29 October 2012 at high tide after taking an unusual path at landfall. The largest storm on record for the Atlantic, Sandy, a combination tropical–extra tropical storm, caused catastrophic damage by overtopping barrier islands and coastal structures with damage throughout the region, including New York City, which experienced a 4.4 m storm surge.

5. VULNERABILITY OF DELTAS AND LOWLAND COASTLINES TO CLIMATE AND ANTHROPOGENIC CHANGES

All the World's major deltas, including the Nile Delta and Louisiana's Mississippi River Delta, developed about seven to eight thousand years ago when Holocene sea-level rise slowed and rivers were able to transport sediment to the coast in volumes sufficient to maintain landforms acted on by marine processes. Deltas develop by lateral movement of channels that spread alluvial sediments while at the same time thick deposits undergo rapid subsidence. Acted on by marine processes over time, deltas become fronted by sandy barrier island shorelines, backed by vegetated wetlands and marshes [10].

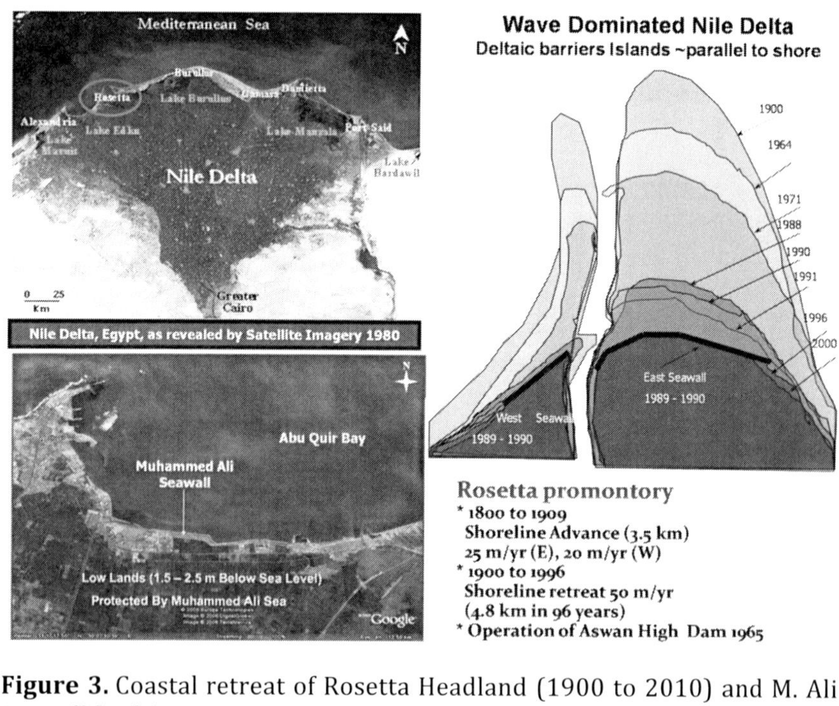

Figure 3. Coastal retreat of Rosetta Headland (1900 to 2010) and M. Ali Seawall [11] from open source maps.

Figure 4. Map image of the Louisiana delta plain, Mississippi River delta, coastal barrier islands and wetlands, and New Orleans just south of Lake Pontchatrain (Google Earth 2014).

Deltas have been subjected to significant coastal changes due to a variety of man-made factors over the past several hundred years, such as reduction in sediment load due to the construction of dams and levees, such as along the Mississippi River and the Nile River. In addition, the reduction of river currents modifies the near shore circulation, induced by wave–current interaction, which maintained the delta morphology for many centuries. Such effects of the combined wave–current flow, on nearshore circulation, are detailed in studies by Ismail [12,13]. For the Nile, the Aswan High Dam built in 1965 has caused significant coastal changes due to sediment depletion, exacerbating subsidence and shoreline erosion (Figure 3). For the USA, Louisiana's delta plain has the highest rates of coastal erosion and wetland loss of any region in the world due to a combination of complex natural processes and anthropogenic actions. Since 1900, about 4900 km^2 of wetlands in coastal Louisiana have been lost (Figure 4).

6. NILE DELTA COAST AND 2010 STORM EFFECTS

The coastal flooding in Alexandria on 12 December 2010 is a striking example of the more progressive severity of events since 2003. Egypt was hit by strong winds, exacerbated by heavy precipitation, up to 60 km/h with 10 h duration. These weather conditions resulted in waves of ~7 m height with a surge of 1.2 m, which forced the closure of Alexandria harbor. The typical significant wave height is 1–1.5 m. Maximum wave height during storm conditions averages 4.5 m. Typical values of storm surge on the delta coast are 40–50 cm. The storm had profound destructive effects on Alexandria, Abu Qir Bay as well as on shorelines between the two river Nile promontories, Rosetta and Damietta.

A comprehensive program has been underway since 1970 to gather data related to coastal erosion along the Nile Delta shore. Review of major coastal problems and general description of the recommended protective measures to address these problems are in the coastal Master Plan for Phase I, as reported by [14]. Plans are underway to prepare Master Plan Phase II for coastal protection of the Nile Delta coast that will further address future climate change impacts and the need for adaptation alternatives (Figure 5).

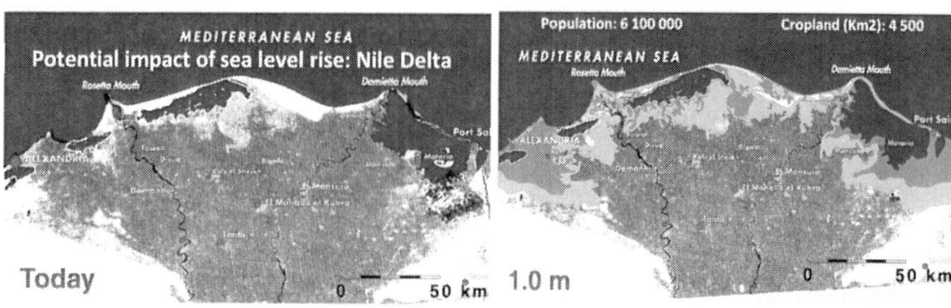

Figure 5. Maps of the Nile Delta showing potential impacts of a 1 m rise in sea level (Maps from open sources).

Mohamed Ali-Seawall Case Study

This project is the design-review and upgrade of Mohamed Ali seawall (revetment) in the partially protected Abu Qir Bay under scenarios of extreme storm conditions for the future 50 years return period. These design conditions include: storm wave height, storm surge, subsidence, and projected LSLR (Figure 6). Rosetta headland with its adjacent Abu Qir Bay form a littoral cell, in which sedimentation is controlled by a combination of waves, coastal currents, tides, and river discharge. The M.A. seawall located in Abu Qir Bay was constructed in 1830 to protect the agriculture lowland against sea flooding (Figure 3). The seawall was repaired and upgraded in 1981 and a beach segment was created for recreation. As in many coastal regions near major urban areas, the coast is used for a variety of purposes including, a power plant and a liquefied natural gas export terminal. The seawall was upgraded again in 2009, one year before the 2010 storm. Damages from the storm showed an urgency for further upgrades, as highlighted by [11].

Hydrodynamic analyses were conducted to provide design recommendations to increase the life of the structure, deter flooding of the lowlands, and increase the stability of the structure to impacts of climate change. Using a future LSLR value of 0.5 m in the next 50 years, resulted in a design wave run-up height of 5.4 m (Figure 6). Two design scenarios were considered to estimate wave height distributions within Abu Qir Bay and along M. Ali seawall.

I. Current morphology with the obtained maximum storm yearly wave condition (H_{max} = 3.5 m, T = 7.5 s, N direction, Water Depth = 14 m).

II. Future morphology (recent bathymetry +0.5 m sea-level rise within the next 50 years) with the maximum yearly wave condition in Abu-Qir Bay (4.8 m).

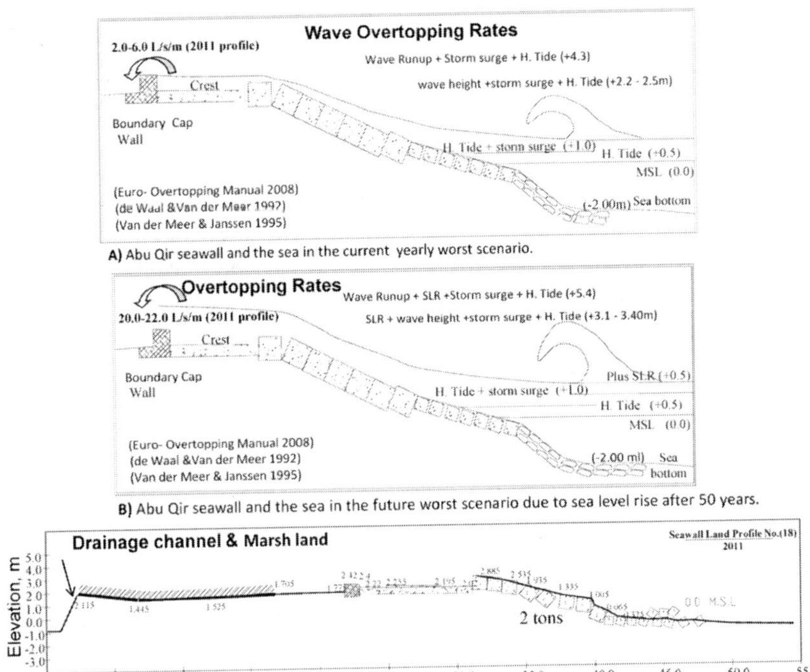

Figure 6. Impact of climate change on the the M.A. design and recommended drainage channel and creation of wetlands. The value of 0.5 m includes projected global sea-level rise and local subsidence [11].

The model results showed that sea-level rise has a noticeable effect for the case of maximum wave condition at the wall. The wave height in front of the coastal structure at the toe will increase by about 25% after 50 years. Further, based on the obtained wave run-up of 5.4 m and subsequent overtopping, recommendations were made to increase the weight of the armor layer from 0.5 t to 2 t, increase the seawall top elevation to 3 m, and reinforce the revetment toe. Further recommendations are given to increase the height of the seawall cap, and to strengthen the beach top and back slope with a facility to drain the overtopped storm water behind the seawall (Figure 6).

7. HURRICANE KATRINA, STORM PROTECTION, AND LOUISIANA COASTAL RESTORATIONS PLANS

Hurricane Katrina (2005) was the fourth most powerful storm to strike Louisiana since 1893 with respect to maximum wind speed and surge elevations at landfall. As Katrina progressed across Breton Sound and Lake Borgne, it generated storm surge of 8.5 m, and 5.5 m wave run-up (total 14 m) on the Mississippi coast, and a 6 m surge southeast of New Orleans, and up to 2 m of additional wave run-up, a total of 8 m. In addition, Katrina dropped 20–30 cm of rain across southeastern Louisiana. In southeastern Louisiana, communities unprotected by levees were flooded; the surge overtopped and destroyed levees protecting eastern New Orleans as well as parishes (*i.e.*, counties) to the south and east, flooding the city and killing an estimated 1400 people. Floodwalls failed along drainage and navigation canals connected to Lakes Pontchartrain and Borgne with significant coastal erosion and wetland loss.

In response to Hurricane Katrina, the Louisiana Coastal Protection and Restoration (LACPR) technical report was prepared by [15]. In collaboration with the State, the USACE developed and analyzed a range of alternatives, based on a number of structural, nonstructural, and coastal restoration measures, to reduce storm surge risk in south Louisiana. As a representation of Category 5 storm risk reduction, the report presents alternatives for the 100-year, 400-year, and 1000-year design levels. The 400-year flood event is an approximation of Hurricane Katrina. The philosophy of integrated defense for reducing risk from hurricane surge was used based on a “Multiple Lines of Defense” strategy that no single measure or approach will be sufficient for achieving the risk reduction objectives. This integrated defense solution has been the Hurricane and Storm Damage Risk Reduction System (HSDRRS), with a budget of $14.5 billon. It includes five parishes and consists of 560 km of levees and floodwalls, 73 non-federal pumping stations, three canal closure structures with pumps, and four gated outlets. One major structural component of the system is the Inner Harbor Navigation Canal—Lake Borgne Surge Barrier (IHNC-LBSB) [16,17,18].

7.1. IHNC Lake Borgne Storm Surge Barrier—Sea Level, Subsidence and Barrier Design Height

The 2.9 km-long IHNC-LBSB is located at the confluence of the Gulf Intracoastal Waterway (GIWW) and the Mississippi River Gulf Outlet (MRGO), about 19 km east of New Orleans (Figure 7). The surge barrier, completed in 2011, works in tandem with the Seabrook Floodgate Complex constructed at the north end of the IHNC at Lake Pontchartrain, (Figure 8) [18,19].

The IHNC-LBSB has three gates that allow vessel passage through the barrier and provides a complete closure of the MRGO navigation channel.

The surge barrier design was based on detailed modeling of wind, surge, waves and rainfall for 152 synthetic storms that were selected to span the historical record of storms that have impacted southeastern Louisiana. Responses from this limited storm set were then interpolated to a much larger storm set to represent the full range of possible storm conditions and to enable recurrence analysis. The final 8 m design height (selected to minimize the impact of a 100-year, 1% chance annual occurrence event and provide resilience to a 500-year, 0.2% chance annual occurrence event) included 0.3 m of LSLR over the 50-year design life [20]. Given the magnitude of the IHNC-LBSB project, the significance of the urban area that is being protected, the global sea level rise predicted by [15], (*i.e.*, [15] Table 4.1, 0.64–1 m by 2100) and the substantial subsidence that has been observed in parts of southern Louisiana, (e.g., Figure 9), one might question why a more conservative design height was not selected (e.g., ~10 m) that would minimize the effects of a 500-year event over a design lifetime reaching until 2100. An alternative means of increasing the design capacity of the IHNC-LBSB is to expand the storage capacity for water that overtops the barrier. Currently there is a modest storage volume inside the IHNC for water to accumulate before exceeding the IHNC floodwalls if it overtops the IHNC-LBSB barrier. This storage volume is critical to providing system resilience to the 500-year event. However, the elevated water levels in the IHNC would bring any vessel that remained inside and broke its mooring to a level where they could directly impact and damage the IHNC floodwalls. The IHNC storage volume could be expanded by more than ten-fold, and the water level minimized, if water is allowed to flow through the Bayou Bienvenue Sector Gate (Figure 8) and enter a marsh area called the Central Wetlands (Figure 8). While the Central Wetlands are actually located inside the HSDRRS, the inhabited areas of Orleans and St. Bernard Parishes are protected by an interior "back" levee system, as shown in Figure 8. We strongly encourage investigation of this option as a critical, long term means of lowering the risk of overtopping or failure of the floodwalls in the IHNC and another catastrophic flooding of New Orleans.

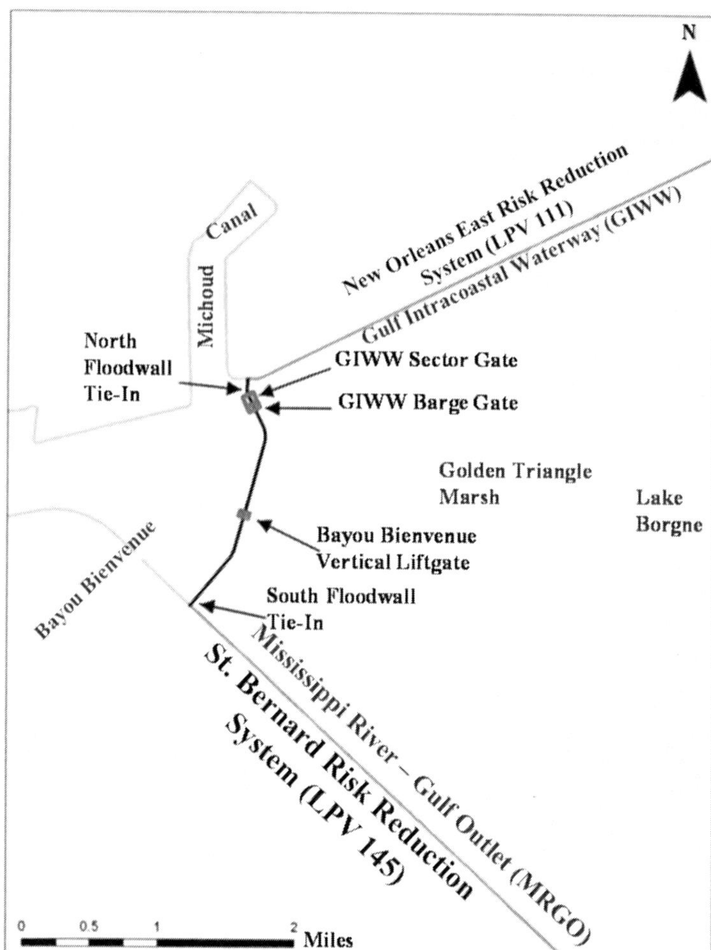

Figure 7. Map from open source of the Lake Borgne surge barrier [15].

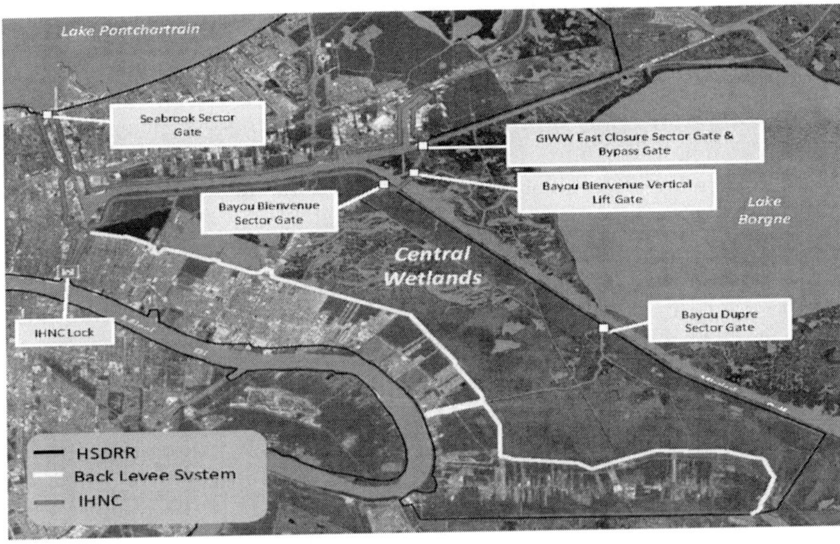

Figure 8. Location map of the IHNC-LBSB, IHNC floodwalls and the Central Wetlands.

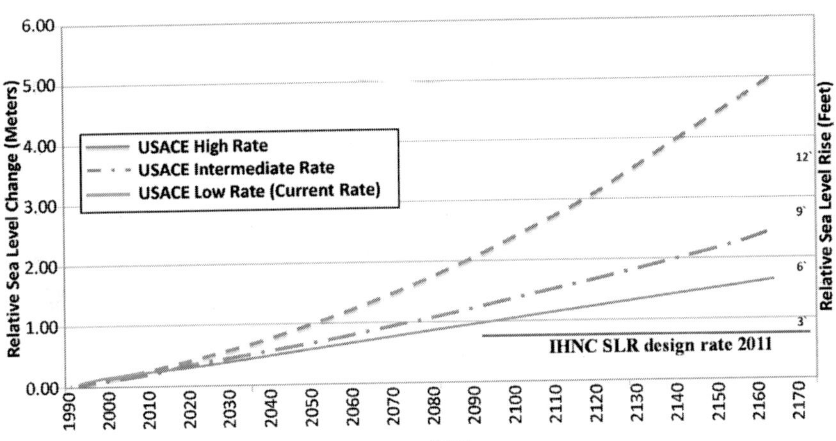

Figure 9. USACE projected sea-level rise curves for Grand Isle, Louisiana. These curves include eustatic (global) sea-level rise values and subsidence rates [7].

7.2. West Closure Pump Station

On the opposite side of the Mississippi River, lies the other main perimeter defense against flooding from storm surge (Figure 10). The West Closure Complex must operate in a fail-safe mode whenever severe storms threaten New Orleans, to prevent interior flooding of the west basin. When the navigable sector gate is closed to block storm surges, catastrophic flooding can be mitigated by pump station operation. Similar concerns to those raised for the Lake Borgne surge barrier exist about the design parameters (including the design storm recurrence and the design lifetime) and the accompanying adequacy of LSLR (subsidence and sea-level rise) assumptions in the design of this facility. In this case, the situation is further exacerbated by the potential impact to pumping operations if the closure is overtopped by surge and waves or experiences an extended power outage that is likely to accompany such an event.

Figure 10. Image of the GIWW—West Closure Pump station.

7.3. Louisiana Comprehensive Master Plan 2012

Over the past 30 years, a series of plans for addressing land loss have been developed, the LACPR is described above, but the most recent and ambitious is the Comprehensive Master Plan completed in 2012 [21]. This plan builds on earlier plans and discusses the importance of trying to deal with land loss by means of a variety of structural, nonstructural, and even "voluntary acquisition". The time frame is limited to 50 years into the future with a projected budget of $50+ billion. Restoration efforts aim at reconnecting the Mississippi River to the deltaic plain. While the objective of the plan is to ultimately have a healthy and sustainable coast, the lack of discussion of conditions likely to prevail beyond the next 50 years is an unfortunate omission.

Future changes to the delta plain over just the next 50 years were addressed in maps and figures showing predicted change, and rather than using the usual terms for sea-level rise projections due to global warming, the plan instead used "moderate environmental scenarios" for the current situation and "less optimistic environmental scenarios" using the GSLR reported values of 0.12 to 0.65 m by 2050. But there is no discussion of climate change and its impacts as the central driver of the less optimistic environmental conditions. Land subsidence, a large component of LSLR (~75%), is highly variable across the delta plain (~2–35 mm/year) and was factored into the reported scenarios. However, discussion is lacking on the natural geologic forces and man-made causes of subsidence such as oil, gas, and ground-water extraction. Since LSLR is already 0.9 m over parts of the coast over the past century [22], the current global average is 3.2 mm/year, and recent climate literature projects GSLR to be 2 m+ by 2100, it appears the CPRA's plan scenarios understate the risk to New Orleans and the delta plain due to continued rapid land subsidence and projected sea-level rise to 2100 and for centuries into the future. More realistic scenario values of LSLR for planning and design might be 0.5 to 1 m by 2050 and 3 m by 2100.

8. HURRICANE SANDY AND NEW YORK CITY COASTAL PROTECTION PLANS

The New York City (NYC) Metropolitan area is the largest of the 25 most densely populated areas in the east coast of the U.S., out of which 23 are highly vulnerable to storms and sea-level rise. Hurricane Sandy made landfall along the New Jersey coast on 29 October 2012. Sandy was the largest extra-tropical cyclone in recorded history for the north Atlantic region. It made landfall near high tide with maximum surge elevation of 4.4 m (NOAA Battery tide gauge station), while the average tide range is 1.4 m. Deaths totaled 72 people ; total damages were ~$50 Billion and New York City sustained $19 Billion in damages, mostly due to surge flooding and fires. Damages were also significant for the northern New Jersey region and extended as far east as Cape Cod, Massachusetts. While Hurricane Sandy was not "caused" by climate change, storm surge elevations were enhanced by LSLR over the past century, which is partly an impact of climate warming. The Battery gauge shows 28 cm of relative rise over the past century and of that total about 12.2 cm is due to land subsidence as a result of rebound of Earth's crust. Moreover, the Sandy Hook, New Jersey gauge located across Lower New York Harbor has even higher rates of 39 cm LSLR over the past century, of which ~22.7 cm is due to subsidence [22].

Integrated elements of plans to rebuild after Sandy include elevating structures, nourishing barrier islands and dunes, and flood-proofing major infrastructure. Another long-term infrastructure adaptation measure considered was installation of storm surge barriers across vulnerable openings to the sea, including the Verrazano Narrows, upper East River and the Arthur Kill channel.

New York City Climate Change Plans for Adaptation

Following Sandy, NYC undertook comprehensive studies of the effects of Sandy on the coast as well as the urban infrastructure and developed plans for making the city more resilient to future storms and the increasing impacts of climate warming [23]. The plans were intermediate in time span, looking to year 2080 (Figure 11). For projecting sea-level rise, two model scenarios were employed. One, using methods from the recent IPCC assessment resulted in GSLR projections of 20.3–58.4 cm by 2080. A second method that included local subsidence and uncertainty values, more rapid rates of ice sheet melt for Greenland and Antarctica and greater atmospheric GHG emissions resulted in LSLR projections of 1.0–1.4 m by 2080, which is in close agreement with [6]. Beyond 2080, GSLR very likely will continue, possibly at even higher rates. Levels of 3 m or higher are possible over the next several hundred years if current high GHG emissions, ice sheet melting and global warming continue.

For many recent U.S. coastal protection plans, sea level change is factored in as a critical element from response to Hurricane Sandy, to the North Atlantic Coast Comprehensive Study, to analyses for projects currently in planning and engineering design. The planning process by NYC will be refined as climate science and predictive tools improve. New Jersey also sustained major flood damage and loss of life from Sandy and restoration is underway, mostly on barrier islands, but except for some floodplain "buy-outs", there appears to be less planning for future conditions and adaptation. Integrated elements of plans to rebuild after Sandy include elevating structures, nourishing barrier islands and dunes, and flood-proofing major infrastructure.

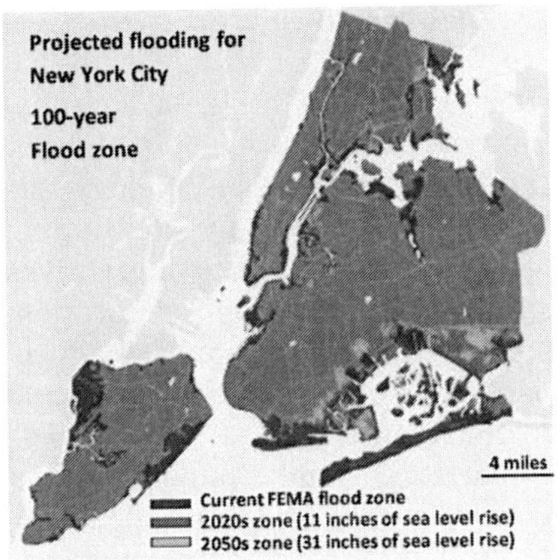

(a)

(b)

Figure 11. (**a**) NYC projected flooded areas for 100-year storm, SLR 79 cm by 2050 [23]. (**b**) Proposed surge barriers [24] and surge barrier with gates [24].

For New York, conceptual design of storm surge barriers was presented by Arcadis [24] across the vulnerable openings to the sea including the Verrazano Narrows, upper East River and Arthur Kill. Due to requirements of maintaining ship navigation, tidal exchange, and river discharge, barrier gates were proposed by Halcrow [24] within a surge barrier to be placed across Ambrose channel to accommodate passage of navigating vessels (Figure 9b). Due to high cost and complexity of such large structures, surge barriers are not in current plans.

9. CONCLUSIONS AND RECOMMENDATIONS

Review of response and planning of coastal projects following major storms in Egypt, Louisiana, and the New York City region are presented. The review considers varying degrees of climate change impacts projected for the next 50–100 years. To reduce risk and vulnerability, [3,8] recommended that adaptation planning using projected GSLR of 0.5–2 m by 2100, plus local geophysical and man-made factors, is advisable. Planning should fit the time frame and degree of protection desired. These boundary conditions demand that the protection of coastal areas requires flexible designs that can be adapted in a phased manner to increasing sea-level rise and increased storm activity over the next 50 to 100 years and beyond [7]. The reality is that it is not economically feasible to protect all coastlines from high rates of sea-level rise and catastrophicsuper-storms. Structures built for short term (~50 years) and modest sea-level rise (~0.5 m) should be designed such that their height can be raised and size increased in the event that even higher sea levels prevail. New planning protocols need to employ cost-effective coastal protection technologies affordable by developing countries. Engineering plans and design need to incorporate alternatives such as beach nourishment, dune stabilization, and wetland restoration, which could serve as buffers protecting mainland areas from full effects of storm surge and wave action as well as provide environmental benefits. High-risk urban areas will likely require hard structures (e.g., seawalls, revetments, and flood barriers) and pump station infrastructure. Pump operations, however, need to be assured during extended storms. Strategic development of new urban planning policies, including relocation, should be considered when risk is great and in-place adaptation is no longer economically feasible.

ACKNOWLEDGMENTS

We would like to thank three anonymous reviewers and the editor for their constructive and helpful edits and suggestions that improved the quality of the paper.

AUTHOR CONTRIBUTIONS

S.J.W. wrote parts of the manuscript on climate change, sea-level rise impacts, Louisiana management plans, and parts of the New York City case study. N. I. had the lead writing on Egypt and parts on the Louisiana Lake Borgne Surge Barrier project.

CONFLICTS OF INTEREST

The authors declare no conflicts of interest.

REFERENCES

1. IPCC. Contribution of Working Group II to the Fourth Assessment Report of Intergovernmental Panel on Climate Change 2007; Parry, M.L., Canziani, O.F.,

Palutikof, J.P., van der Linden, P.J., Hanson, C.E., Eds.; Cambridge University Press: Cambridge, UK; New York, NY, USA, 2007; p. 976.

2. Sweet, W.V.; Park, J. From the extreme to the mean: Acceleration and tipping points of coastal inundation from sea-level rise. Earth's Future 2014, 2, 579–600.
3. Williams, S.J. Sea-level rise implications for coastal regions. In Understanding and Predicting Change in the Coastal Ecosystems of the Northern Gulf of Mexico; Brock, J.C., Barras, J.A., Williams, S.J., Eds.; Coastal Education and Research Foundation, Inc.: Coconut Creek, FL, USA, 2013; pp. 184–196.
4. Parris, A.; Bromirskiet, P.; Burkett, V.; Cayan, D.; Culver, M.; Hall, J.; Horton, R.; Knuuti, K.; Moss, R.; Obeysekera, J.; et al. Global Sea-Level Rise Scenarios for the U.S. National Climate Assessment; NOAA Tech Memo OAR CPO-1; Climate Program Office (CPO): Silver Spring, MD, USA, 2012; p. 37.
5. IPCC. Summary for Policy Makers, Climate Change: 5th Assessment Report of the Intergovernmental Panel on Climate Change 2013; Stocker, T.F., Qin, D., Plattner, G.-K., Tignor, M.M.B., Allen, S.K., Boschung, J., Nauels, A., Xia, Y., Bex, V., Midgley, P.M., Eds.; Cambridge University Press: Cambridge, UK; p. 28.
6. Kopp, R.E.; Horton, R.M.; Little, C.M.; Mitrovica, J.X.; Oppenheimer, M.; Rasmussen, D.J.; Strauss, B.H.; Tebaldi, C. Probabilistic 21st and 22nd century sea-level projections at a global network of tide-gauge sites. Earth's Future 2014, 2, 383–406.
7. U.S. Army Corps of Engineers (USACE). Procedures to Evaluate Sea Level Change: Impacts, Responses, and Adaptation, ETL 1100-2-1; USACE: Washington, DC, USA, 2014; p. 254.
8. Moser, S.C.; Williams, S.J.; Boesch, D.F. Wicked Challenges at Land's End: Managing Coastal Vulnerability under Climate Change. Annual. Rev. Environ. Res. 2012, 37, 51–78.
9. Rignot, E.; Mouginot, J.; Morlighem, M.; Seroussi, H.; Scheuchl, B. Widespread, rapid grounding line retreat of Pine Island, Thwaites, Smith, and Kohler glaciers, West Antarctica, from 1992 to 2011. Geophys. Res. Lett. 2014, 41, 3502–3509.
10. Williams, S.J.; Kulp, M.; Penland, S.; Kindinger, J.L.; Flocks, J.G. Mississippi River Delta Plain, Louisiana coast and inner shelf: Holocene geologic framework and processes, and resources. In Gulf of Mexico Origin, Water and Biota: Vol. 3, Geology; Buster, N.A., Holmes, C.W., Eds.; Texas A & M University Press: College Station, TX, USA, 2011; pp. 175–193.
11. Ismail, N.; Iskander, M.; El-Sayed, W. Assessment of coastal flooding at southern Mediterranean with global outlook for lowland coastal zones. In Proceeding of International Conference on Coastal Engineering, ASCE, Santander, Spain, 1–6 July 2012.
12. Ismail, N.M. Effect of wave-current interaction on littoral drifts. J. Am. Shore Beach Preserv. Assoc. 1982, 50, 35–38.
13. Ismail, N.M.; Wiegel, R.L. Opposing waves effect on momentum jets spreading rate. J. Waterw. Port. Coast. Ocean. Div.1983, 109, 465–483.
14. Kadib, A.L.; Shak, A.T.; Mazen, A.; Nadar, M.K. Shore protection plan for the Nile Delta coastline. In Proceeding of 20th International Conference on Coastal Engineering, Taipei, Taiwan, 9–14 November 1986; Volume III, pp. 2530–2544.
15. U.S. Army Corps of Engineers (USACE). Louisiana Coastal Protection and Restoration (LACPR), Final Technical Report; USACE: New Orleans District, MVD, USA, 2009; p. 293.
16. Tetra Tech. Inner Harbor Navigation Canal (IHNC) Hurricane Protection Barrier. Avaiable online: http://www.incainc.com/flood-control/inner-harbor-navigation-canal-ihnc-hurricane-protection-barrier.html (accessed on 20 January 2015).
17. U.S. Army Corps of Engineers (USACE). Climate change adaptation plan. Available online: http://www.iwr.usace.army.mil/Media/NewsStories/tabid/11418/Article/494304/procedures-to-evaluate-sea-level-change-impacts-responses-and-adaptation.aspx (accessed on 15 April 2015).

18. Huntsman, S.R. Design and construction of the lake Borgne surge barrier in response to Hurricane Katrina. In Proceeding of Conference on Coastal Engineering Practice 2011, San Diego, CA, USA, 21–24 August 2011.
19. DeSoto-Duncan, A.; Hess, C.; O'Sullivan, M. "The Great Wall of Louisiana" Protecting the Coastline from Extreme Storm Surge and Sea Level Rise. In Proceeding of the 2011 Solutions to Coastal Disasters Conference, Anchorage, AK, USA, 26–29 June 2011; pp. 690–701.
20. Grieshaber, J.; USACE/Hurricane Protection Office. Inner Harbor Navigation Canal (IHNC) Basin 1% (100-year) and 0.2% (500-year) Surge and Wave Event Water Levels; SE Louisiana Flood Protection Authority: New Orleans, LA, USA, 2010.
21. Coastal Protection and Restoration Authority of Louisiana (CPRA). Louisiana's Comprehensive Master Plan for a Sustainable Coast; Coastal Protection and Restoration Authority of Louisiana: Baton Rouge, LA, USA, 2012; p. 190.
22. Zervas, C.; Gill, S.; Sweet, W. Estimating Vertical Land Motion from Long-Term Tide Gauge Records; Tech Report NOS CO-OPS 065; NOAA: Silver Spring, MD, USA, 2013; p. 22.
23. NYCPCC. Climate risk information 2013: Observations, climate change projections, and maps. In New York City Panel on Climate Change; Rosenzweig, C., Solecki, W., Eds.; NYCPCC: New York, NY, USA, 2013; p. 38.
24. Hill, D.; Bowman, M.J.; Khinda, J.S. Storm surge barriers to protect New York against the deluge. In Proceeding of COPRI Committee Report, ASCE Conference, Brooklyn, NY, USA, 30–31 March 2009.

CHAPTER 4

The Impact of Uncertainties in Ice Sheet Dynamics on Sea-Level Allowances at Tide Gauge Locations

Aimée B. A. Slangen [1,2,*], Roderik S. W. van de Wal [2], Thomas J. Reerink [2], Renske C. de Winter [3], John R. Hunter [4], Philip L. Woodworth [5] and Tamsin Edwards [6]

[1]*Royal Netherlands Institute for Sea Research (NIOZ), Department of Estuarine and Delta Systems (EDS), Utrecht University, Korringaweg 7, Yerseke 4401 NT, The Netherlands*

[2]*Institute for Marine and Atmospheric research Utrecht (IMAU), Princetonplein 5, Utrecht 3584 CC, The Netherlands*

[3]*Department of Physical Geography, Utrecht University, Heidelberglaan 2, Utrecht 3584 CS, The Netherlands*

[4]*Antarctic Climate & Ecosystems Cooperative Research Centre, 20 Castray Esplanade, Hobart TAS 7000, Australia*

[5]*National Oceanography Centre, 6 Brownlow Street, Liverpool L3 5DA, UK*

[6]*School of Environment, Earth & Ecosystem Sciences, Faculty of Science, Technology, Engineering & Mathematics, The Open University, Milton Keynes MK7 6AA, UK*

ABSTRACT

Sea level is projected to rise in the coming centuries as a result of a changing climate. One of the major uncertainties is the projected contribution of the ice sheets in Greenland and Antarctica to sea-level rise (SLR). Here, we study the impact of different shapes of uncertainty distributions of the ice sheets on so-called sea-level allowances. An allowance indicates the height a coastal structure needs to be elevated to keep the same frequency and likelihood of sea-level extremes under a projected amount of mean SLR. Allowances are always larger than the projected SLR. Their magnitude depends on several factors, such as projection uncertainty and the typical variability of the extreme events at a location. Our results show that allowances increase significantly for ice sheet dynamics' uncertainty distributions that are more skewed (more than twice, compared to Gaussian uncertainty distributions), due to the increased probability of a much larger ice sheet contribution to SLR. The allowances are largest in regions where a relatively small observed variability in the extremes is paired with relatively large magnitude and/or large uncertainty in the projected SLR, typically around the equator. Under the RCP8.5 (Representative Concentration Pathway) projections of SLR, the likelihood of extremes increases more than a factor 104 at more than 50–87% of the tide gauges.

Keywords: sea-level rise; allowances; sea-level extremes

1. INTRODUCTION

Sea level is projected to rise in the coming century and beyond as a result of a warming climate, with major contributions from warming of the ocean water and increasing land ice mass loss [1]. To protect against or mitigate the effects of sea-level rise (SLR) in highly populated coastal zones, it is important to understand the processes causing SLR, and to reduce the uncertainties for each of the projected contributions. Sea-level rise here is *relative* sea-level change, i.e., the difference as measured by tide gauges attached to the Earth's surface. Presently, one of the main uncertainties in SLR is the contribution of the Antarctic and Greenland ice sheets [2,3]. Particularly the timing and magnitude of dynamical changes associated with the ice sheets (e.g., marine ice sheet instability) are still uncertain. This is therefore an active area of research.

In the Fifth Assessment Report (AR5) of the Intergovernmental Panel on Climate Change (IPCC), the projected ice dynamical contribution to SLR was based on observations and a range of scenarios from models ([1], Table 13.5). Around the same time, Bamber and Aspinall [4] presented an estimate based on expert judgement, which had a strongly skewed uncertainty distribution rather than the symmetric distribution presented in IPCC. With the same data but a different way to process the expert opinions, De Vries and Van de Wal [5] presented an estimate with a lower high-end estimate than Bamber and Aspinall [4], though still a strongly skewed uncertainty distribution. Although expert elicitations document the level of consensus on a certain topic, they are sensitive to, for instance, the level of expertise of the interviewees and the way experts are selected. Therefore, process-based estimates are generally a preferable method. Until recently, process-based estimates were only available for specific parts of the ice sheets (e.g., Pine Island Glacier [6] or Thwaites Glacier [7]). However, Ritz et al. [8] recently presented an estimate for the Antarctic ice dynamical contribution to SLR in the 21st century using a process-based approach. They too found a non-normal distribution of the uncertainties, though less skewed than the De Vries and Van de Wal [5] distribution.

In this study, we will compare and contrast three different types of uncertainty distributions for the mass loss contribution to SLR due to ice dynamics [1,5,8]. We will specifically look at the effect on sea-level allowances rather than focusing on the sea-level projections. A sea-level allowance indicates the height a coastal structure needs to be elevated to maintain the same frequency and likelihood of extreme sea-level events under future SLR [9,10]. An allowance gives more information than just a 90% or 95% confidence level of projected mean SLR because it also accounts for the local variability in the sea-level extremes. An important assumption for the allowances in general is that the variability in the extremes does not change under the influence of climate change. For the 20th century, this is a reasonable assumption, as Menéndez and Woodworth [11] found that the historical change in extremes is mostly due to changes in mean sea level rather than changes in the storminess contributions to the extremes. For the 21st century, model projections of changes in the wave climate and storm surges are subject to large uncertainties, and it is therefore difficult to say if the assumption will hold in the future [1].

We use the regional relative sea-level projections of Slangen et al. [12], which describe the change of the ocean surface *relative* to the ocean floor. The projections follow the high-end RCP8.5 (Representative Concentration Pathway [13]) climate change scenario based on CMIP5 model output (5th phase of the Climate Model Intercomparison Project [14]). While these projections assume a normal (Gaussian) uncertainty distribution, more recent publications have incorporated skewed ice sheet dynamics distributions into their projections to provide a more complete uncertainty distribution (e.g., [15,16,17,18]) which can be used for risk-assessment and policy purposes. We follow this line and combine the Slangen et al. [12] sea-level projections with three different ice sheet dynamics scenarios (Section 2.1) using the framework of De Winter et al. [18]. Whereas De Winter et al. [18] discussed the impact and sensitivity of the high-end regional sea-level projections to different ice sheet uncertainty distributions, here we focus on the effect of the shape of the skewed ice sheet uncertainty distributions on the projected allowances. De Winter et al. [18] showed that the 90th, 95th and 97.5th percentiles of regional SLR projections are highly sensitive to the uncertainty distribution of ice sheet dynamics. In contrast to De Winter et al. [18], all the ice sheet dynamics scenarios will be shifted to the same median (as described in Section 2.1) to focus on the effects of the different shapes rather than differences in their magnitude. Previously, regional allowances were mainly computed using normal (Gaussian) uncertainty distributions for the projected SLR [10]. Buchanan et al. [19] used full probability distributions in combination with allowances to compute an "Average Annual Design-Life Level" for the United States coastline. Here, we present allowances for skewed uncertainty distributions for tide gauge stations across the world and compare the different shapes of the distributions (Section 2.3).

There are various ways to express sea-level extreme events statistically, depending on the application and the quality of the available data. Methods such as fitting Generalised Extreme Value distributions (of which the Gumbel distribution is a special case) to annual maxima in water levels and fitting Generalised Pareto distributions to peaks over a threshold [20] are often used to represent the statistics of extreme water levels. There is no consensus as to a universal best approach [21], and the method chosen depends on the purpose of the analysis and the data availability and quality. Here, we use a Gumbel parameterisation, which will allow us to compute the allowances following Hunter [9] and Hunter et al. [10]. We use the GESLA-2 tide gauge dataset (Global Extreme Sea Level Analysis Version 2 [22,23], Section 2.2), which significantly increases the number of locations with respect to previous studies of allowances [9,10]. Finally, we show the increase in the frequency of sea level extremes under the RCP8.5 scenario if the allowances are not applied (Section 3.3).

2. DATA AND METHODOLOGY

2.1. Uncertainty Distributions of Sea-Level Change

We use the regional sea-level projections for the RCP8.5 scenario based on CMIP5 model output as presented in Slangen et al. [12], for the period 2010–2100. These projections are similar to the results presented in IPCC AR5 [1] and include the contributions of steric/dynamic change, glacier and ice sheet mass change, groundwater extraction and Glacial Isostatic Adjustment. The global mean projected change for 2010–2100 is 0.72 ± 0.28 m, of which the Antarctic dynamics mass loss contribution is 0.08 ± 0.06 m and the Greenland dynamics mass loss contribution 0.05 ± 0.06 m. All uncertainty distributions in Slangen et al. [12] are assumed to be normal, with means and standard deviations as provided in their Table 1.

The two dynamical ice sheet contributions will be removed from the total projections of Slangen et al. [12], and replaced by three different scenarios for the ice sheet dynamically-driven mass loss. For clarity, this means that the ice sheet surface mass balance (SMB) contribution to SLR remains unaltered with respect to Slangen et al. [12]. The three scenarios used are:

(1) The IPCC scenario (Figure 1, green, from Church et al. [1]). We use the cumulative ice sheet dynamical contributions for the Antarctic Ice Sheet (AIS) and Greenland Ice Sheet (GRIS) as presented in IPCC AR5 Table 13.5 (0.05 ± 0.03 m for Greenland, 0.07 ± 0.08 for Antarctica), assuming a normal distribution. As Church et al. [1] present the change for 1986–2005 to 2081–2100, we correct the values linearly to estimate the change for 2010 to 2100. The Antarctic contribution is split into 11% for the East Antarctic Ice Sheet (EAIS) and 89% for the West Antarctic Ice Sheet (WAIS), which is the split for the year 2010 as presented in the IMBIE project (Ice sheet Mass Balance Inter-comparison Exercise, [24]).

(2) The VW15 scenario (Figure 1, blue) is from De Vries and Van de Wal [5]. We use the skewed uncertainty distribution of rates of SLR in 2100 for WAIS, EAIS and GRIS contributions, based on expert elicitation estimates. VW15 presented a reanalysis of the data in Bamber and Aspinall [4], which more rigorously accounts for the lack of consensus in the expert estimates. To convert mm/year in 2100 into cumulative SLR for 2010–2100, we assume a rate of change in 2010 of 0.6 mm/year for WAIS, 0.0 mm/year for EAIS and 0.3 mm/year for GRIS following VW15 and Bamber and Aspinall [4], and assume a linear change in the rate for the period 2010–2100.

(3) The R15 scenario for the Antarctic dynamical ice sheet contribution (Figure 1, black) is from Ritz et al. [8]. This is an observation and model-based estimate of the ice dynamical mass loss for 2010–2100. The shape of the Antarctic distributions differs slightly from the original R15 study due to the use of a narrower bandwidth (less smoothing) in the kernel density estimation. For Greenland dynamics, this scenario is the same as the IPCC scenario, as R15 did not calculate the Greenland contribution.

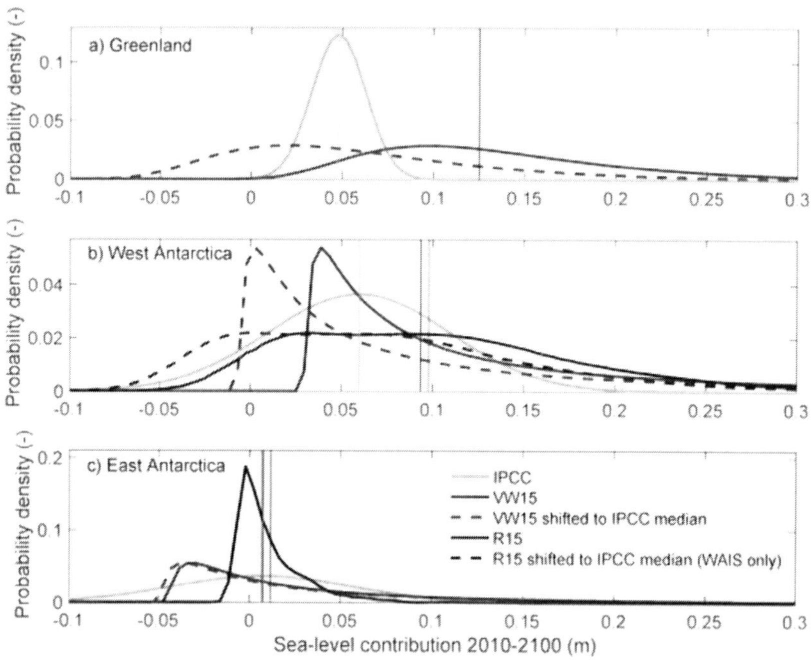

Figure 1. The three scenarios for ice sheet dynamics contributions to sea-level rise used in this study, for (**a**) Greenland, (**b**) West Antarctica and (**c**) East Antarctica (cumulative m sea-level change between 2010–2100). Original uncertainty distributions in solid lines, medians in vertical lines, shifted uncertainty distributions (where all medians match the IPCC medians) in dashed lines.

All the ice dynamics scenarios are shifted to match the median values of the IPCC scenario (Figure 1, dashed lines), in order to make sure that the differences in the allowances are only due to changes in the *shape* of the distribution rather than the *magnitude*. We use the median (i.e., the point where the cumulative density function is 0.5) for this shift, rather than the mean, as the median is less sensitive to extremely large values in the non-normal distributions. Although the published VW15 estimates are for SMB and ice sheet dynamics combined, we here assume that the skewed *shape* of the distribution is primarily a result of the uncertainty in the dynamical processes and correct for the differences in *magnitude* by the aforementioned shift in the median.

The regional SLR for each of the ice sheet dynamics contributions is computed by multiplying each contribution as presented in Figure 1 (Greenland, East Antarctica, West Antarctica) with its respective sea-level "fingerprint" ratio (e.g., [25]). These fingerprints are computed with a gravitationally consistent sea-level model. The resulting projected ice dynamical contribution to the relative regional SLR was shown in Slangen et al. [12] in their Figure 1e. Typically, sea level falls near the sources of ice mass loss, and rises above average in the far field. The individual fingerprint ratios of each ice sheet may be scaled with the global mean SLR contribution of each ice sheet if the location of the mass change remains the same as the location used to compute the fingerprint. This assumption is most significant in proximity to the ice sheet, where gravitational gradients decrease relatively sharply from the centres of mass loss.

We use the SEAWISE model [18] to combine the skewed SLR distributions for Greenland, West-Antarctica and East-Antarctica one by one with the projected non-ice sheet dynamics SLR from Slangen et al. [12] (for which a normal distribution is assumed) at each tide gauge location (Appendix, Figure A1). SEAWISE first scales the ice dynamic uncertainty distributions with the individual fingerprints to obtain the ice sheet uncertainty distribution at each location. Then, SEAWISE combines multiple SLR distributions, e.g., for West-Antarctica and Greenland (these can be normal or skewed). The model can combine independent distributions or assign a correlation coefficient to assume a dependency between distributions. Here, all distributions are assumed to be independent. Details of the SEAWISE methodology are presented in the Appendix and in De Winter et al. [18].

2.2. The Statistics of Sea-Level Extremes

In addition to the regional sea-level projections, we require location-specific information on the statistics of sea-level extremes resulting from tides, storm surges and other (high-frequency) sea-level variability. For this, we use Gumbel scale parameters. The Gumbel scale parameter for a certain location can be derived from historical records of high-frequency sea-level measurements (see Figure 2 for four examples). Here, we use data from the GESLA-2 database (Global Extreme Sea Level Analysis Version 2 [22,23]), which contains high-frequency records (at least hourly) from 1300 tide gauge stations as of February 2016. Gumbel distributions were fitted to the annual maxima (expressed relative to the linear trend in mean sea level over the record) of each of the 658 tide gauges, using the Matlab® *evfit* (Mathworks, Natick, MA, USA) function, as described in Hunter et al. [26]. In a comparison of the Gumbel scale parameters obtained from the *evfit* method to the *ismev* method as used in Hunter [9] (both methods are extensively described in Hunter et al. [26]), we found that the results are very similar to the extent required for the present analysis and we therefore use the values from the Matlab® *evfit* method only. We select all tide gauge records that contain at least 20 years of data and where each year with data is more than 75% complete (Figure 3), which yields a data set of 658 records. Following Hunter et al. [26], we reject four stations that show significant non-Gumbel behaviour, and also leave out two Hudson Bay tide gauges (Canada) as we do not have sea-level projections available there. This leaves 652 records, of which 448 records have ≥30 years of data available, 319 records ≥40 years, 164 records ≥50 years and 94 records ≥60 years.

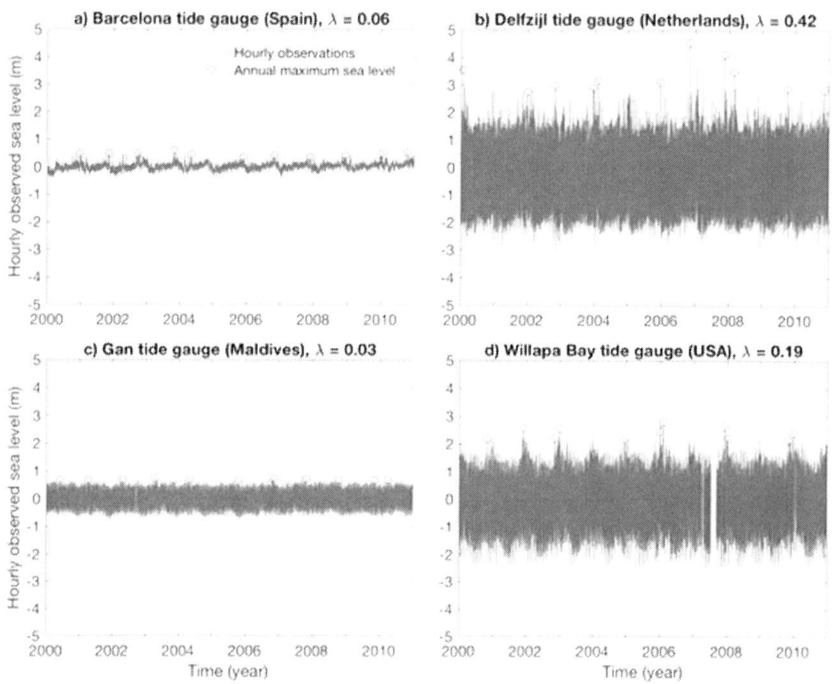

Figure 2. Four examples of tide gauge records: (**a**) Barcelona, (**b**) Delfzijl, (**c**) Gan, (**d**) Willapa Bay. Hourly data (in blue) and their annual maximum values (red) for 2000–2010 (m), where the mean of each time series has been removed. λ= the Gumbel scale parameter (m). Locations are indicated in Figure 5a by their first letter.

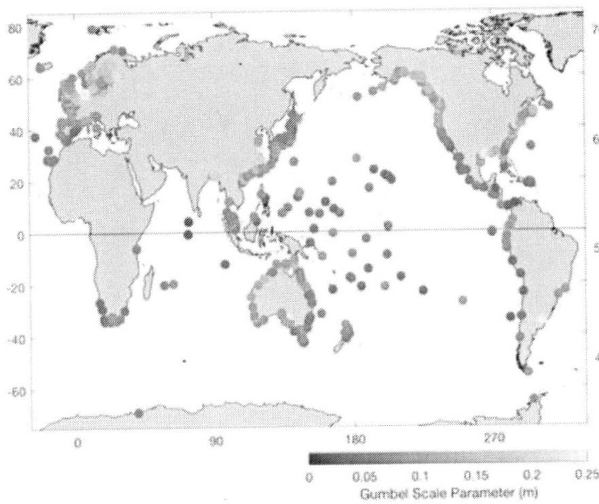

Figure 3. (**left**) Gumbel scale parameters (m) at 652 tide gauge stations and (**right**) a zoom on the European region.

The Gumbel scale parameter describes the variability of the extremes; i.e., a smaller parameter means that the *variability in the extremes* is small (e.g., Figure 2a,c) — this does not necessarily mean that the average *magnitudes* of the extremes are small. A larger scale parameter indicates larger variability in the extremes (e.g., Figure 2b,d). Sea-level extremes can be a combination of high tide and extreme storm surge, but also a medium tide and an extreme surge, or a high tide and a medium surge. Other factors such as interannual or seasonal variability can have significant effects as well: Barcelona (Figure 2a), for instance, has a strong seasonal cycle, leading to a more regular occurrence of the annual maximum than for instance in Delfzijl (Figure 2b). For a detailed discussion of contributing factors to extreme sea levels, we refer the reader to Merrifield et al. [27].

Historically, there have been more tide gauge stations in the Northern Hemisphere, resulting in a larger number of estimates of Gumbel scale parameters in the Northern Hemisphere than in the Southern Hemisphere. The Gumbel scale parameters are typically in the range of tenths of meters (0.12 ± 0.06 m, mean ± 1σ). Values are generally lower around the equator, increasing towards the mid-latitudes (specifically in the Northern Hemisphere), with the highest values between 40o and 70o N (Figure 3), where both the tidal range and the storm surge activity tend to be larger [27]. For instance, Gumbel scale parameters are high at the Dutch coast (e.g., Figure 2b, Delfzijl tide gauge), due to a large tidal range and a large variability in storm surge heights. This is the result of

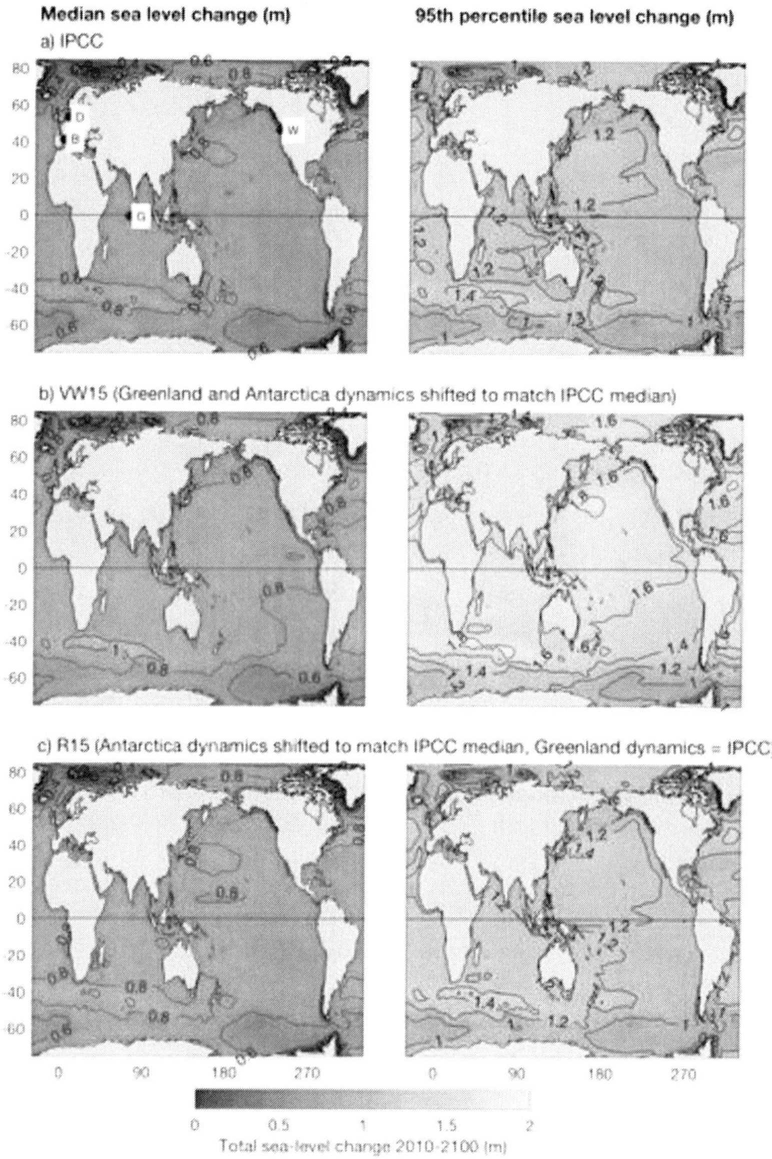

Figure 5. Projected cumulative sea-level change (m) for 2010–2100; median (left column) and 95th percentile (right column). All projections have the same median for the projected individual dynamical ice sheet contributions to SLR but a different shape of the uncertainty distribution. (**a**) IPCC; (**b**) VW15; (**c**) R15. The locations of the tide gauges in Figure 2 are indicated in (**a**) by their first letter.

the geometry of the semi-enclosed and relatively shallow North Sea basin, resulting in the development of larger storm surges under north western storm wind conditions, while the storm surge levels are considerably lower for other wind directions (e.g., [28]). This in contrast to the Barcelona tide gauge (Figure 2a), which has a much smaller tidal range since it is located in the Mediterranean and experiences less extreme storm surge heights.

2.3. Allowances Methodology

We assume a projected sea-level rise by an amount z′, where z′ has an uncertainty distribution function P(z′). If a coastal structure is raised by an allowance *a* to preserve the frequency of extreme (flooding) events in the future, *a* is given by

$$a = \lambda \ln\left(\int_{-\infty}^{\infty} P(z')\exp\left(\frac{z'}{\lambda}\right)dz'\right) \tag{1}$$

where λ is the scale parameter of the Gumbel distribution (Section 2.2), which describes the sea level extreme events and which is assumed to remain unchanged under sea-level rise. For the full derivation of the allowance, the reader is referred to Hunter [9] and Hunter et al. [10]. Note that, in these earlier papers, z′ was defined relative to a projected mean sea level, whereas here we use z′ as the total projected sea level to simplify the equations.

For a normal distribution, the allowance can be reduced to a simple analytical form:

$$a_{normal} = \mu + \frac{\sigma^2}{2\lambda} \tag{2}$$

(Equation viii in Supplementary Material of Hunter [9]), where μ is the mean SLR and σ the standard deviation of SLR when P(z′)follows a normal distribution.

However, Equation (2) is only valid for simple normal distributions of SLR. As we want to test different types of (non-normal) uncertainty distributions, the approach needs to be modified. We first redefine the skewed distributions of the projected SLR (from Section 2.1) in such a way that they can be used in the allowance computation. This is done by using a nonlinear fitting procedure (Powell's method, Press et al. [29]), which fits up to four normal distributions (see Figure 4 for an example) to optimally represent the (skewed) sea-level distribution. Then, the equation for the allowances only needs to be modified such that the allowance represents this group of normal distributions rather than just one normal distribution.

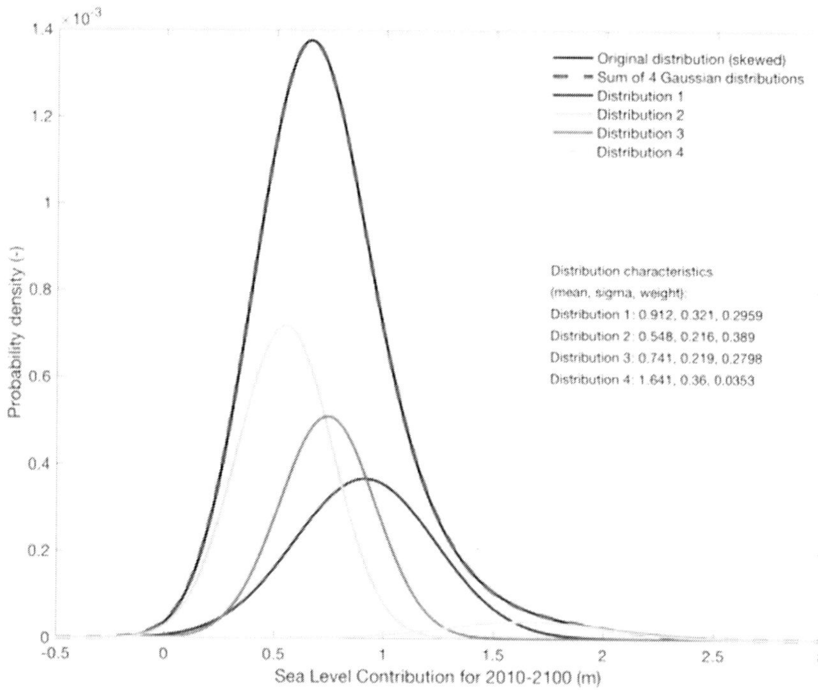

Figure 4. Example of a skewed sea-level uncertainty distribution (black), broken down into a set of four normal distributions (blue), which together describe the original distribution best (red dash).

We start from the description of a single normal distribution with a zero mean and standard deviation σ, where P(z′) is expressed as:

$$P(z') = \frac{1}{\sigma\sqrt{2\pi}} \exp\left(-\frac{(z')^2}{2\sigma^2}\right) \tag{3}$$

(Eq. ii from Supplementary Material of Hunter [9]). Then, the distribution P(z′) is modified such that it represents a group of normal distributions rather than one normal distribution:

$$P(z') = \frac{1}{\sqrt{2\pi}} \sum_{i=1}^{n} \frac{w_i}{\sigma_i} \exp\left(-\frac{(z'-\mu_i)^2}{2\sigma_i^2}\right) \tag{4}$$

where each distribution i in the group of normals has a weight wi, a standard deviation σi, and is centered on μi, such that P(z′) is described by a group of n normal distributions. The sum of all weights w equals 1, where individual weights may be negative. Now, Equation (4) is inserted in Equation (1):

$$\int_{-\infty}^{\infty} P(z')\exp\left(\frac{z'}{\lambda}\right)dz' = \sum_{i=1}^{n} w_i \exp\left(\frac{\mu_i}{\lambda} + \frac{\sigma_i^2}{2\lambda^2}\right) \tag{5}$$

This results in the following definition of the allowance for a group of normal distributions:

$$a_{skewed} = \lambda \ln\left(\sum_{i=1}^{n} w_i \exp\left(\frac{\mu_i}{\lambda} + \frac{\sigma_i^2}{2\lambda^2}\right)\right) \tag{6}$$

This equation will be used to compute allowances for non-normal distributions in the remainder of the paper. For one normal distribution (n = 1) with weight 1, Equation (6) reduces again to Equation (2).

3. RESULTS

3.1. Sea-Level Projections for 2010–2100

Sea-level projections are computed for each grid point of the RCP8.5 projections of Slangen et al. [12] in conjunction with the regional patterns of the three scenarios for ice sheet dynamics (Section 2.1), which leads to three sea-level projection scenarios (Figure 5). In the three scenarios, the median values (i.e., the point where the cumulative density function reaches 0.5) of the uncertainty distributions of WAIS, EAIS and GRIS are set to match the IPCC median values (Figure 1, dashed lines) in order to study only the effect of the change in the shape of the uncertainty distributions, rather than the shifts in magnitude. This is a different approach compared to the work presented in De Winter et al. [18], who did not shift the medians, as the purpose of that study was to look at the sensitivity of the high-end percentiles of regional sea-level projections to different ice sheet scenarios, considering both the shape and the magnitude of the distributions.

As a result of the differences in the shape of the ice sheet dynamics distributions, the final median SLR is not exactly the same for the three scenarios, even though the medians of the individual contributions are the same. Therefore, the VW15 scenario has a higher median (over the global ocean area) of 0.81 m compared to 0.71 m for IPCC and 0.73 m for R15 (Figure 5, left column). The maps show low or negative SLR around the glaciers and ice sheets. As a result of projected changes in the ocean heat content and circulation, there is a dipolar band associated with a shift in the Antarctic Circumpolar Current stretching south of South Africa and Australia [30] and generally higher values in a wide band around the equator in the Indian and Pacific Oceans.

The 95th percentile figures of the sea-level projections (Figure 5, right column) show significantly larger values for the VW15 scenario compared to the other two scenarios, reaching projected SLR over 1.75 m. The global mean 95th percentile of the VW15 scenario is 1.56 m, which is higher than the other two scenarios (1.16 m IPCC and 1.18 m R15). This is a result of the skewness of the VW15 dynamical ice sheet uncertainty distribution, which has a heavier tail that leads to much higher values for the 95th percentile of all three dynamical ice sheet contributions in the VW15 scenario (Figure 1). Although the R15 scenario has a higher 95th percentile contribution for the West-Antarctic ice dynamics (+0.12 m compared to IPCC) and a slightly longer tail to higher values, the shape of the distribution is less skewed than that of the VW15 scenario. The R15 East Antarctic 95th percentile contribution is smaller (−0.04 m compared to IPCC) due to the distribution being much sharper than the IPCC distribution. Combined, the 95th percentile total SLR of R15 is only marginally larger than that of IPCC (+0.02 m).

3.2. Allowances for Different Uncertainty Distributions

From the sea-level projections and the Gumbel scale parameters, we compute sea-level allowances for the year 2100 (with respect to 2010) as outlined in Section 2.3. The allowances can only be computed at tide gauge locations, where Gumbel scale parameters are available. The projected SLR for each location is taken from the regional sea-level maps using a nearest neighbour approach. In some regions with multiple tide gauge stations, this means that the same sea-level projection grid point will be used for different stations.

Similar to the 95th percentile sea-level projections (Figure 5), the allowances show a distinct effect from the skewed distribution of the VW15 scenario, and less so from the more mildly skewed R15 distribution (Figure 6, left column). On average, the VW15 allowances are more than twice as large as the allowances in the other two scenarios (2.19 m vs. 1.01 (IPCC) and 1.03 (R15) m). For the normal and close to normal distributions, it appears that the allowance is slightly less than the 95th percentile SLR at the tide gauge locations (which is 1.08 m for IPCC and 1.12 m for R15). For the skewed VW15 distribution, the allowance is much larger than the 95th percentile SLR at the tide gauges (1.50 m), due to the heavy upper tails of the uncertainty distributions of the dynamical ice sheet contributions.

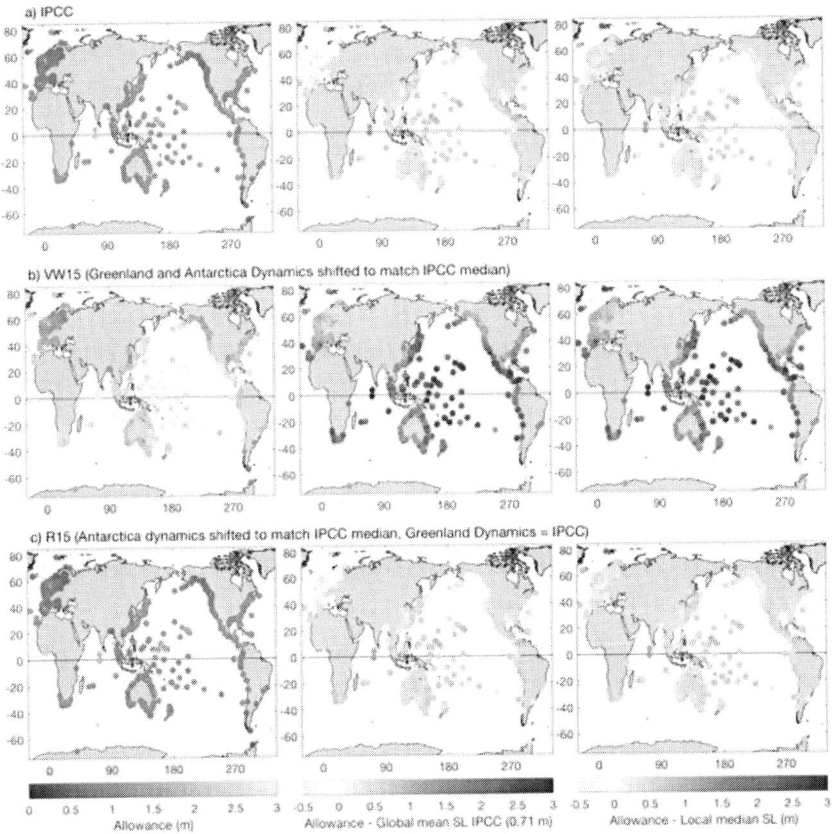

Figure 6. Allowances (m) for 2100 (left column), Allowances minus global mean SLR (middle column, m) and Allowances minus local SLR (right column, m), using 2010–2100 sea-level change projections, all projections have the same median for the projected dynamical ice sheet contributions to SLR but a different shape of the uncertainty distribution. (**a**) IPCC; (**b**) VW15; (**c**) R15.

When we compare the allowance to the global mean SLR (Figure 6, middle column) and to the local projected SLR (Figure 6, right column), this shows that an allowance is more than a simple addition of the change in the mean sea level: it accounts for the local distribution of extremes through the Gumbel scale parameter as well. In some regions, the projected local SLR is less than the global mean change in the IPCC and R15 scenarios, for instance in the Baltic Sea and the northwest of the United States. Although the allowance may be less than the *global mean* SLR in some places, it is always larger than the *local* SLR (compare middle to right columns, Figure 6). The largest differences between the projected SLR (global and regional) and the allowances are in and along the Central Pacific ocean basin, where Gumbel scale parameters (Figure 3) tend to be lower (indicating smaller variability in the extremes) and projected SLR and its uncertainty tend to be higher (due to large distance from ice mass loss sources and large steric/dynamic changes).

A focus on Europe (Figure 7) shows that the allowances in this region fall between 0.5 and 1 m for the IPCC and R15 scenarios, while the allowances are larger for the VW15 scenario, being at least 1 m and as much as 2.5 m. The allowances pattern generally is opposite to the Gumbel scale parameter (Figure 3): regions with higher Gumbel scale parameters tend to have lower allowances and vice versa. Due to Glacial Isostatic Adjustment (ongoing response of the solid Earth following ice sheet unloading after the Last Glacial Maximum [31]), the projected SLR is lower in the Baltic, leading to lower allowances in all three scenarios. Around the Mediterranean basin, projected SLR is substantial and the present sea-level variability is relatively small (small Gumbel scale parameters), which leads to the highest allowances in Europe (Figure 2a).

There is no immediately obvious relation between the Gumbel scale parameter and the projected mean SLR, apart from a tendency for locations with a smaller Gumbel scale parameter (<0.1 m) to be in locations with an average to above-average SLR projection, for all three scenarios (Figure 8). However, if the allowance is included (colours in Figure 8), a pattern emerges in the allowance that results from the combination of the projections and the scale parameters. It shows that tide gauges with smaller Gumbel scale parameters and larger projected SLR (uncertainties) require larger allowances to maintain the same likelihood of flooding (following Equation (6)).

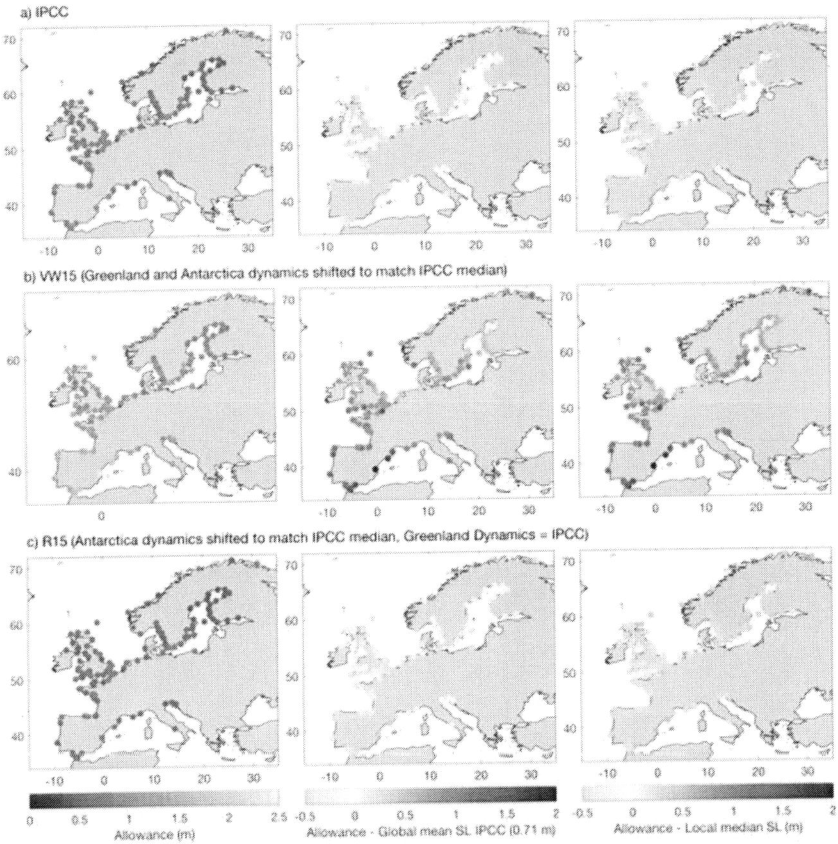

Figure 7. Same as Figure 6, but zoom on European region. Scenarios (**a**) IPCC; (**b**) VW15; (**c**) R15.

3.3. Changes in the Frequency of Extreme Events

The frequency of sea-level extremes at any given location will change when mean sea level changes. Looking at Figure 2, the same amount of SLR would have a larger effect on the number of extreme events in places with small variability (both in sense of total range of variability and in the small variability in the extremes, such as Barcelona and Gan, Figure 2a,c). In these places, even 0.5 m of SLR would bring the mean sea level outside the range of current sea-level variability. In places such as Delfzijl and Willapa Bay (Figure 2b,d), which have a larger range already, a mean SLR of 0.5 m would not be out of the ordinary range of variability. However, it would mean that the extremes, which are already of considerable magnitude, would become even larger.

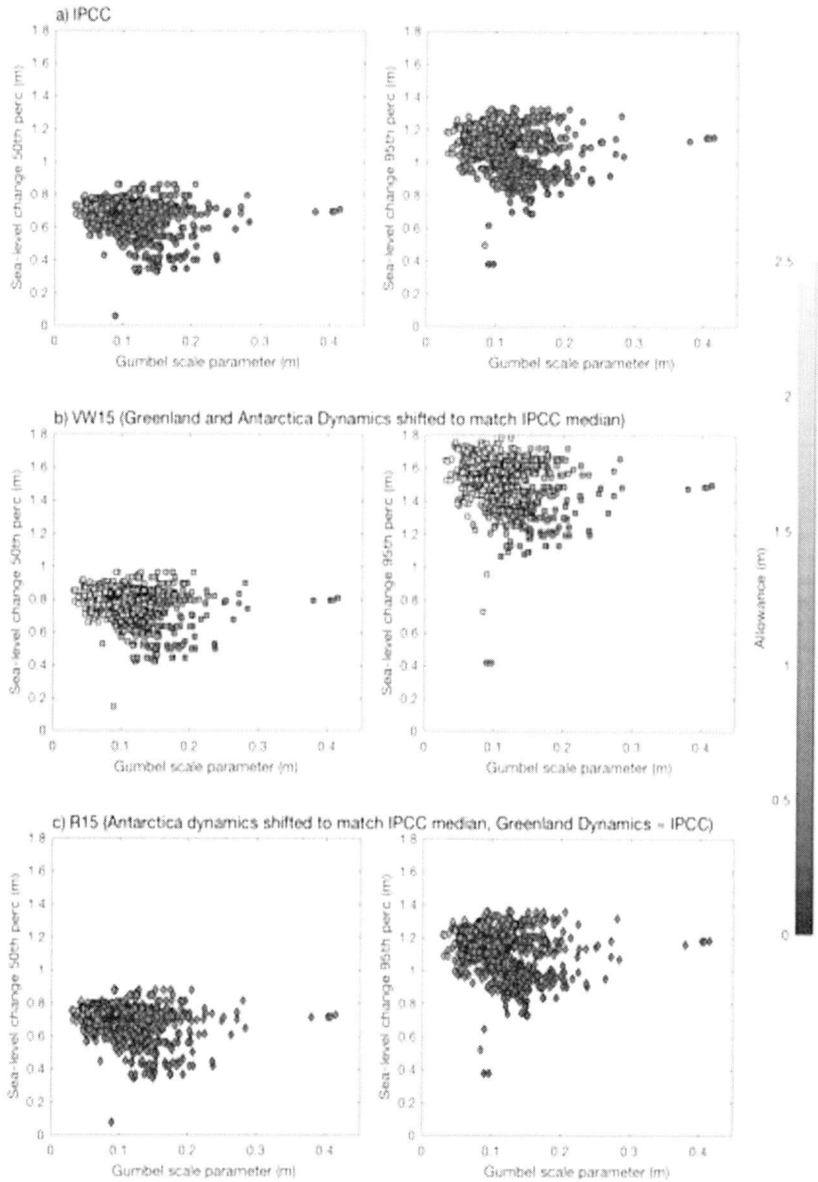

Figure 8. Scatter of 2010–2100 change in sea level (m) median (left column) and 95th percentile (right column) vs. the Gumbel scale parameter (m), colourscale indicating allowance (m), for the three scenarios (**a**) IPCC; (**b**) VW15; (**c**) R15.

The sea-level allowance (Equation (6)) is designed such that a location maintains the same frequency of extremes when the infrastructure is raised by the amount of the allowance. To quantify the effect of not applying the allowance (i.e., the infrastructure is not moved), we compute the change in the frequency of extremes *N* following

$$N = \exp\left(a/\lambda\right) \tag{7}$$

where *a* is the (skewed or normal) allowance and λ the Gumbel scale parameter. If the frequency of the extremes increases by a factor 500 this means that a 1 in 10,000 year extreme event will by 2100 occur once every 20 years. For a factor 5, the 1 in 10,000 year event would become a 1 in 2000 year event. It is therefore important to consider regional sea-level projections and their uncertainties, and sea-level extreme statistics, which all differ significantly between different locations.

The changing frequencies of extreme events for the IPCC and R15 scenarios (Figure 9), computed using Equation (7), show larger increases than in (Church et al. [1] their Figure 13.25), due to the use of the RCP8.5 scenario here rather than the RCP4.5 scenario in Figure 13.25b. Another difference with respect to Church et al. [1] is that there are more stations with updated time series included in Figure 9, which can lead to small variations in the Gumbel scale parameter λ. Since the allowances for the VW15 scenario are on average twice as large as the allowances in the other two scenarios due to the skewed uncertainty distribution, the increase in frequencies is even larger due to the exponent in Equation (7) (Figure 9, Table 1). Overall, the largest increases in frequencies are projected in the Equatorial Pacific and along the Southern and Eastern coasts of the U.S., where projected SLR and uncertainties are relatively large and Gumbel scale parameters are small. If Equation (7) results in ratios of more than 104, this means that a 1 in 10,000 year event would occur at least once every year by 2100, and a 1 in 100 year event would occur at least 100 times per year by 2100. Factors over 104 therefore effectively mean that a coastal structure, designed for any normal range of return periods (i.e., 100–10,000 years), would be lost unless the allowance is applied. For IPCC and R15, we find factors over 104 for 50% and 52% of the tide gauge locations, respectively, while, for VW15, this increases to 87% due to the skewed ice sheet contribution (Table 1).

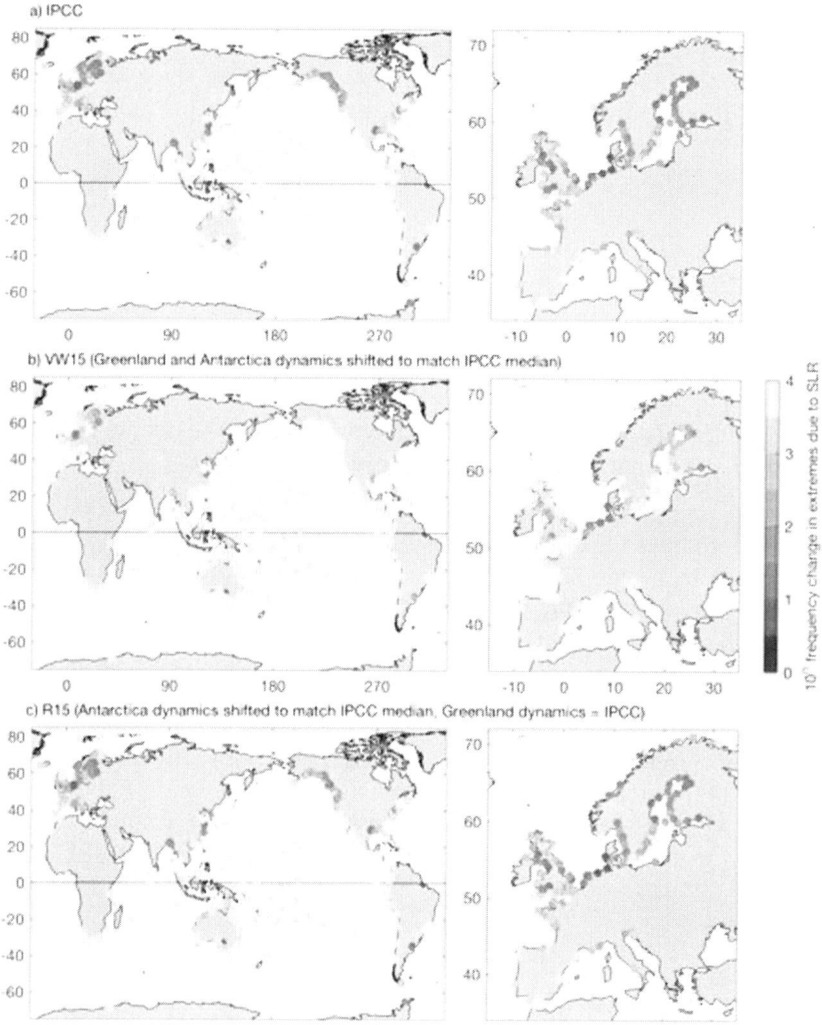

Figure 9. Ratio of the change in the frequency of extreme events (2100 vs. 2010) if allowances are not applied, based on the (**a**) IPCC (**b**) VW15 and (**c**) R15 sea-level change scenario. Note the logarithmic scale. A limit of 104 has been imposed, as frequency increases above this level mean that a coastal structure effectively would be lost if no allowance is applied.

Table 1. Percentage (number) of tide gauge stations for every 10n increase in the frequency of extreme events by 2100 per ice sheet scenario IPCC, VW15 and R15).

10^n	IPCC	VW15	R15
0–1	1% (6)	0% (2)	1% (6)
1–2	11% (73)	1% (9)	10% (67)
2–3	21% (139)	6% (37)	21% (136)
3–4	17% (109)	6% (39)	16% (105)
>4	50% (325)	87% (565)	52% (338)

4. DISCUSSION

In computing the allowances, several choices and assumptions were made for the projected mean SLR and for the representation of extreme sea levels.

Regarding the projections, we used the RCP8.5 scenario as presented in Slangen et al. [12]. The RCP8.5 scenario is chosen because it is currently the more realistic option among the available scenarios, which is therefore relevant for coastal protection. Compared to the projections presented in [1], there are small differences (for the period 1986–2005 vs. 2080–2100, RCP8.5 scenario) in the individual contributions: glaciers (+6 cm), Greenland SMB (−1 cm), Antarctic SMB (−4 cm), Greenland Dynamics (+1 cm), Antarctic Dynamics (+2 cm) and groundwater depletion (+4 cm). The main reason is that Slangen et al. [12] used one model or approach for each contribution to sea level, while the IPCC AR5 projections are based on an assessment of multiple estimates for each contribution. As a result, the global mean SLR under the RCP8.5 scenario is 71 cm in Slangen et al. [12], compared to 63 cm in IPCC AR5 (for 2081–2100 relative to 1986–2005). However, the methodology used for computing regional patterns is the same for the IPCC AR5 and the Slangen et al. [12] projections, and therefore any differences are mainly in the magnitude of the signal rather than the spatial distribution.

For the contributions of the ice sheet dynamics, we used three different shapes for the uncertainty distributions, from three different sources. The IPCC estimate is a consensus estimate, based on a multitude of models and publications, the VW15 estimate is based on an expert elicitation, and the R15 estimate is based on physical and statistical models. We chose to use these published estimates of the actual ice sheet dynamical contributions rather than completely artificial uncertainty distributions, which probably would have not included a distribution with the shape of the R15 distribution. A systematic sensitivity study could be part of future work. Nevertheless, the current set of distributions shows that a highly skewed distribution leads to much higher estimates of the allowances. It is therefore very important to better understand the shape of the uncertainty distribution using physical understanding and modelling, as the uncertainties are critical to estimating future SLR and to decide on the necessary coastal protection.

In order to focus on the influence of the shape of the uncertainty distributions rather than the actual magnitude, we shifted all distributions such that they matched the median values of IPCC. If we compare the differences in allowances before and after the shift of the medians, we find that for the VW15 scenario the shift in the median SLR projections is −12 cm, but the impact on the allowances (comparing allowances with and without median shift) is −8 cm. For the R15 scenarios, the medians were shifted by only −3 cm, and the impact on the allowances is −4 cm.

With regard to the use of the Gumbel scale parameter, which is based on annual maximum sea levels, it is considered to be a method that is relatively straightforward, but perhaps not the ideal representation of sea-level extremes. It is relatively robust for temporal and spatial variations and therefore widely used for global applications [32], but, on the other hand, does not use all the information that is available. Some other potential methods, such as the Generalised Extreme Value distribution or a Generalised Pareto distribution, include more information, as they not only provide a slope, but also a shape of the extreme value distribution [33]. However, these more complicated distributions cannot (yet) be used to compute allowances. In a potential follow-up of this work, it would be useful to examine the effect of different types of extreme value distributions on sea-level allowances, but here we chose to focus on the sensitivity to the shape of the sea-level projection uncertainties alone.

Generally, the longer the time period used, the better the estimate of the extreme sea-level variability [34]. Here, we use a minimum of 20 years of data in the tide gauge record as a constraint, but find that the average allowance only varies with a maximum of 0.06 m for the VW15 scenario when selecting longer time series (which results in less tide gauge records): 2.19 m for ≥20 years vs. 2.13 for records ≥50 years. For IPCC and R15, the differences are even smaller and the selection of longer time series has a negligible effect on the allowances (respectively, 1.01 and 1.03 m for ≥20 years and, respectively, 1.00 and 1.02 m for ≥50 years).

One caveat for all extreme value statistics is that they depend on the occurrence of extreme events in the observational period, which becomes a more relevant restriction for longer return periods (e.g., 1 in 10,000 years) [35]. As a result, the statistics may change significantly when new extreme events are included in the analysis and it is therefore important to keep them up to date.

A final assumption regarding the statistics of the extremes is that they do not change over time, i.e., that both the magnitude with respect to the mean sea level and the frequency not change. This assumption is supported by observations and analysis of past extreme sea levels, which show that the observed changes in the frequency of flooding events is mainly caused by the increase in mean sea levels (e.g., [11]). However, the statistics of storms, and therefore of sea-level extremes, could well change significantly in the future. It is important to keep in mind that sea level can also rise due to local land subsidence, for instance due to groundwater extraction. In some locations, the subsidence even exceeds the climate-driven sea-level rise and therefore amplifies the flooding frequency [36] and changes the location of the coastline due to submergence of the coast. However, information on vertical land motion (VLM, measured with GPS) is not available all over the world and projections are difficult since they directly depend on human decisions. Therefore, VLM and changing coastlines could not yet be taken into account in the sea-level projections.

5. CONCLUSIONS

We have compared and contrasted three different scenarios for the projected dynamical ice sheet contribution to SLR. For each scenario, we combined information on sea-level extremes at more than 650 tide gauge stations with regional SLR projections in 2100 with respect to 2010, to compute so-called sea-level allowances. An allowance is a guide for the expected height a coastal structure needs to be elevated to keep the same flooding frequency. Allowances depend not only on the mean SLR, but also on the uncertainty in the mean sea-level projection and on the variability and magnitude of sea-level extremes. Each of the three sea-level projection scenarios has a different shape for the uncertainty distributions in the ice sheet dynamics contributions, leading to differences in the allowances.

We found that the allowances change significantly for our skewed scenario, doubling their values compared to allowances based on a normal distribution of the ice sheet dynamics. This is due to a heavier upper tail associated with higher contributions from ice sheet dynamics (the low-probability/high-impact events). The IPCC-based 95% SLR at the tide gauge stations is 1.08 m, with an average allowance of 1.01 m, while the skewed VW15 scenario leads to a projected 95% SLR of 1.50 m at the tide gauges and a much larger allowance of 2.19 m. For the R15 scenario, which has a less skewed but still non-normal distribution, the 95% SLR at the tide gauges is 1.12 m and the average allowance is 1.03 m. This sensitivity to the skewness of the uncertainty distribution means that it is very important to improve estimates of ice sheet dynamics contributions to SLR, and in particular their uncertainties. As we focused on the shape of the uncertainty distribution here rather than the magnitude, computations of allowances should be repeated with improved ice sheet uncertainty distributions to deliver allowances that are suitable for decision making and coastal protection.

The sea-level allowances are largest in regions with a relatively small variability in sea-level extremes and SLR projections of large magnitude and/or uncertainty. These are also the regions where frequencies in extremes will increase most if the allowances are not applied (i.e., coastal structure is not elevated), especially for the skewed probabilities in the VW15 scenario. Most of these vulnerable locations are in the tropics, where the largest regional sea-level changes (and uncertainties) are projected due to increasing thermal expansion and above-average regional contributions due to land ice mass loss [12]. This means that these regions would also be particularly vulnerable to larger contributions from the ice sheets.

Here, the computation of allowances could only be done at tide gauge locations, as this is where high-resolution observational time series are available for reasonable lengths of time. It would, however, be interesting to try and fill in the gaps for the entire coastline, using modelling approaches. Recently, Muis et al. [32] published a global set of extreme sea levels and storm surges (the GTSR data set), including Gumbel scale parameters along the global coastline, which showed good agreement between observations and the model data set. This set could possibly be used in future work to provide sea-level allowances at all coastal locations, which would be helpful for coastal planning purposes.

ACKNOWLEDGMENTS

We gratefully acknowledge the contributors to the GESLA-2 database which are listed on www.gesla.org. A.B.A.S. and R.C.d.W. were funded by the NWO Netherlands Polar Programme, and T.J.R. was funded by the Netherlands Earth Science Center. P.L.W. was funded partly by the Leverhulme Trust.

AUTHOR CONTRIBUTIONS

A.B.A.S. and R.S.W.v.d.W. conceived and designed the experiments; A.B.A.S. performed the experiments, analysed the data and wrote the paper; and all authors contributed to the writing of the paper. R.S.W.v.d.W., R.C.d.W. and T.J.R. designed the SEAWISE model and T.J.R. developed the code. J.R.H., P.L.W. and T.E. contributed with data and analysis tools.

CONFLICTS OF INTEREST

The authors declare no conflict of interest.

REFERENCES

1. Church, J.; Clark, P.; Cazenave, A.; Gregory, J.; Jevrejeva, S.; Levermann, A.; Merrifield, M.; Milne, G.; Nerem, R.; Nunn, P.; et al. Sea Level Change. In Climate Change 2013: The Physical Science Basis. Contribution of Working Group I to the Fifth Assessment Report of the Intergovernmental Panel on Climate Change; Stocker, T., Qin, D., Plattner, G.K., Tignor, M., Allen, S., Boschung, J., Nauels, A., Xia, Y., Bex, V., Midgley, P., Eds.; Cambridge University Press: Cambridge, UK; New York, NY, USA, 2013. []
2. Clark, P.U.; Church, J.A.; Gregory, J.M.; Payne, A.J. Recent Progress in Understanding and Projecting Regional and Global Mean Sea Level Change. Curr. Clim. Chang. Rep. 2015, 1, 224–246.
3. Slangen, A.B.A.; Adloff, F.; Jevrejeva, S.; Leclercq, P.W.; Marzeion, B.; Wada, Y.; Winkelmann, R. A Review of Recent Updates of Sea-Level Projections at Global and Regional Scales. Surv. Geophys. 2016, 38, 385–406.
4. Bamber, J.L.; Aspinall, W.P. An expert judgement assessment of future sea level rise from the ice sheets. Nat. Clim. Chang.2013, 3, 424–427.
5. De Vries, H.; van de Wal, R.S.W. How to interpret expert judgment assessments of 21st century sea-level rise. Clim. Chang.2015, 130, 87–100.
6. Favier, L.; Durand, G.; Cornford, S.L.; Gudmundsson, G.H.; Gagliardini, O.; Gillet-Chaulet, F.; Zwinger, T.; Payne, A.J.; le Brocq, A.M. Retreat of Pine Island Glacier controlled by marine ice-sheet instability. Nat. Clim. Chang. 2014, 4, 117–121.
7. Joughin, I.; Smith, B.E.; Medley, B. Marine Ice Sheet Collapse Potentially Under Way for the Thwaites Glacier Basin, West Antarctica. Science 2014, 344, 735–738.
8. Ritz, C.; Edwards, T.L.; Durand, G.; Payne, A.J.; Peyaud, V.; Hindmarsh, R.C.A. Potential sea-level rise from Antarctic ice-sheet instability constrained by observations. Nature 2015, 528, 115–118.
9. Hunter, J. A simple technique for estimating an allowance for uncertain sea-level rise. Clim. Chang. 2012, 113, 239–252.
10. Hunter, J.R.; Church, J.A.; White, N.J.; Zhang, X. Towards a global regionally varying allowance for sea-level rise. Ocean Eng.2013, 71, 17–27.
11. Menéndez, M.; Woodworth, P.L. Changes in extreme high water levels based on a quasi-global tide-gauge dataset. J. Geophys. Res. 2010, 115, C10011.

12. Slangen, A.B.A.; Carson, M.; Katsman, C.; van de Wal, R.; Koehl, A.; Vermeersen, L.; Stammer, D. Projecting twenty-first century regional sea-level changes. Clim. Chang. 2014, 124, 317–332.
13. Moss, R.H.; Edmonds, J.A.; Hibbard, K.A.; Manning, M.R.; Rose, S.K.; van Vuuren, D.P.; Carter, T.R.; Emori, S.; Kainuma, M.; Kram, T.; et al. The next generation of scenarios for climate change research and assessment. Nature 2010, 463, 747–756.
14. Taylor, K.; Stouffer, R.J.; Meehl, G.A. An overview of CMIP5 and the experiment design. Bull. Am. Meteorol. Soc. 2012, 93, 485–498.
15. Kopp, R.E.; Horton, R.M.; Little, C.M.; Mitrovica, J.X.; Oppenheimer, M.; Rasmussen, D.J.; Strauss, B.H.; Tebaldi, C. Probabilistic 21st and 22nd century sea-level projections at a global network of tide-gauge sites. Earth Future 2014, 2, 383–406.
16. Jackson, L.P.; Jevrejeva, S. A probabilistic approach to 21st century regional sea-level projections using RCP and High-end scenarios. Glob. Planet. Chang. 2016, 146, 179–189.
17. Jevrejeva, S.; Jackson, L.P.; Riva, R.E.M.; Grinsted, A.; Moore, J.C. Coastal sea level rise with warming above 2 °C. Proc. Natl. Acad. Sci. USA 2016, 113, 13342–13347.
18. De Winter, R.; Reerink, T.J.; de Vries, H.; Slangen, A.B.A.; van de Wal, R.S.W. Impact of asymmetric uncertainties in ice sheet dynamics on regional sea level projections. Nat. Hazards Earth Syst. Sci. Discuss. 2017. In Review.
19. Buchanan, M.K.; Kopp, R.E.; Oppenheimer, M.; Tebaldi, C. Allowances for evolving coastal flood risk under uncertain local sea-level rise. Clim. Chang. 2016, 137, 347–362.
20. Coles, S. An Introduction to Statistical Modeling of Extreme Values; Springer: London, UK, 2001. []
21. Arns, A.; Wahl, T.; Haigh, I.D.; Jensen, J.; Pattiaratchi, C. Estimating extreme water level probabilities: A comparison of the direct methods and recommendations for best practise. Coast. Eng. 2013, 81, 51–66.
22. GESLA. Available online: www.gesla.org (accessed on 23 May 2017).
23. Woodworth, P.; Hunter, J.; Marcos, M.; Caldwell, P.; Menendez, M.; Haigh, I. Towards a global higher-frequency sea level data set. Geosci. Data J. 2016, 3, 50–59.
24. Shepherd, A.; Ivins, E.R.; Geruo, A.; Barletta, V.R.; Bentley, M.J.; Bettadpur, S.; Briggs, K.H.; Bromwich, D.H.; Forsberg, R.; Galin, N.; et al. A Reconciled Estimate of Ice-Sheet Mass Balance. Science 2012, 338, 1183–1189.
25. Mitrovica, J.X.; Tamisiea, M.E.; Davis, J.L.; Milne, G.A. Recent mass balance of polar ice sheets inferred from patterns of global sea-level change. Nature 2001, 409, 1026–1029.
26. Hunter, J.R.; Woodworth, P.L.; Wahl, T.; Nicholls, R.J. Using Global Tide Gauge Data to Validate and Improve the Representation of Extreme Sea Levels in Flood Impact Studies. Glob. Planet. Chang. 2017. Under Review. []
27. Merrifield, M.A.; Genz, A.S.; Kontoes, C.P.; Marra, J.J. Annual maximum water levels from tide gauges: Contributing factors and geographic patterns. J. Geophys. Res. Oceans 2013, 118, 2535–2546.
28. De Winter, R.C.; Sterl, A.; Ruessink, B.G. Wind extremes in the North Sea Basin under climate change: An ensemble study of 12 CMIP5 GCMs. J. Geophys. Res. Atmos. 2013, 118, 1601–1612.
29. Press, W.; Teukolsky, S.A.; Vetterling, W.T.; Flannery, B.P. Numerical Recipes 3rd Edition: The Art of Scientific Computing, 3rd ed.; Cambridge University Press: New York, NY, USA, 2007. []
30. Slangen, A.B.A.; Church, J.A.; Zhang, X.; Monselesan, D. The sea-level response to external forcings in CMIP5 climate models. J. Clim. 2015, 28, 8521–8539.
31. Tamisiea, M.E.; Mitrovica, J.X. The moving boundaries of sea level change: Understanding the origins of geographic variability. Oceanography 2011, 24, 24–39.
32. Muis, S.; Verlaan, M.; Winsemius, H.C.; Aers, J.C.J.H.; Ward, P.J. A global reanalysis of storm surges and extreme sea levels. Nat. Commun. 2016, 7, 11969.

33. Wahl, T.; Haigh, I.; Nicholls, R.; Arns, A.; Dangendorf, S.; Hinkel, J.; Slangen, A. Understanding extreme sea levels for broad-scale coastal impact and adaptation analysis. Nat. Commun. 2017. Accepted. []
34. Haigh, I.D.; Nicholls, R.; Wells, N. A comparison of the main methods for estimating probabilities of extreme still water levels. Coast. Eng. 2010, 57, 838–849.
35. Dangendorf, S.; Arns, A.; Pinto, J.G.; Ludwig, P.; Jensen, J. The exceptional influence of storm 'Xaver' on design water levels in the German Bight. Environ. Res. Lett. 2016, 11, 045001.
36. Erkens, G.; Bucx, T.; Dam, R.; de Lange, G.; Lambert, J.L. Sinking coastal cities. Proc. IAHS 2015, 372, 189–198.

CHAPTER 5

Morphodynamics at the Coastal Zone in the Laizhou Bay, Bohai Sea

Junjie Deng[1 2], Jan Harff[3], Yanfang Li[4] Yan Zhao[4], Hua Zhang[4]

[1]*Institute of Marine and Coastal Research, University of Szczecin, Szczecin, Poland*

[2]*School of Earth and Environmental Sciences, University of Wollongong, Australia*

[3]*Institute of Marine and Coastal Research, University of Szczecin, Szczecin, Poland*

[4]*Yantai Institute of Coastal Zone Research, Chinese Academy of Sciences, Yantai, Shandong 264003, P. R. China*

ABSTRACT

Deng, J. J.; Harff, J.; Li, Y. F., Zhao, Y., and Zhang, H., 2016. Morphodynamics at the coastal zone in the Laizhou Bay, Bohai Sea.

Morphogenetic processes of the Yellow River Delta and the Laizhou Bay, Bohai Sea, China, have to be described by the interrelation of riverine sediment supply, relative sea level change and the effects of wind driven waves and nearshore currents. The research area is regarded a natural laboratory of the development of river dominated coastal zone with interfering natural and anthropogenic forcing factors. As well for the historical hindcast as for future projections of the morphogenetic coastal processes the Dynamic Equilibrium Shore Model (DESM) can be applied. This model generalizes the standard Bruun rule model and generates Digital Elevation Models (DEMs) scenarios on decadal to centennial time scales for the geological past and future. The basic concept is a dynamic equilibrium coastal profile evolution in adaption to the sediment budget in a spatially three-dimensional domain. By adding parameters to account for sediment mass contribution from riverine sediment flux, the DESM can be used to explore coastal morphological equilibrium states responding to sediment budget changes for river-dominated coastal zones. For the parameterization historical maps of the southern Bohai Sea have been applied to reconstruct paleo-coastlines for the 19th century for the comparison with a modern DEM. Gauge measurements provided the data for an estimation of trends in sea level change for the Laizhou Bay on the decadal scale. Modern sea level rise together with reduced riverine sediment supply caused by anthropogenic activities such as damming up-streams may change the depositional environment at the Yellow River Delta and related areas of the Laizhou Bay from river dominated (progradational) to wave dominated (regressive) environment.

Keywords: Coastal morphodynamics, Laizhou Bay, Dynamic Equilibrium Shore Model, river sediment flux, relative sea level change.

INTRODUCTION

Coastal morphodynamic evolutions are subject to changing climate such as accelerated rising sea level and increasing frequency of storm events (IPCC, 2013). Furthermore, coasts are more vulnerable at the glacial-isostatiscal subsiding coastal areas than the uplifting coasts (Harff *et al.*, 2007). During the last century, most deltas in the world are also vulnerable to subsidence because of decreasing riverine sediment sources or/and sediment compaction that are largely influenced by anthropogenic activities (Saito *et al.*, 2007; Syvitski *et al.*, 2009; Syvitski and Kettner, 2011). Special attentions should be given to these subsiding deltas when considering impacts of rising sea level on the coasts (Nicholls and Cazenave, 2010).

River mouth systems in the Bohai Sea (Yellow River mouth) are in the focus of environmental research because of the enormous economic importance of the area so that for sustainable planning and management functional models of the environmental system are needed. These models allow the generation of scenarios not only for recent processes, but also for the geological development of the area and its future projection. The comparison of coastal and river mouth environments of different marginal seas will lead to a deeper understanding of the processes within the land-ocean transition zones. In the past, comparative research studies have been already conducted along the northern shelf of the South China Sea, in particular in the Pearl River Estuary and the Beibu Gulf (Harff *et al.*, 2010; 2013; Yao, 2009). In both cases the studies benefit from two preceding research projects run along the southern coast of the Baltic Sea:

"Dynamics of Natural and Anthropogenic Sedimentation-DYNAS" (Harff *et al.*, 2009) and "Sinking Coasts-Geosphere, Ecosphere and Anthroposphere of the Holocene Southern Baltic Sea-SINCOS" (Harff and Lüth, 2011). Because of its steep gradients of oceanographic and geological parameters, such as salinity, sea level change, and vertical displacement of the Earth's crust (Glacio-Isostatic Adjustment-GIA), the Baltic Sea serves as a model ocean where coastal and shelf sea processes can be studied exemplarily in an exceptional manner. So, special function-models have been developed within the frame of the SINCOS and DYNAS projects allowing as well the historical hindcast as future projection of the development of coastal areas as a result of interference of geological, climatic, oceanographic driving forces with anthropogenic activities (Harff and Lüth, 2011; Harff and Meyer, 2011; Zhang *et al.*, 2012). Within the frame of a research project "Coastal Changes at the Southern Baltic Sea-Past and Future Projection- CoPaF" the SINCOS approach has been specified in particular to morphodynamic changes of the coastal zone affected by the climatically controlled sea level rise (Deng *et al.*, 2014). The level of conceptual generalization was set for any case to let scientists apply the corresponding models not only to the Baltic Sea, but also to other marginal seas even in different climatic zones and different oceanographic conditions.

Parameterization of a long-term multi-scale morphodynamic model to generate future scenarios has to base on the information derived from the reconstruction of paleo-coastal geomorphological changes with similar duration to the future scenarios (*e.g.* Deng *et al.*, 2014; Dissanayake *et al.*, 2012; Zhang *et al.*, 2011). This fact highlights the importance of modelling paleo-coastal geomorphology at the corresponding time span to future projection (i. e. ca. 100 years according to common climate modelling periods). A model to generate paleo-digital elevation scenarios has been elaborated for the wave-dominated sandy coast (i. e. the Pomeranian Bay, southern Baltic Sea) by Deng *et al.* (2014). This so-called Dynamic Equilibrium Shore Model (DESM) generates paleo-Digital Elevation Models (DEMs) on decadal to centennial time scales. The basic concept is a dynamic equilibrium coastal profile evolution in adaption to the sediment budget in a spatially three-dimensional domain. This concept is generalised from the standard Bruun rule model (Bruun, 1962; 1988). Our main objective is to generalize the DESM by adding parameters to account for sediment mass contribution from riverine sediment flux, and to use the model to explore coastal morphological equilibrium states responding to sediment budget changes.

At the Laizhou Bay, located at the south of Bohai Sea, fluviatile sources supply most of the sediments accumulated at sinks along the coast. Among numerous rivers merging the southern Bohai Sea, the Yellow River dominates fluviatile sediment sources. The main sediment sink area coincides with the subsiding zone of the western Laizhou Bay, whereas the eastern Laizhou Bay coast belongs to the uplifting Shangdong Peninsula (Li *et al.*, 1991). It is notable that besides natural forces, during the last decades anthropogenic factors influence increasingly coastal morphogenesis. Dams and reservoirs erected upstream diminish continuously the suspended matter load of Yellow River delivered to the receiving marine basins (Wang *et al.*, 2007). This decrease makes sea level change and wind-wave dynamics become increasingly important for the coastal development compared to fluviatile factors. At the same time the tide plays the dominant role in transporting sediments out of the sheltered embayment system to deeper shelf and the Yellow Sea (Bian *et al.*, 2013; Lu *et al.*, 2013). Numerical studies confirm the interpretation and estimation of Yang and Liu (2007) that along shelf sediment transport delivered by Yellow River formed clinoform deposits along the eastern tip of Shandong Peninsula in the Yellow Sea during the last 7000 years. These studies providing qualitative and quantitative information of sediment source to sink transport make the Laizhou Bay and the adjacent Yellow River Delta as an ideal case for generalising the DESM model by applying sediment mass balancing.

ENVIRONMENTAL SETTING

The Bohai Sea (Figure 1), a marginal inland sea of China, serves as a key area to study the interrelation of different factors influencing the relative sea level and coastline change. As a semi-enclosed marginal Sea the shallow Bohai Sea (average depth of 18 m) is separated in the Southeast from the Yellow Sea by the Shandong Peninsula. Through the 90 km wide Bohai Strait the Bohai Sea is connected with the Yellow Sea. Tidal dynamics and wind waves are main processes ruling morphodynamics.

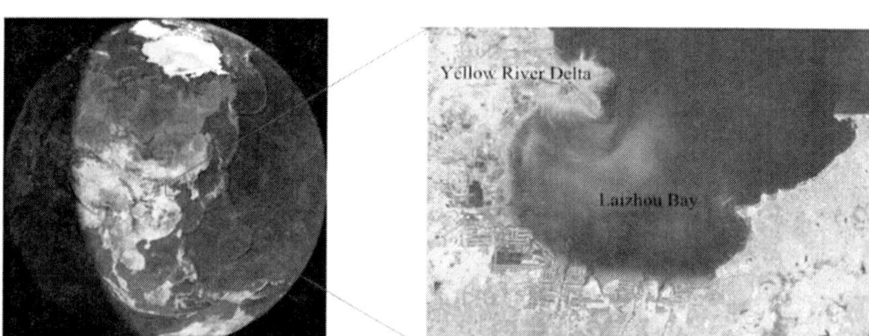

Figure 1. Study area: Laizhou Bay, Southern Bohai Sea, China

Maps of Asia and Bohai Sea from http://www.ngdc.noaa. Gov /mgg/global/global.html; Laizhou Bay data from Landsat–5 TM satellite image, July, 2009

The general development of the Bohai Sea is determined by the postglacial Sea level rise. Regionally, the general trend of coastline advance (regression) and retreat (transgression) varies due to tectonically induced differentiated basin subsidence and the formation of river deltas such as the Yellow River, Haihe River, Liaohe River Delta, and natural, anthropogenically induced change in riverine sediment discharge to the Bohai Sea.

In the Laizhou Bay, southern Bohai Sea, these processes can be studied exemplarily. The Bay is subdivided into three parts: Yellow River Delta (progradational sediment wedge), western and southern Laizhou Bay (muddy coastal embayment), and eastern Laizhou Bay (sandy to rocky coast).

The western-subsiding-part is dominated by the Yellow River delta and the discharge of the riverine particulate matter. By a counterclockwise current system fine-grain sediments are transported to the shallow coastal areas of the southern Laizhou Bay where also smaller rivers deliver their load to the Bay. The subsiding western basin-dominated by the delta-is separated from the eastern part of the Shandong Peninsula by the NNE striking Tan-Lu Fault. The eastern coast of the bay belongs to the East China uplifting belt (Li *et al.*, 1991) and is determined by Mesozoic granites and sandstones. Here besides river load the coastal sediments descend from the erosion of Mesozoic coastal rocks.

The coastal development in the Bohai Sea is determined by the Postglacial sea level rise having caused a transgression which reached its maximum during the Holocene Climate Optimum about 6 ka BP. After the maximal transgression a drop in the relative sea level most likely related to the Neoglacial cooling is recorded by proxy data in several key areas of the eastern and southern area of Chinese marginal seas (Pirazzoli, 1991). Figure 2 shows the coastline displacement in the southern Bohai Sea since the maximum transgression of 7000 a BP. The coastlines from 7000 a BP until 1128 AD are reconstructed from geological proxies. The 1891 AD coastline was derived from a historical nautical chart. This paleo-coastline information reflects dynamic coastal evolution varying at different time intervals, which provide essential input data for the model implementaion. River course shifting of Yellow River, and significant increase of Yellow River sediment discharge after 11 AD (Figure 3) may explain the phenomena of coastline displacement at the Laizhou Bay.

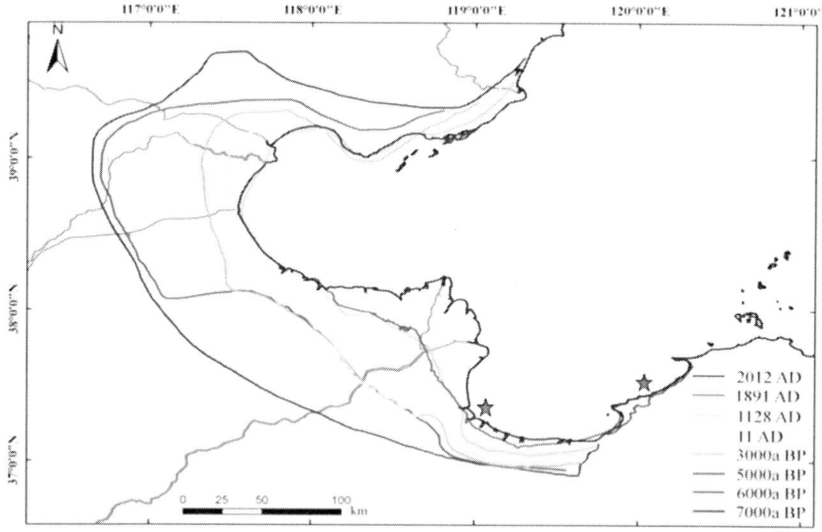

Figure 2. Ancient and recent coastlines of West Bohai Sea

2012 AD: Northwest Bohai Sea: nautical chart of Qinhuangdao Gang to Qihe Kou (Anonymous (1959–2009)); Southwest Bohai Sea: nautical chart of Qihe Kou to Longkou Gang (Anonymous (1958–2011)) 1891 AD: R. C. Carrington of the Hydrographic Office, 1891;7000a BP, 5000a BP–1128 AD: Xue (2009), 6000a BP: Zhuang (1987) blue asterisk: Yangjiaogou gauge, red asterisk: Longkou Gang gauge

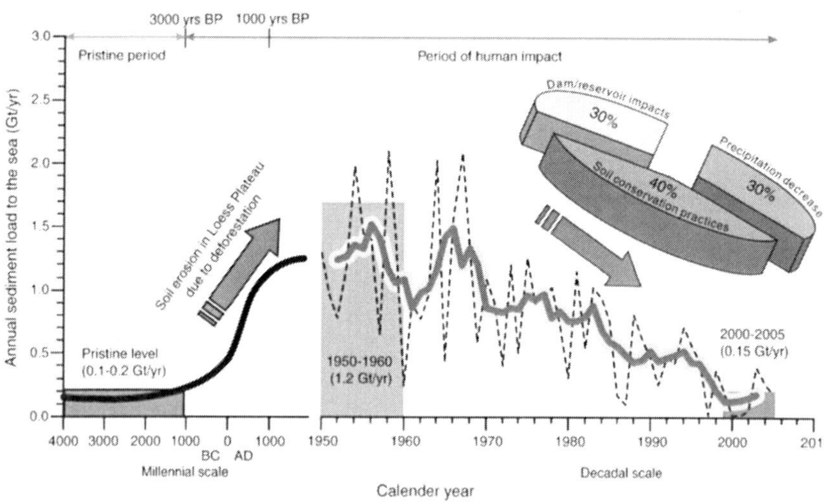

Figure 3. Past and present sediment load discharged from the Huanghe (Yellow River) to the sea (Wang *et al.*, 2007)

The Holocene regression after 7000 a BP reflects besides sea level drop the increase in the Yellow River sediment discharge. Figure 3 shows the paleo- and present sediment load supplied by the Yellow River to the Bohai Sea according to Wang *et al.* (2007). According to the data, the Holocene pristine sediment discharge measures about 0.1–0.2 Gt/yr. During the Zhou Dynasty due to deforestation the sediment discharge increased, and a high level of discharge was maintained by the Mid–20th century. The curve shows clearly the from about 1950 AD a decrease of sediment supply from 1.5 Gt/yr in 1950 to 0.2 Gt/yr in 2000 AD because of dam reservoir impacts, soil conservation practices and (natural) precipitation decrease. Also, the shifts in the sub-recent outlets are a main effect of hydro-engineering.

Recent relative sea level changes are recorded by gauges' time series data from the western and the eastern Laizhou Bay from different geological settings. The mean relative sea level rise is about 2.1 mm/yr from 1950AD to 1994AD at the western subsiding coast (Yangjiaogou gauge in Figure 2), while there is a mean sea level rise of 1.7 mm/yr from 1950AD to 2010 AD at the eastern coast (Longkou Gang gauge in Figure 2). The reason for minor relative sea level rise values in the East of the Laizhou Bay is explained here by the uplifting tendency of the Shandong Peninsula (Li *et al.*, 1991). The shift in the tendency of sea level change from a decrease during the Late Holocene to the last decades along with diminished sediment supply by the Yellow River explains increasing erosional effects at the eastern delta area and in the Laizhou Bay.

METHODS

General concept of morphodynamic modeling

The model is utilized to describe the change of a reference digital elevation model DEM_t during a time span Δ_t. It is taking into account the relative sea level change ($RSL_{\Delta t}$) and the change in sediment thickness because of accumulation and erosion of sediments ($SED_{\Delta t}$):

$$DEM_{t+\Delta t} = DEM_t + RSL_{\Delta t} + SED_{\Delta t}. \qquad (1)$$

This model has been applied successfully by Qi (2013) to reconstruct a paleogeographic scenario of the southwestern Bohai Sea. In the model the relation describing the relative sea level change $RSL_{\Delta t}$ plays a central role

$$RSL_{\Delta t} = EC_{\Delta t} + GIA_{\Delta t} + \sum_i E^i_{\Delta t} \qquad (2)$$

Symbols in † and † represent 2D fields covering an area of investigation **R** with values allocated to vectors *r* marking any location in the area **R** (for instance $dem_t(r) \in DEM_t$, $\forall\, r \in$ **R**).

The relative sea level change is composed of different components, the eustatic (climatically driven) sea level change ($EC_{\Delta t)}$, the glacial-isostatic adjustment ($GIA_{\Delta t)}$ and minor effects $E'_{\Delta t}$ as gravitational effect, compaction and others. For the historical reconstruction of paleo-digital elevation models, paleo-coastlines and *RSL*-data can be derived from interpolation of geological proxy-data pointing spatially to the position of ancient coasts or-for the younger history-from historical maps.

For a quantification of the $SED_{\Delta t}$ component in † the so-called Dynamic Equilibrium Shore Model-DESM (Deng *et al.*, 2014) is applied. First, the area of research (model domain) is to be defined. For optimal parameterization it should be ensured that the area can be regarded semi-enclosed. Regarding the parameters, there are three of them of special importance: the relative sea level change during the time span of interest, the supply of sediments from external sources and the redistribution of sediments (budgeting) within the area of investigation.

Generalised description of the model DESM at the wave-dominated sandy coast

At the wave-dominated sandy coast, the model domain is a semi-enclosed area in terms of sediment budget, when lateral sediment in-flux $Q_{LST,in}$ (m$^{3)}$ and out-flux $Q_{LST,out}$ (m$^{3)}$ at the domain's lateral boundary is zero (i. e. $Q_{LST,in} = Q_{LST,out} = 0$) in Figure 4. The model domain of DESM consists of retreating and advancing coast segments. The whole model domain is thus discretized alongshore into $n+m$ zones, each zone represented by one cross-shore profile (Figure 4). Based on recent DEM, for any time point $t \leqslant 0$, the depth of cross-shore profile representing each zone is described by the exponential function with the origin at the shoreline and the end point at the closure depth. Coastal bathymetrical profiles at the coastal erosion and accretion zones are described respectively as follow:

$$y = a(1 - e^{(const1 * b_0)(x+\Delta c_t)}) - s_t, \; t \leqslant 0$$
$$y = a(1 - e^{(const2 * b_0)(x+\Delta c_t)}) - s_t, \; t \leqslant 0 \qquad (3)$$

where y is the water depth (m) under the mean sea level and x is the offshore distance from the shoreline (m); a is the exponential limit of the profile; b_0 is the curvature coefficient of the present cross-shore bathymetrical profile; Δc_t is the known distance (m) between historical and present coastlines; and S_t is known relative sea level changes (m) compared with recent time 2000 AD ($t = 0$). The *const*1 and *const*2 are curvature coefficients. When $t = 0$, † represents the recent coastal submarine profile.

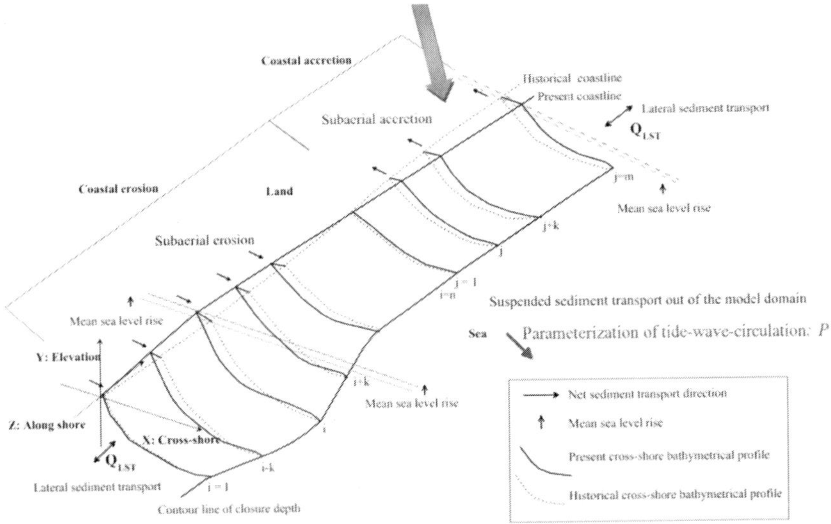

Figure 4. Sketch of a three-dimensional concept of the DESM model for coasts with river and tide influences

At a given time t, the subaerial mass volumes can be estimated on a basis of the modern DEM and the data of Δc_t and s_t, as is given in Deng *et al.* (2014). At the semi-enclosed coast in terms of sediment budget for a given time t, the mass balanced mathematical expression can be given below:

$$\left|\sum_{i=1}^{m} V_{erosion,i}(const1, \Delta c_{t,i}, s_{t,i}, b_{0,i}, a_i)\right| = \left|\sum_{j=1}^{n} V_{accretion,j}(const2, \Delta c_{t,j}, s_{t,j}, b_{0,j}, a_j)\right|, \ t \leqslant 0 \quad (4)$$

$V_{erosion}$ and $V_{accretion}$ are respectively erosion and accretion sediment mass volume (m^3), and subscript "i" or "j" denote the profile number in Figure 4. For a given time t, the data of a modern Digital Elevation Model, historical coastline changes, and relative sea level records can provide the information of parameters of Δc_t, s_t, b_0 and a for each profile i or j, hence, the only unknown parameters are *const*1 and *const*2. † can be thus simplified as:

$$\left|\sum_{i=1}^{m} V_{erosion,i}(const1)\right| = \left|\sum_{j=1}^{n} V_{accretion,j}(const2)\right| \quad (5)$$

In this equation, *const*1 is the uniform ratio between paleo -and present-curvature coefficients of all cross-shore profiles at the retreating coastal zones, and *const*2 is the ratio at the advancing coastal zones. Hence, by applying sediment mass balancing, the value of *const*1 can be determined by iterative inverse procedure when *const*2 is set to 1, or *const*2 can be calculated by iterative inverse procedure when *const*1 is set to 1. For the detail of this inverse modelling method, please see Deng *et al.* (2014).

Generalization of the model DESM to the river dominated muddy coast with influence of tide

To generalize the model at the muddy coast with the river mouth system, the seaward boundary is determined by the closure depth as well and at the alongshore direction it is bounded by the location of transition from coastal accretion to erosion. The sub-aerial mass volume can be estimated by using parameters of sub-aerial terrain heights, coastal retreat and advance distances and relative sea level changes, as like the examples given in Deng *et al.* (2014). In addition, the tidal flat, riverine sediment sources and the amount of them deposited at the model domain needs to be parameterized. This parameterization has been preliminarily described in Deng *et al.* (2014) in an abstract form. Here, a complete description is provided.

As like the model DESM, the whole coastal domain at the underwater part is discretized alongshore into m zones, each zone represented by one cross-shore profile (Figure 5). Based on the recent DEM, for any time point $t \leqslant 0$, the depth of cross-shore profile representing each zone is described by the exponential function with the origin at the shoreline and the end point at the closure depth. Coastal bathymetrical profiles at the coastal accretion zones are described as follow:

$$
\begin{aligned}
y &= a\left(1 - e^{(const2 * b_0)(x + \Delta c_t - \Delta l_{tf})}\right) - s_t,\ t \leqslant 0,\ x \geqslant \Delta l_{tf} \\
y &= -\frac{h_{tf}}{l_{tf}}(x + \Delta c_t) - s_t,\ t \leqslant 0, \Delta c_t \leqslant x \leqslant \Delta l_{tf} \qquad (6)
\end{aligned}
$$

where y is the elevation (m) above the mean sea level and x is the distance (m) away from the shoreline where the elevation is zero; a is the exponential limit of the profile that remain unchanged with the time and Δl_{tf} and h_{tf} are respectively the width (m) and the deepest depth (m) of tidal flat that are assumed to be independent of the time; b_0 is the curvature coefficient of the modern cross-shore bathymetrical profile; Δc_t is the known distance (m) between historical and present coastlines; and S_t is known relative sea level changes (m) compared with recent time 2000 AD (t= 0). When t = 0, † describes the recent submarine coastal profile. The *const*2 are the curvature coefficient reflecting the geomorphological changes.

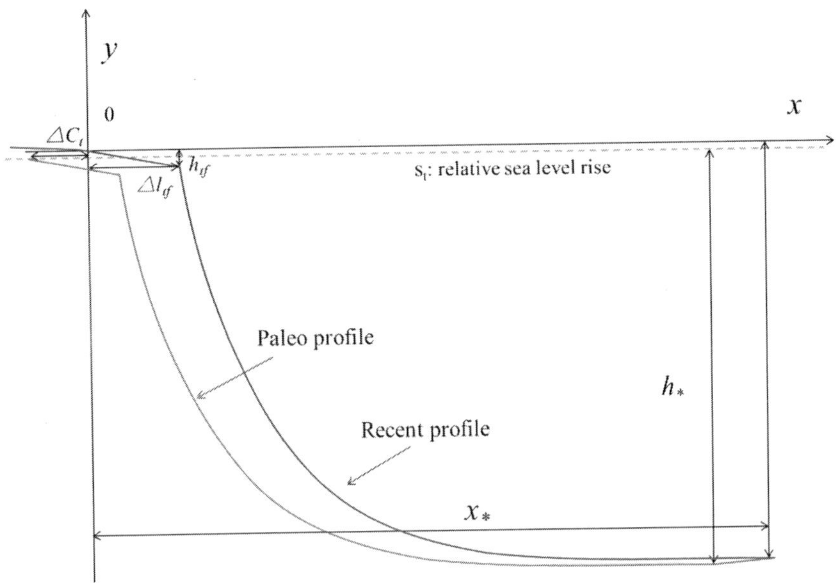

Figure 5. Sketch showing parameters for modeling coastal profile shapes (accretion type)

The mass balanced mathematical equation can be expressed as follow:

$$P * V_{River} = \left| \sum_{j=1}^{n} V_{accretion,j}(const2, \Delta c_{t,j}, s_{t,j}, b_{0,j}, a_j) \right|, \quad t \leq 0 \qquad (7)$$

In this equation, V_{River} is the sediment flux (m^3) from the river and P stands for the percentage of this flux controlling deposition at the model domain. P is a function of joint effects of wave, tide and current, which is derived from the process-based modelling experiments. For a given time t, the data of a modern Digital Elevation Model, historical coastline changes, and relative sea level records can provide the information of parameters of Δc_t, s_t, b_0 and a, hence, the only unknown parameters are *const*1 and *const*2. † can be thus simplified as:

$$P * V_{River} = \left| \sum_{j=1}^{n} V_{accretion,j}(const2) \right| \qquad (8)$$

*Const*2 is the uniform ratio between paleo and present curvature coefficients of all cross-shore profiles at the accreted coastal zones. By using the information of a modern DEM, historical coastline changes, and relative sea level records, †describes how the coastal profile shape (i. e. *const*2) is determined by riverine sediment source V_{River} and hydrodynamic force P at the research area.

When the alongshore boundary is determined by the nodal points where lateral sediment flux is zero and the whole model domain includes the erosion and accretion coastal segments, then † should be re-written as follow:

$$P * V_{River} = \left| \sum_{j=1}^{n} V_{accretion,j}(const2, \Delta c_{t,j}, s_{t,j}, b_{0,j}, a_j) \right| - \left| \sum_{i=1}^{m} V_{erosion,i}(const1, \Delta c_{t,i}, s_{t,i}, b_{0,i}, a_i) \right|, \; t \leq 0 \quad (9)$$

In this case, either *const*1 or *const*2 has to be assumed to equal to 1 so as to obtain the solution of †. The calibration procedure will be needed to decide which shape parameter needs to be set to 1. Variation of the riverine sediment sources V_{River} and hydrodynamic (i. e. wave, current and tide) force P induce a dynamic equilibrium evolution of coastal profiles, as is shown in Figure 6.

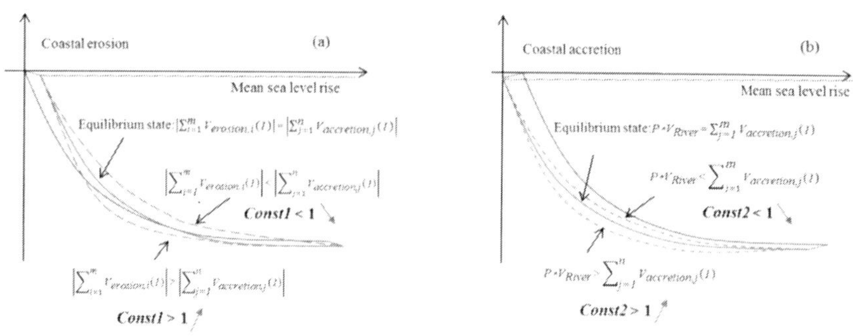

Figure 6. Examples showing how sediment sources and sinks affect coastal bathymetrical profile morphology in the DESM model for (a) semi-enclosed wave dominated sandy coast; (b) river-dominated coastal accretion area with influence of tide (Blue curve: present coastal profile; Red curve: paleo-coastal profile) (Modified from Deng *et al.*, 2014)

MODEL APPLICATIONS

Model setting

Historical coastline changes

For the historical coastline changes, the British map (R. C. Carrington of the Hydrographic Office, 1891) covering the whole Bohai Sea and north Yellow Sea (Figure 7) is used for the comparison. As can be seen in Figure 7, this map does not include, not proved any human constructions that can be used as the control points. The coordinates given in the map are regarded as control points. The projected map coordinate system is UTM zone 50 on WGS84 spheroid. As this map covers a large area, only local study area Laizhou Bay is geo-referenced by using first-order polynomial transformation method. The Root Mean Square Error (RMSE) is about 300 m. The lack of control points in the map makes the accuracy assessment and precise rectification rather difficult. But, the alternative assessment was conducted by measuring the width of coastline line (that is about 200 m) in the geo-referenced map.

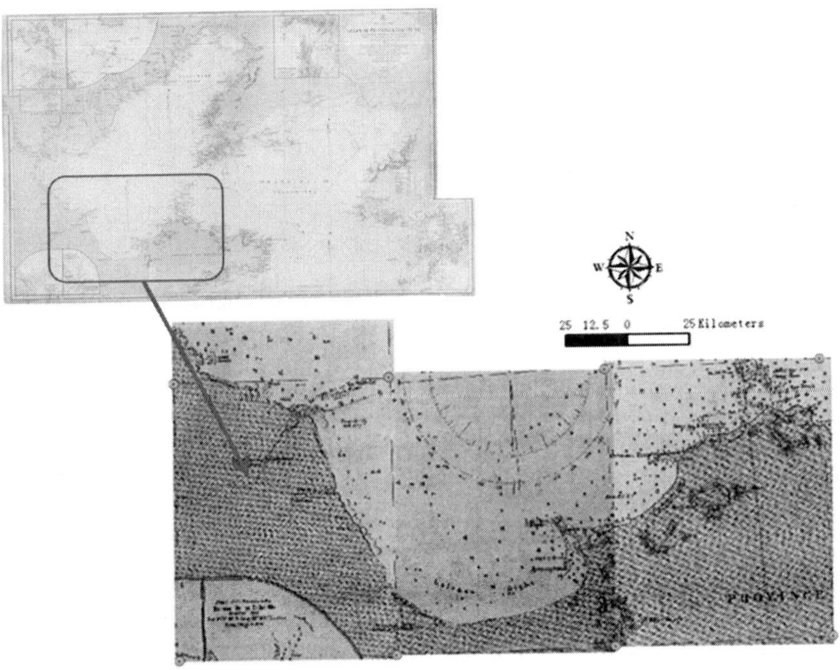

Figure 7. British map covering the whole Bohai Sea and north Yellow Sea (Control points marked as solid circle)

The result of the comparison of coastlines in Figure 8 shows a rapid progradation of Yellow River Delta, and a relocation of Yellow River Mouth. The accuracy error is definitely not comparable to this magnitude of coastline changes, and this allows the application of the DESM model to quantify the sediment mass balancing. The model domain thus includes the Yellow River Delta and only the muddy coastal part.

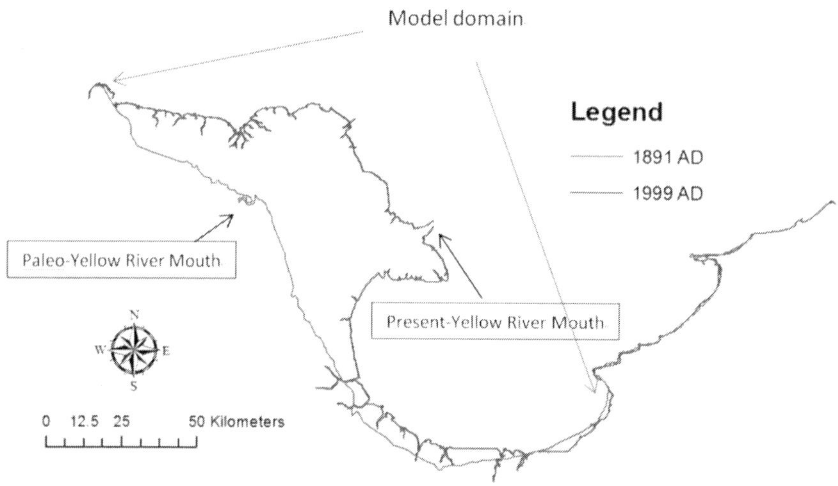

Figure 8. Comparison of coastlines between 1891 AD and 1999 AD

Recent DEM

Figure 9 shows the recent DEM in 2000s that comprises the nautical chart data mainly measured in 2002 AD, the Shuttle Radar Topography Mission (SRTM) data for the land relief elevation, and the water line extracted from the Landsat satellite images in 1999 AD. According to the tidal table (http://app.cnss.com.cn/tide_search.php), the water line is at the mean water line that is regarded as the coastline in 1999 AD. Hereafter, the recent DEM is called 2000 AD DEM.

The whole model domain depicted in Figure 8 is discretized into 242 zones alongshore with a uniform spacing of 1000 m. This long coastal section of 242 km length needs a changing value of closure depth along the coast, as the alongshore similarity of hydrodynamic forces is unlikely to remain. This is particular the case at the prograding River Mouth System where the closure depth should be the largest in front of the River Mouth due to the direct input of river sediment flux. Based on the morphological feature from the recent DEM (Figure 9), the 20 m isobath in front of the paleo-Yellow River Mouth seems to be the seaward limit of sediment progradation at the relatively flat sea bottom surface. Accordingly, the closure depth decreases eastward to 12 m (Figure 9a). Therefore, the artificial selection of alongshore variation of the closure depth to some extent follows the feature of sediment dynamic at the research area. The precise determination is not able to be done according to the current knowledge.

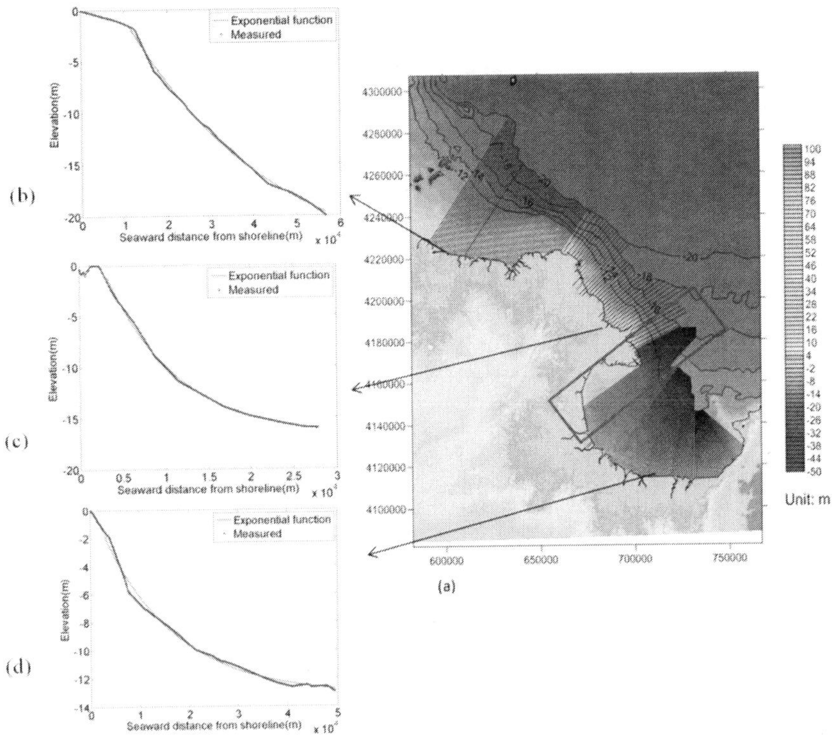

Figure 9. Discretization of model domain (DEM) and the three examples showing the fitness between exponential function and the bathymetric al profiles

The fitness between the exponential function and the measured bathymetrical profiles below the tidal flat is tested by using R^2. The average R^2 of the 242 profiles is 0.96. There are three examples are given in Figure 9b, c and d. The tidal flat with the width up to several kilometers is assumed to be a linear surface whose boundary is defined as the 1.5 m isobaths that is almost able to distinguish tidal flat at all profiles based on the modern DEM in 2000 AD. For some examples of submarine profiles, please see Figure 9.

Sea level and river sediment discharge

Recent relative sea level changes are recorded by gauges' time series data from the western and the eastern Laizhou Bay from different geological settings. In Figure 10, gauge data of the western Laizhou Bay (Yangjiaogou) and eastern Laizhou Bay are depicted. The mean relative sea level rise is about 2.1 mm/yr from 1950 AD to 1994 AD at the western subsiding coast, while there is only 1.7 mm/yr from 1950 AD to 2010 AD at the eastern coast. The reason for minor relative sea level rise values in the East of the Laizhou Bay is explained here by the uplifting tendency of the Shandong Peninsula (Li *et al.*, 1991).

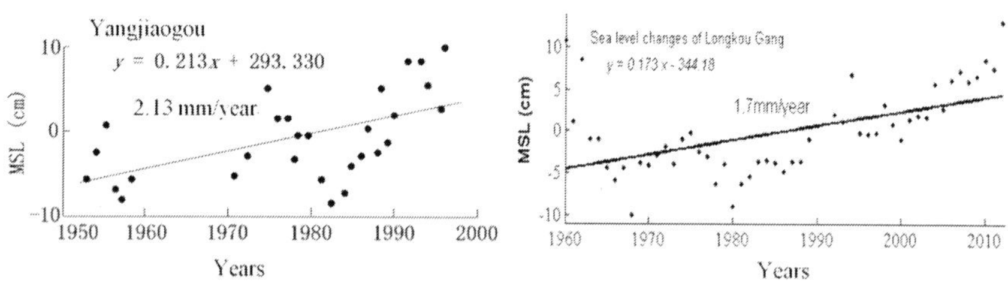

Figure 10. Sea level changes from gauge data of Yangjiaogou (SW Laizhou Bay) and Longkou Gang (E Laizhou Bay). For the location, please see Figure 2

The Yellow River sediment flux decreases from up to $20*10^8$ t/yr in 1950s AD to ca. $0.2*10^8$ t/yr in 2005 AD at the Lijin station due to mainly anthropogenic influences (Figure 3). However, for the time span from 1891 AD to 2000 AD, the annual sediment load of 749 Mt/yr is adopted and the density of sediment is 1.36×10^3 kg/m^3 that is used for conversion to the volumetric unit. About 70% of this sediment load from the river deposited in the Bohai Sea, according to the wave-tide-circulation modelling study at the Bohai Sea and Yellow Sea that was carried out by Lu *et al.* (2013).

Model results

After the iterative inverse procedure, the *const*2 in the † is determined as 0.78. The output DEM in 1819 AD is compared with 2000 AD DEM in Figure 11. In this comparison, it shows a progradation process between 1891 AD and 2000 AD. The outcome of the model indicates a tendency direction of a dynamic equilibrium evolution of coastal sea bottom morphology. This tendency of coastal profile evolution coincides with the research by Liu *et al.* (2011) on the measured data of estuarine morphology.

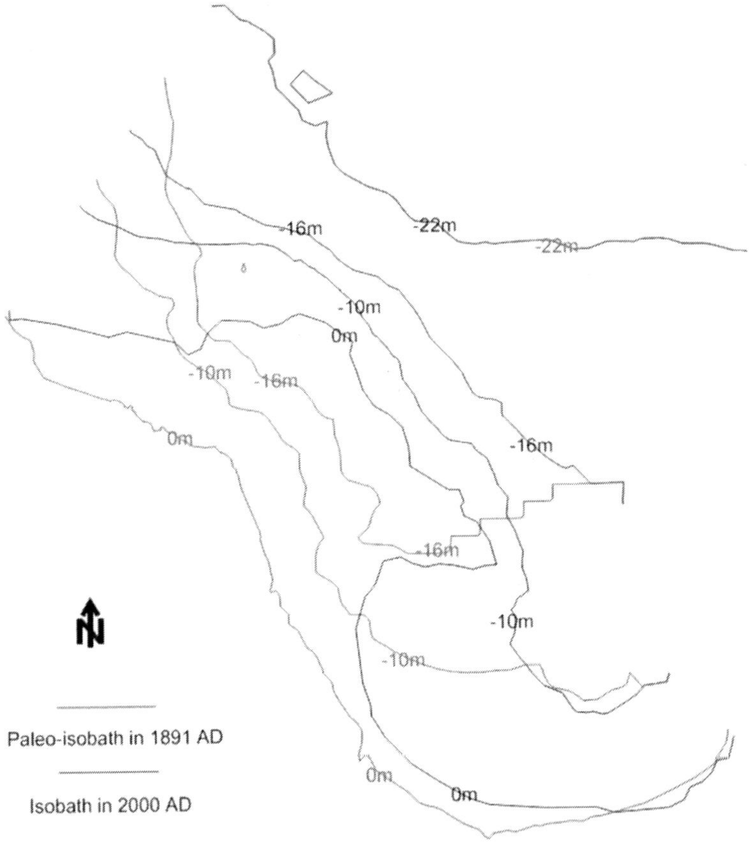

Figure 11. Isobaths comparison between 1819 AD DEM and 2000 AD DEM

We tested how changing river sediment flux influences the coastal morphological stability when the parameters of coastline changes stay unchanged. If the river sediment flux increases by ~40%,

$$P * V_{River} = \left| \sum_{j=1}^{n} V_{accretion,j} (1) \right|$$

that means the coastal morphology is stable. Under this condition, the riverine sediment flux determines this morphological evolution. Figure 6 illustrates how river sediment flux affects dynamic equilibrium states of coastal profiles. If the river sediment flux is lower than the given value in the model setting, the value of *const*2 will be even smaller than 0.78. The smaller *const*2 means the present coastal profile shape would become more steeper than the paleo one, and vice versa (Figure 6). We also tested the sensitivity of the parameter of relative sea level rise. The double rate of sea level rise from 1891AD to 2000 AD can only accommodate additional 5% of the total sediment volume. In order to balance the sediment budget, the DESM model has induced a smaller *const*2 of 0.73 than 0.78. This smaller *const*2 also means a more steeper present coastal profile relative to the paleo one.

Yellow River sediment flux tends from Figure 3 to decrease in the future if anthropogenic activities continue in the upstream of the river, and the sea level rise is likely to be accelerated due to climate change and delta subsidence. The above model sensitive tests show that both boundary conditions of the sea level and river sediment flux are main driving factors for coastal morphology development. When assuming constant coastline changes in the future, it can be anticipated that the current tendency of river sediment flux and sea level change will not induce an equilibrium state of coastal morphodynamic evolution. On the contrary, it is likely that the coastal morphology will continue this tendency and will not have sufficient sediments to fill the increasing accommodation space because of the rising sea level. At the south and east of the Laizhou bay that is relatively far away from the Yellow River Delta, there are less sediments coming from Yellow River than the adjacent area of the river delta. The decreasing river sediment flux and rising sea level would make this area more vulnerable to erosion than the west Laizhou Bay that is just next to the Yellow River Delta. As the model can only reflect the aggregated effects, the model cannot describe the intermediate processes between 1891 AD and 2000 AD, such as river route shifts that may temporally influence the sediment transport pattern.

DISCUSSION AND SUMMARY

River mouth systems play a key role for the generation of source-to-sink models describing the pathway of particulate matter discharged by the rivers to the ocean. The transformation of matter and energy in these zones of transition from the continent to the receiving marine basins depends on the dominance of the interfering main sources of energy governing the sediment dynamics: river (fluvial), tides and waves (Milliman and Farnsworth, 2013). During the younger Holocene the Yellow River Delta and the coastal zones within the vicinity are regarded a typical river dominated system with a prograding delta and a regressive sea in the easterly adjacent Laizhou Bay. The morphodynamic development of the river mouth system is well recorded by paleo-coastlines and for the younger history by historical maps. The progradation of the sedimentary system was even amplified by the Holocene sea regression.

But, during the last decades a shift in the development of the coastal area is being recorded. The Yellow River mouth starves of sediment load since damming up-streams reduces the discharge of suspended matter to the Bohai Sea. So, the influence of wave and tide dynamics is replacing fluvial influences increasingly. Additionally sea level gauge data records a rise of the sea level simultaneously. Our investigation of waterline shifts supports the hypothesis of increasing coastal retreat in Laizhou Bay (Yan *et al.*, 2014). Moreover, our DESM model results have also demonstrated the future tendency of coastal retreat is likely to happen particularly at the south and east of Laizhou Bay.

According to this result, we have to recognize that after a period of Holocene regression the area undergoes probably a new period of sea transgression. This transgression is most probably due to the recent sea level rise as a result of modern global warming. A second reason has to be searched in the continuous decrease of riverine sediment supply. For the Yellow River delta we can formulate the hypothesis of a shift in the environment as depicted in Figure 12. This envisaged development would cause quite different conditions for the management of the delta and the coasts in its vicinity. In order to get closer to a realistic future projection new concepts in morphogenetic modelling have to be set up. Besides data about sea level change and river discharge variability, meteorological information about wind and wave climate and storm frequencies have to be involved into the analysis. Different data categories have to be made available. The scale spans from future climate modeling results to planning information about actions upstream the Yellow River valley steering the discharge of water and sediment supply. The DESM model (Deng *et al.*, 2014) has proved to provide reasonable scenarios as future projections of coastal developments of marginal seas. It is suggested to continue the study of coastal development of the Bohai Sea by applying the DESM model.

The complexity of the system to be studied requires interdisciplinary and international co-operation. On the other side, the changing environmental conditions of river mouth systems turn out to become of global importance. So, comparative studies, in particular with marginal seas in low and high latitudes will help to understand functioning of river mouth and coastal systems.

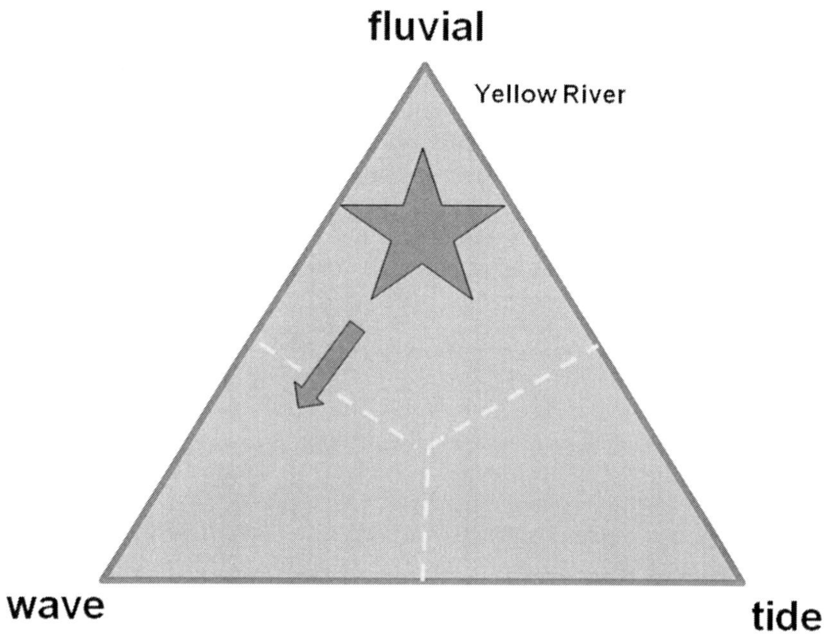

Figure 12. River mouth classification system by acting factors and current positions of the Yellow River Delta. The arrow marks a hypothetical anthropogenically triggered change of the natural environments

ACKNOWLEDGMENTS

This study is carried out in the frame of the bilateral co-operation between University of Szczecin, Poland and Yantai Institute of Coastal Zone Research, CAS, China. The authors express their thanks to the Polish National Science Centre (Narodowe Centrum Nauki: NCN) for the financial support with the No. DEC–2011/01/N/ST10/07531 as well as to the Chinese Academy of Sciences for financing the exchange of scientists.

LITERATURE CITED

1. Anonymous: Qinhuangdao Gang to Qihe Kou, Scale 1 : 250000, The Navigation Guarantee Department of the Chinese Navy Headquarters, Tianjin (1959–2009).
2. Anonymous: Qihe Kou to Longkou Gang, Scale 1 : 250000, The Navigation Guarantee Department of the Chinese Navy Headquarters, Tianjin (1958–2011).

3. Bian, C.; Jiang, W., and Greatbatch, R. J., 2013. An exploratory model study of sediment transport sources and deposits in the Bohai Sea, Yellow Sea, and East China Sea. *Journal of Geophysical Research*, 118, 5908–5923.
4. Bruun, P., 1962. Sea-level rise as a cause of shore erosion. *Journal Waterways and Harbours Division*, 88(1–3), 117–130.
5. Bruun, P., 1988. The Bruun Rule of Erosion by Sea-Level Rise: A Discussion on Large-Scale Two- and Three-Dimensional Usages. *Journal of Coastal Research*, 4(4), 627–648.
6. Deng, J.; Zhang, W.; Harff, J.; Schneider, R.; Dudzinskanowak, J.; Terefenko, P., and Furmańczyk, K., 2014. A numerical approach for approximating the historical morphology of wave-dominated coasts—A case study of the Pomeranian Bight, southern Baltic Sea. *Geomorphology*, 204, 425–443.
7. Deng, J.; Harff, J.; Li, Y., and Zhang, W., 2014. Modelling the Paleo-geomorphology of wave-dominated sandy and tide-influenced muddy coastal embayments on the decadal to centennial time scale-a comparative study. In: Raju, N. J. (eds.), Geostatistical and Geospatial Approaches for the Characterization of Natural Resources in the Environment: Challenges, Processes and Strategies. Springer-Verlag, Heidelberg, 548–551.
8. Dissanayake, D. M. P. K.; Ranasinghe, R., and Roelvink, J. A., 2012. The morphological response of large tidal inlet/basin systems to Relative Sea Level Rise. *Climatic Change*, 113, 253–276.
9. Harff, J. and Meyer, M., 2011. Coastlines of the Baltic Sea—zones of competition between geological processes anda changing climate: examples from the Southern Baltic—Springer. In: Harff, J., Björck, S., Hoth, P. (eds.), *The Baltic Sea Basin*. Springer-Verlag, Heidelberg. pp. 149–164.
10. Harff, J. and Lüth, F., 2011. SINCOS II – Sinking Coasts: Geosphere Ecosphere and Anthroposphere of the Holocene Southern Baltic Sea. – *Ber. Röm. – Germ. Komm.*, 92, 1–380.
11. Harff, J.; Meyer, M.; Zhang, W.; Barthel, A., and Naumann, M., 2011. Holocene sediment dynamics at the southern Baltic Sea. – *Ber. Röm. – Germ. Komm.*, 92, 41–76.
12. Harff, J.; Leipe, T., and Zhou, D., 2010. Pearl River Estuary related sediments as response to Holocene climate change and anthropogenic impact (PECAI). *Journal of Marine Systems*, 82(1), S1–S89.
13. Harff, J.; Leipe, T.; Waniek, J., and Zhou, D., 2013. Depositional environments and multiple forcing factors at the South China Sea's northern shelf. *Journal of Coastal Research*, SI66, 90 p.
14. Harff, J.; Graf, G., and Bobertz, B., 2009. Dynamics of natural and anthropogenic sedimentation (DYNAS). *Journal of Marine Systems*, 75, 315–451.
15. IPCC, 2013. The Physical Science Basis. Contribution of Working Group I to the Fifth Assessment Report of the Intergovernmental Panel on Climate Change. In: Stocker, T. F.; Qin, D.; Plattner, G. – K.; Tignor, M.; Allen, S. K.; Boschung, J.; Nauels, A.; Xia, Y.; Bex, V., and Midgley, P. M. (eds.), *Climate Change* 2013 (1535). Cambridge University Press, Cambridge, United Kingdom and New York, NY, USA.
16. Li, C.; Chen, G.; Yao, M., and Wang, P., 1991. The influences of suspended load on the sedimentation in the coastal zones and continental shelves of China. *Marine Geology*, 96, 341–352.
17. Liu, F.; Chen, S.; Peng, J., and Chen, G., 2011. Temporal variability of water discharge and sediment load of the Yellow River into the sea during 1950–2008. *Journal of Geographical Sciences*, 21(6), 1047–1061.
18. Lu, J.; Qiao, F.; Wang, X.; Teng, Y.; Jung, K. T., and Liu, Y., 2013. Modeling the Yellow River sediment flux and its deposition patterns under climatological conditions. *Ocean Dynamics*, 63, 709–722.
19. Pirazzoli, P. A. and Pluet, J., 1991. World atlas of Holocene sea-level changes. *Elsevier Science*, 121 p.
20. R. C. Carrington of the Hydrographic Office: China–East Coast: Bohai and Noth Yellow Sea. *United Kingdom of Britain*, 1891.

21. Saito, Y.; Chaimanee, N.; Jarupongsakul, T., and Syvitski, J. P. M., 2007. Shrinking megadeltas in Asia: Sea-level rise and sediment reductionimpacts from case study of the Chao Phraya Delta, Inprint: newsletter of the IGBP/IHDP Land Ocean Interaction in the Coastal Zone, pp. 3–9.
22. Syvitski, J. P. M.; Kettner, A. J.; Overeem, I.; Hutton, E. W. H.; Hannon, M. T.; Brakenridge, G. R.; Day, J.; Vörösmarty, C.; Saito, Y.; Giosan, L., and Nicholls, R. J., 2009. Sinking deltas due to human activities. *Nature Geoscience*, 2, 681–686
23. Wang, H.; Yang, Z.; Saito, Y.; Liu, J.; Sun, X., and Wang, Y., 2007. Stepwise decreases of the Huanghe (Yellow River) sediment load (1950–2005): Impacts of climate change and human. activities. *Global and Planetary Change*, 57, 331–354.
24. Xue, C., 2009. Historical changes of coastlines on west and south coasts of Bohai Sea since 7 000 a B. P. *Scientia Geographica Sinica*, 29(2), 217–222.
25. Zhang, J. and Wang, J., 1999. Combined impacts of MSL rise and the enlarged tidal range on the engineering design standard in the areas around the Huanghe River Mouth in Chinese. *Marine Science Bulletin*, 18(5), 1–9.
26. Zhang, W. Y.; Harff, J.; Schneider, R., and Wu, C. Y., 2011. A multi-scale centennial morphodynamic model for the southern Baltic coast. *Journal of Coastal Research*, 27, 890–917
27. Zhang, W. Y.; Schneider, R., and Harff, J., 2012. A multiscale hybrid long-term morphodynamic model for wave-dominated coasts. *Geomorphology*, 149–150, 49–61.
28. Zhang, Z.; Li, J.; Qiu, S. H.; Xian, Z. Q.; Bao, G. C., and Wang, Y. J., 1987. Holocene transgression and its layers in the east coast of Laizhou Gulf, Bohai. *Transactions of Oceanology and Limnology*, 2, 31–39.
29. Zhao, Y.; Li, Y.; Deng, J; Harff, J.; Tang, C., and Zhang, H., 2014. Sea level and coastline change in the Bohai Seanatural and anthropogenic impacts, poster presentation–*"The conference" Our Climate-Our Future, "Regional perspectives on a global challenge"* (Berlin, Germany).

CHAPTER 6

Coastal Flood Assessment Based on Field Debris Measurements and Wave Runup Empirical Model4

David Didier *, Pascal Bernatchez, Geneviève Boucher-Brossard, Adrien Lambert, Christian Fraser, Robert L. Barnett and Stefanie Van-Wierts

Coastal Geoscience, Centre for Northern Studies, University of Quebec in Rimouski, Rimouski, QC G5L3A1, Canada

ABSTRACT

On 6 December 2010, an extra-tropical storm reached Atlantic Canada, causing coastal flooding due to high water levels being driven toward the north shore of Chaleur Bay. The extent of flooding was identified in the field along the coastline at Maria using DGPS. Using the assumption that the maximum elevation of flooded areas represents the combination of astronomical tide, storm surge and wave runup, which is the maximum elevation reached by the breaking waves on the beach, all flood limits were identified. A flood-zone delineation was performed using GIS and LiDAR data. An empirical formula was used to estimate runup elevation during the flood event. A coastal flood map of the 6 December flood event was made using empirical data and runup calculations according to offshore wave climate simulations. Along the natural beach, results show that estimating runup based on offshore wave data and upper foreshore beach slope represents well the observed flood extent. Where a seawall occupies the beach, wave breaking occurs at the toe of the structure and wave height needs to be considered independently of runup. In both cases (artificial and natural), flood risk is underestimated if storm surge height alone is considered. There is a need to incorporate wave characteristics in order to adequately model potential flood extent. A coastal flooding projection is proposed for Pointe Verte based on total water levels estimated according to wave climate simulation return periods and relative sea-level rise for the Chaleur Bay.

Keywords: storm surge; wave runup; coastal flood mapping; coastal defense

1. INTRODUCTION

Sea-level rise, as a consequence of climate change, is a primary concern to coastal communities due to the increased threat of storm surges and coastal flooding [1,2]. The rate of global mean sea-level rise increased from 1.9 mm/year to 3.2 mm/year between the periods 1961 to 2009 and 1993 to 2011 [3,4]. This acceleration, combined with future projections of sea level and extreme water level change [5,6,7,8,9,10,11] has led the scientific community to pay greater attention to the development of vulnerability assessment methods for coastal communities and infrastructure to flooding phenomenon [12]. Previously, the vulnerability of coastal communities to future sea-level rise has been estimated by applying hypothetical sea-level increases (e.g. often a hypothetical 1 m rise is used, [13,14,15]) or predictions from SRES and IPCC scenarios [16,17] to topographic maps or digital elevation models at national or greater scales [18,19]. However, adaptation approaches to coastal flooding require the development of local to regional level assessments [20]. In this context, the regional distribution and amplitude of sea-level rise remains a significant uncertainty when forecasting future trends [6]. In addition, the isostatic component (vertical land motion) is still rarely accounted for in coastal flooding assessment methods. It is evidently essential, therefore, to assess future *relative* sea-level rise [21], especially in areas experiencing crustal subsidence due to postglacial isostasy [22,23].

Coastal flooding occurs when water levels overtop the first natural ridge or flood defense crest landward of the beach, generating landward flow and sediment transport [24]. Alongshore water level variability, relative to the crest elevation, further affects overwash potential [25] and can lead to an increase in the risk of flooding. During coastal storms, resulting water level maxima are a product of astronomical tide, barometric pressure and wind-induced surge [26,27]. In addition, landward propagating waves in shallow waters are affected by beach morphology due to the effects of, e.g., shoaling, wave breaking and energy dissipation [28]. Such interactions can increase water levels in the onshore direction through the process of wave runup following breaking in the surf zone [29]. Runup (*R*) is defined as the vertical elevation difference between the maximum shoreward location of water and still water level (predicted tide + surge) [30,31,32] and includes two components: the setup and swash [33]. Setup is defined as a mean water surface elevation in response to wave breaking, which is a superelevation, whereas swash refers to oscillations across the water-land interface around setup. The sum of tide prediction (, surge and runup on natural beaches correspond to total water levels (TWL) [34,35]. Runup is generally calculated in terms of the two percent exceedance value of wave runup ($R_{2\%}$) for the TWL [34].

Runup (*R*) on a natural beach mainly depends on offshore significant wave height (*H0*) and wavelength (; where g is the acceleration due to gravity (981m/s), *T* is the wave period (s)) and the beach slope [29,36,37,38] grouped under the non-dimensional Iribarren number . The Iribarren number, also referred to surf zone similarity parameter, is widely used in the computation of runup and overtopping discharge [37] and in near-shore process assessments [28]. Beach slope is therefore an important parameter for determining maximum wave runup height along coasts. However, beach surfaces used to calculate slopes, which can differ significantly between case studies, are not always clearly identified or described in the literature [29,32,34,39].

Wave runup on structures is typically calculated using a design wave for an expected water level in order to determine the required crest elevation for defense structures [40]. Types of structures range from low sloping coastal defenses to vertical seawalls. The main factors that influence flooding over sloping structures are wave runup and overtopping [41,42]. However, for vertical seawalls where , wave runup is minimal and is limited to a theoretical value of [43]. In such cases, where waves frequently impact the structure, wave height is usually twice the incident wave amplitude and can produce a vertical plume of *circa* 5.5 times the wave amplitude [44]. On sloping coastal defenses, such as dykes, wave overtopping occurs when wave runup exceeds the height of defense structures [45].

It is a challenge to accurately model the landward flow of water following overtopping of defense structures due to complex interactions between wave morphology and the structures themselves [46]. Studies on flow characteristics and discharge rates following wave overtopping have highlighted the importance of the roughness factor and permeability associated with different defense structures [47,48]. Whereas overtopping associated with different coastal structure types has been evaluated [47,49], less is known on the spatial propagation of water levels in backshore zones following overtopping of the structures [46], which is of key relevance to coastal managers responsible for land use planning and flood risk management [50]. Moreover, the establishment of rigid defense structures in response to ongoing erosion processes often results in decreasing beach and intertidal zone widths [51]. This can lead to increased wave overtopping of defense structures during storms [52,53]. Such modifications to natural beach states can increase flood risk [51,54]. The alongshore variability in flood elevation as a result of human intervention in response to coastal hazard represents a threat to coastal communities.

Studies on runup have typically been carried out along stretches of natural beaches (e.g. [29,38,55,56,57]] or in areas of armoured coastline (e.g. [41,42]). In reality, coastal zones prone to flood risk will contain a combination of both natural and artificial sections, which highlights the need for a local to regional approach when considering coastal flooding vulnerability.

To date, the majority of coastal flooding studies have been carried out along coastlines exposed to ocean swell with long fetch and rarely on sheltered embayments [58]. In sheltered coastal environments, it is generally considered that the waves have less influence on coastal flooding and the tide and storm surge are the main drivers [58]. In Eastern Canada, the entire area of the Chaleur Bay (Figure 1) has been hit by two major and destructive storms recorded within a five year interval (December 2005 to December 2010), despite the fact that the Acadian peninsula partially protect the bay to wave agitation coming from the Gulf of St. Lawrence. However, the recent storm events have raised questions with regards to the influence of wave effects along coastlines in sheltered bays with short fetch. There is a critical need to understand wave influence in these environments in order to adapt and improve coastal zone management in Eastern Canada.

Mapping coastal flood risk areas is an important step for risk management planning and for guiding adaptation procedures necessary to reduce vulnerability to flooding. The classical method of flood mapping, which typically integrates extreme water levels, derived from historic tide gauge data, with a digital elevation model ("bathtub model") to identify areas at risk [31,59,60,61], has been criticized in recent years [62]. Recent storm events along North American (e.g. Katrina, Wilma, Sandy) and European (e.g. Xynthia) coastlines have highlighted the need to understand the effects of beach geomorphology and hydrodynamic conditions close to the coast in order to accurately determine backshore zones at risk of flooding [20,31,34]. Due to a paucity of wave data at local to regional scales, wave runup is not always included in coastal flood mapping. This is especially prevalent in Atlantic Canada [63,64]. Another challenge to flood risk mapping derives from forecasting extreme water level return periods based only on short tide gauge series [65,66]. It is also important to integrate the probability of extreme tidal water levels with runup data [34] for future coastal management and robust flood hazard mapping. Recent studies have shown that airborne LiDAR data, widely used for digital elevation models in flood hazard mapping, are not precise enough to accurately identify crest elevations of the first line of defense and to represent adequately elevation changes of flood defenses (embankment crests and walls) or beach berms at the coastal front or along river banks, which leads to over- or under-estimating of risk areas [67,68,69,70].

The development of methods for flood hazard mapping remains a challenge, especially at the local to regional scale, in order to be applicable and easy to use by local authorities responsible for land use and risk management. Therefore, only few approaches integrate the different components required to effectively map the exposed areas to flood hazards for the future.

The aim of this paper is to develop a coastal flood mapping approach based on *in situ* measurements, modelled wave characteristics, sea-level rise, estimated extreme water levels return periods, and wave runup estimations using empirical formula. This method is compared and validated from field surveys during and immediately after the 6 December 2010 storm event that caused extensive damage to the coast of Eastern Canada. We compare the altitude reached by the total water level during the storm based on the presence or not of coastal defenses and assess the importance of the waves in the flood pattern for sheltered coasts. Since slope determination is often of primary concern in runup parameterization, a focus was made on beach morphology with a mobile terrestrial LiDAR system (MTLS) to calculate beach slopes. The use of terrestrial LiDAR system also aims to solve the problem of identifying the elevation crest of the first coastal defense line. Finally, the proposed approach addresses future coastal flood scenarios at the community scale that can be easy to apply for coastal managers.

2. STUDY AREA

In Atlantic Canada, common extra-tropical cyclones typically generate northwestern and northeastern winds toward the shores of the Gulf of St. Lawrence, creating storm surges and high waves [71,72]. As a result of these low-pressure systems, coastal flooding and erosion have increased in recent years throughout the Chaleur Bay [73]. The study area is located in the municipality of Maria (Figure 1) on the north shore of the Chaleur Bay in the province of Quebec, Eastern Canada. Maria represents a typical coastal community for Eastern Canada, with a population of 2500 and an established mix of infrastructure (such as residential, recreational, commercial and tourism). The national road, 132, which links coastal towns throughout the Gaspé Peninsula with eastern Quebec, is particularly vulnerable to flooding at Maria [74]. Despite damage caused by several flooding events [75], there is currently no coastal flood hazard mapping in the province of Quebec.

Flood hazard has been assessed for the southeast part of Maria where there are more than 40 houses built on the sandy low-lying coast (Figure 1). This area was flooded both on 2 December 2005 and 6 December 2010. This stretch of coastline contains a uniform sandy and gravel beach to the east (~22 m width), which is in a quasi-natural state (Figure 2a) and a protected zone to the west, which contains a wooden seawall protecting different properties [52]. The eastern section contains some low lying coastal defenses, which are situated landward of the beach or on the upper beach, with a structure toe above HAT. We therefore assume an alongshore and natural wave energy dissipation across the beach in this area, where runup can occur on the sloping beach, as underlined by the Mase *et al.* [41] study on wave runup and overtopping.

At the western zone of the study site, almost the entire beach has been eroded as a result of wave reflection and scouring at the wall toe [51,52] (Figure 2b). Beach erosion has exposed a substrate of pebbles and gravel (Figure 2b). Backshore elevation in this area is higher than in the eastern part of the site. The region experiences a semi-diurnal and meso-tidal regime, with a mean sea level (MSL) of 1.33 m relative to Chart Datum [76] (Table 1). During highest astronomical tide (HAT) (1.66 m), the seawall toe is submerged and oscillating water levels occur on the seaward face of the wall. As observed during the 2005 flood event [52], continuous wave overtopping can occur during periods of high water levels and large waves.

Table 1. Mean estimated tidal values in 2010 according to chart and geodetic datum, Belledune station (NB) (Canadian Hydrographic Service).

Water Levels	Mean Estimated Value (2010) Chart Datum (CD)	Canadian Geodetic Vertical Datum 1928 (CGVD28)
Extreme level (tide + storm surge)	3.64	2.46
Highest Astronomical Tide (HAT)	2.84	1.66
Mean Sea Level (MSL)	1.33	0.15
Lowest Astronomical tide (LAT)	0.1	−1.08

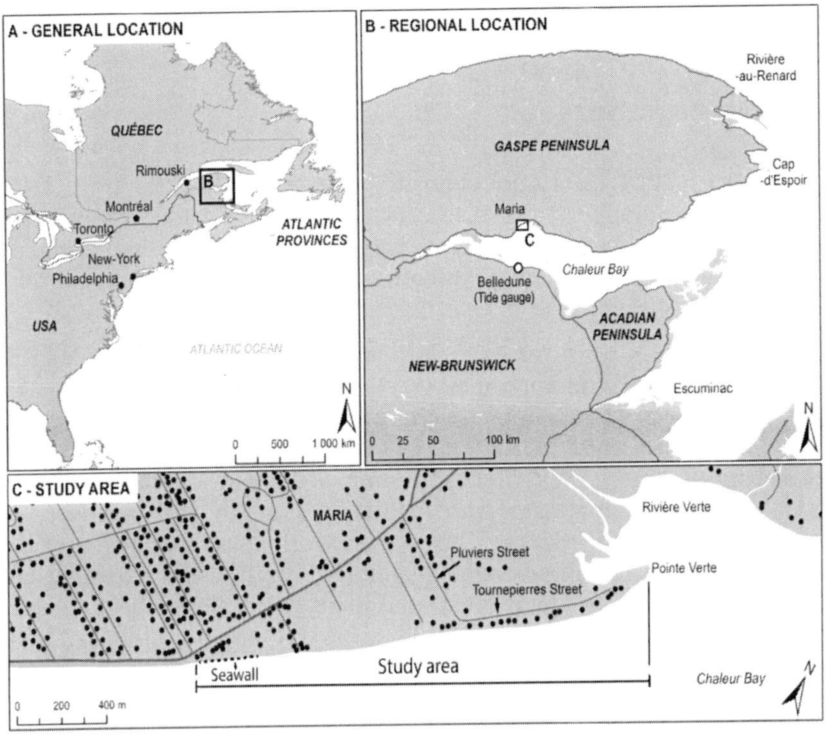

Figure 1. Location map of the study area in Maria (Quebec) on the north shore of the Chaleur Bay.

Figure 2. The beach of the study area is mainly natural on the eastern part (**a**) and characterized by a seawall on the western coastline (**b**).

3. METHODOLOGY

3.1. Post-Storm Surveys

Storm debris [77,78] and tidal deposits [79] can be used as water level limit and maximum runup extent indicators. A field survey was conducted on the day of, and immediately following, the 6 December 2010 flood event, in order to identify maximum flood water levels reached at Maria (Figure 3a).

Storm debris lines were located *in situ* using a DGPS Thales ProMark3 system (Figure 3b), with horizontal (x, y) and vertical (z) uncertainties of ± 1 cm. A total of 188 data points were collected for the 2010 event along debris lines, which were located along 2.2 km of coastline. A database of georeferenced points was then superimposed over a 2007 airborne LiDAR raster elevation data map (spatial resolution of 1 m, vertical accuracy of 20 cm) in ArcMAP. The points were connected together to form a continuous line indicating the landward limits of the water during the storm. Points were then generated every 5 m along the line and elevation values were extracted from the raster providing mean flood elevations for the (protected) western and (unprotected) eastern zones, relative to geodetic datum CGVD28.

3.2. Mobile Terrestrial LiDAR Data Acquisition

The alongshore and cross-shore variability in beach morphology at Maria has been studied by Bernatchez *et al.* [52] using DGPS profiles. Even if the natural beach is mainly homogenous in terms of sediments and morphological patterns (width, volume), morphological variability exists between areas with defense structures (west zone) and those in a quasi-natural state (east zone). We initially used airborne LiDAR data to map the first coastal defense line, but the spatial resolution and accuracy in z does not allow us to determine with sufficient precision the height of the crest elevation. We therefore used mobile terrestrial LiDAR system (MTLS) on the entire beach of the study area on 28 September 2011 (Figure 4a) to improve on spatial resolution and vertical precision of previous studies [51,52]. The survey was conducted at low tide and covered the entire foreshore to a minimum seaward elevation of 1 m below CGVD28 (close to LAT).

Figure 3. Field survey after the flood showing (**a**) flood generated overwash fan and (**b**) the debris line delineation. Limits were located using DGPS on the day following the storm event (Maria, 7 December 2010).

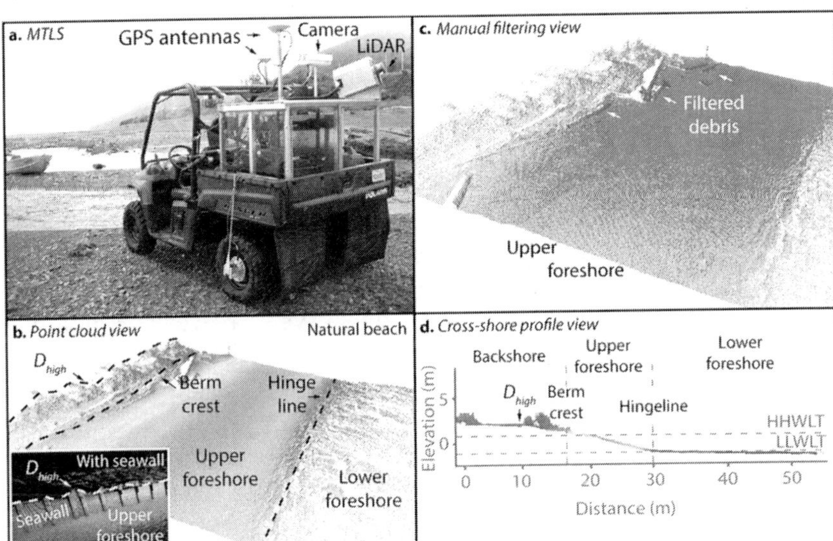

Figure 4. Mobile terrestrial LiDAR system (**a**) with point cloud and filtering view examples (**b**,**c**). Resulting data are used in cross-shore view to localize the first ridge on the natural beach or the coastal structure crest elevation (D_{high}), the berm crest and the hinge line (**d**).

The MTLS is a multi-sensor system mounted on a utility side by side vehicle (Figure 4). The multi-sensor system includes a laser scanner Riegl VQ-250, a camera PointGrey Grasshopper GRAS-50S5C-C and a GPS inertial navigation system Applanix POS-LV 220 (Figure 4a) [80]. Post-processing was performed using PosPAC and Trimble Trident Analyst software. The projection used was NAD83 MTM5 with geoid HT2.0. The LiDAR point cloud has a spatial resolution of less than 0.05 m × 0.05 m (Figure 4b). Different types of beach debris were filtered manually in LP360-QCoherent software (Figure 4c). A grid elevation with a spatial resolution of 0.5 m was created in LP360-QCoherent software in order to identify the beach slope with high accuracy.

The first line of defense (D_{high}), corresponding to the defense structure crest in the west zone, or to the crest elevation of the first ridge landward of the berm crest in the east zone, was manually digitized in ArcMAP. To calculate the accuracy of the laser data, a total of 23 control points were measured using a D-GPS Trimble R8 over the entire study site. The error was calculated by the root-mean-square (RMS) of the difference between control points and the adjacent laser points. Calculated horizontal (x,y) and vertical (z) errors are ± 0.03 m (RMS).

3.3. Hydrodynamic Data Analysis

3.3.1. Wave Dataset

No measured wave data are available for the Chaleur Bay. We therefore used a modelled wave dataset from Environment Canada's (EC) operational wave forecasting model based on WAM cycle 4 MW3 "PROMISE" [81,82,83,84]. Archived outputs cover a 10-year period, from 1 January 2003 to 31 December 2013 with a data hiatus for 2006. Time series are available at three-hour increments.

EC's implementation of WAM extends throughout the St. Lawrence Gulf and Estuary, from Montmagny to Cabot Strait and Belle-Isle Strait, at a spatial grid resolution of 0.04° latitude and 0.06° longitude (approximatively 5 by 5 km at 45° north). Selected source terms include formulation for exponential wave growth [85], formulation for linear wave generation [86], formulations for deep and intermediate water non-linear wave–wave interactions (quadruplets) and dissipation due to whitecapping [81], JONSWAP formulation for dissipation due to bottom friction [87], and formulation for dissipation due to depth-induced breaking [88].

Wave propagation schemes include refraction over bottom and non-stationary ambient currents. Non-stationary boundary conditions are based on 40 m tri-hourly winds from EC's Global Environmental

Multiscale Model (GEM) and surface currents (at 5 m depth) from EC's oceanographic model of the Estuary and Gulf of St. Lawrence (MOR). Output wave parameters derived from spectra are significant spectral wave height (H_{m0}), mean spectral wave period (T_{m02}) and mean wave direction (*Dir*).Wave data quality has been assessed by Jacob *et al.* [84] and Lambert *et al.* [89,90,91] for the Estuary and Gulf of St. Lawrence domain. Selected data for this study were extracted from EC's model at grid point i = 67, j = 69 (−65.950 O, 48.1 N), located 7 km east south-eastward from Maria's shoreline, to a depth of −18.2 m. Given the limited fetch length in front of Maria (70 km), most simulated wave periods were under 5 s. We therefore selected a position for wave parameter extraction close to the shore yet remaining in deep-water conditions ($d > 1/2\ L_0$) for most wave periods. Selected wave records were linearly interpolated to an hourly resolution in order to match temporal resolution with the available tidal dataset.

3.3.2. Determination of Near-Shore Wave Parameters

Runup values, as a result of wave setup and uprush, are determined by deep-water significant wave conditions (H_0 and L_0) and beach slope with the Stockdon *et al.* equation [29]. Wave shoaling and refraction over coastal waters can be empirically determined using weighting coefficients rather than physical computations. We will refer to H_0 in the paper as a reference to H_{m0}. The depth of wave breaking and the location of the break point on the sloping beach or relatively to a defense structure is an important factor contributing to erosion and flooding. Depth of breaking d_b is computed from (1) [92]:

$$d_b = \frac{1}{g^{1/5}\,\gamma^{4/5}}\left(\frac{H_0^2 C_0 \cos\theta_0}{2}\right)^{2/5} \tag{1}$$

where = 0.8, = deep-water wave angle relative to shore normal direction (incident wave = 0°), and deep-water phase velocity.

3.4. Wave Runup Estimation and Coastal Morphology

The flood assessment applied in this paper estimates wave runup with the empirical formulation proposed by Stockdon *et al.* [29] based on *in situ* measured runup values via extensive research conducted on ten sandy field experiments in the USA and Netherlands. The following equation has been applied because it is valid for a broad range of reflective to dissipative beaches:

$$R_{2\%} = 1.1\left(0.35\tan\beta(H_0L_0)^{1/2} + \frac{[H_0L_0(0.563\tan\beta^2 + 0.004)]^{1/2}}{2}\right) \quad (2)$$

where is the elevation above the still water level that is exceeded by 2% of the wave runups [53] and is the beach slope. This empirical runup formulation (RMS error = 0.38 m), hereinafter referred to as the RS06 model, is considered a good predictor of runup [25,33,93] and is often used in coastal flood and vulnerability assessment [67,94,95,96]. The static flood model applied in this paper consists of a summation of observed tidal level and storm surge, wave characteristics and wave runup according to the RS06 empirical model. It is based on the comparison of total water level (TWL) against ground elevation assuming an instant and constant flood surface on the area [67]. Despite its simplicity, the RS06 formulation requires a beach morphology assessment. In natural settings, the beach slope is not a straightforward calculation [32,39,97]. Some authors suggest using the foreshore beach slope in the computation of the empirical formula [29,30,63,96]. Cariolet and Suanez [32] obtained a good fit relationship between measured and predicted runup using the slope of the active section of the beach in a macrotidal environment, which corresponds to the upper foreshore slope. This is in accordance with Stockdon *et al.* [25] who used the mean beach slope between the berm crest and mean high water level to define $\tan\beta$ in storm conditions where the swash is moved up the beach. Two beach slopes on the natural beach were compared in our approach (Figure 5).

Beach slope at high tide during the flood event corresponded to the slope between the point of maximum runup elevation on the beach (R_{max}) and $T_{obs}-1.5H_0$ [43], where T_{obs} is the still water level without the wave component. This slope, referred to as $\tan\beta_{1.5H0}$, was compared to the upper foreshore slope ($\tan\beta_{ufs}$). The upper foreshore slope is relatively located upward of mean sea level (CGVD = 0.15 m) and is located between the point of maximum runup elevation and the hinge line. Considering the maximum flood height along the coast on 6 December 2010, R_{max} is assumed to be equal to the berm crest elevation in both situations. Both berm crest and hinge line were manually delineated for the entire beach as a line feature in LP360 using a 3D visualization window (Figure 4b,d). The width of the beach Y_1 was calculated between beach berm crest and $T_{obs}-1.5H_0$. Using the MTLS derived raster, a contour line equal to $T_{obs}-1.5H_0$ was automatically digitised in ArcMAP. The mean beach slope, $\tan\beta_{1.5H0}$, was then defined $(Z_{crest}-Z_{Tobs-1.5H0})/Y_1$. The mean beach slope, $\tan\beta_{ufs}$, given a beach width Y_2 between the berm crest and the hinge line, equals to $(Z_{crest}-Z_{Hinge\ line}/Y_2)$. Beach slopes were calculated every 50 meters along the coast, for a total of 28 cross-shore profiles.

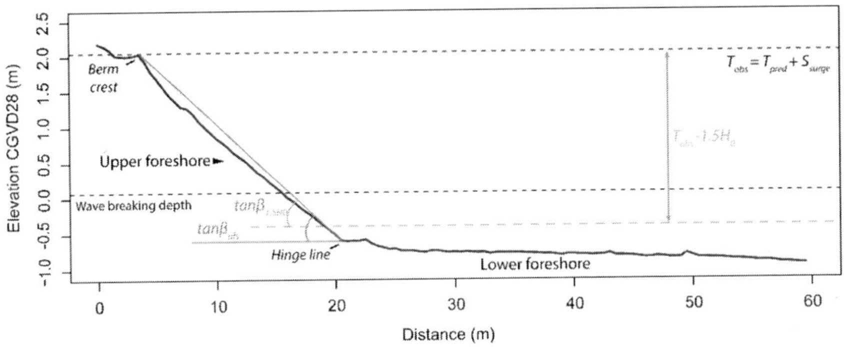

Figure 5. Natural beach slopes definitions used in the empirical runup formula. The slope $\tan\beta_{1.5H0}$ is located between the point of maximum runup elevation on the beach (R_{max}) and $T_{obs}-1.5H_0$. The upper foreshore slope ($\tan\beta_{ufs}$) is located between the point of maximum runup elevation and the hinge line. See the profile location in Figure 10, on the beach only (black line).

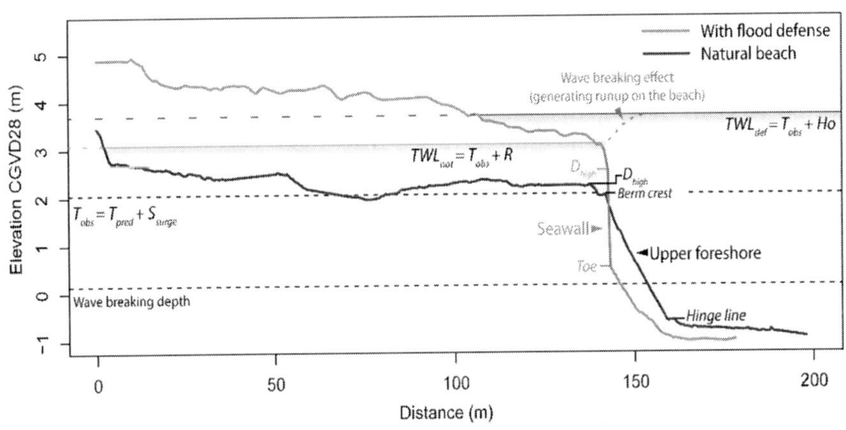

Figure 6. Sketch of the morphological difference between the defended (red) and natural (black) coastline. Modelled TWL are shown for both the natural beach (light blue) and defended coastline (dark blue). Profile locations are shown in Figure 10.

In the present study, where high water levels and wave conditions led to overtopping and flooding, bottom induced wave breaking was located close to the seawall. Since radiation stress induced wave set-down is at its maximum at the breaking point and static wave setup starts at the breaking point [98], we can assume that there is no contribution of wave setup to the total water level on the wall. As a consequence, neither runup nor setup were calculated in front of the seawall in the west zone. During the 6 December event, the cumulative offshore wave height plus storm water level elevation overtopped the defense crest by several tens centimeters. Thus, an approach using only the wave height as a criterion for flood assessment has been proposed (Figure 6).

3.5. Sea-Level Rise Projection and Flood Mapping

3.5.1. Sea-Level Analysis

Relative sea-level rise, as a result of regional glacio-isostatic adjustment and eustatic contributions, is a concern along Atlantic Canada in terms of future coastal flood risk assessment [59,63,71,99]. In the southern Gulf of St. Lawrence, the isostatic crustal movement ranges from −1 to −4 mm/year [22,23]. In Maria, the rate is estimated to be −1.78 mm/year) [100]. The nearby tide gauge at Belledune (Figure 1) provides a local relative sea-level trend of 4.06 mm/year between 1964 and 2014 for the study site, which is located 36 km northwest of the tide gauge. This tide gauge recorded data in 1964 and between 2000 and 2014. Hourly observed water levels were imported from the Canadian Tides and Water Levels Data Archives, available online [101]. The hourly predicted water levels were provided by the Canadian Hydrographic Service and were used to calculate the magnitude of surges [102]. Original tidal data are expressed relative to Chart Datum, which is 1.18 m below geodetic datum (CGVD28). All tidal and elevation references are expressed relative to CGVD28. Hourly data were smoothed to obtain monthly mean tidal levels from which the sea-level rise (SLR) trend was extrapolated. The averaged seasonal cycle and months containing less than 90% of data were removed following Daigle [103]. The SLR trend was obtained using linear regression following [104]:

$$Y = b_0 + b_1 X + \varepsilon \tag{3}$$

where b_1 is the regression coefficient estimating the SLR trend. An estimation of future water levels with the associated confidence intervals was also performed following the approach by Boon [104].

3.5.2. Estimation of Return Periods for Extreme Water Levels

For the natural beach at Pointe Verte, hourly wave runup time series were combined with measured sea-level (T_{pred} + surge) providing total water level (TWL_{nat}), following Ruggiero *et al.* [34]. For coastal flooding assessment, extreme water levels are most relevant. Different approaches based on observed water levels exist to calculated return periods [65,105,106,107] but all of them except the peak over threshold model [106] are time-series length dependent. Since the joint times series of wave runup data and water level data is relatively short (eight years), the peak over threshold model with the generalized Pareto distribution was used to compute return periods associated with extreme water levels [106]. This method is widely used elsewhere in the northern Atlantic for extreme water level predictions [108,109,110]. The analysis was performed on daily maximum levels (high water tide + sea level rise + surge + wave runup) rather than hourly values to avoid temporal dependence. A flood map assessment was performed for the 2010 event based on identified field debris and the RS06 model. The associated return period of this event was calculated with the Pareto analysis. Considering the relatively short time series of tidal data available from Belledune station, we limit the time horizon for flood mapping to 2030. Adding the sea level projection to tidal, barometric surge and runup components gives a future scenario of the same return period for any given year. For the natural beach:

$$TWL = T_{pred} + S_{surge} + R + SLR \tag{4}$$

where SLR is the sea-level rise between 2010 (T_1) and any future year T_2. For the artificially protected area, the runup component is replaced by H_0 (see Section 3.4) thence, for defense structures:

$$TWL = T_{pred} + S_{surge} + H_0 + SLR \tag{5}$$

4. RESULTS

4.1. Coastal Flood Mapping for the 6 December Event

4.1.1. Flood Event Analysis

During the day of 6 December 2010, the Belledune tide station recorded a high water level of 2.05 m at 4:00 p.m. local time (20:00 p.m. UTC), which included a storm surge of 0.63 m combined with an astronomical tide of 1.42 m (Figure 7a,b). Despite the fact that this water level was below the beach crest along most of the coastline (Figure 8), a significant area of the municipality of Maria was flooded by extreme water levels (Figure 9).

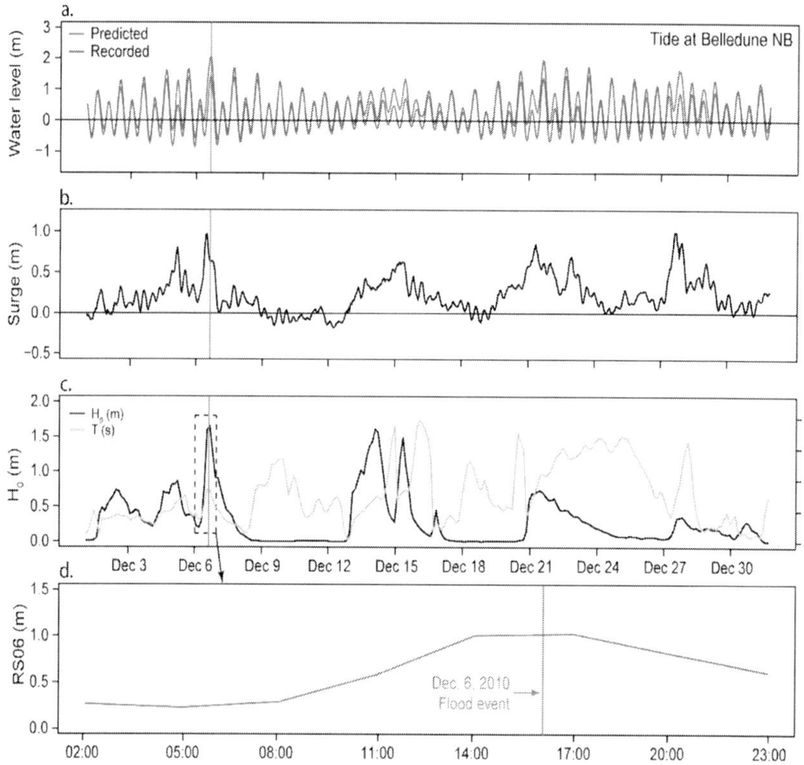

Figure 7. Tidal and hydrodynamic data on 6 December 2010. (**a**) Represents predicted and observed tide (m) at Belledune tide gauge and (**b**) The residual surge (m). The storm surge occurred during high waves reaching 1.63 m (**c**). (**c,d**) show the flooding event on 6 December, where a 1.02 m wave runup was estimated at 4:00 p.m. local time (8:00 p.m. UTC).

Mean observed flood limits reached 3.68 m and 3.07 m for areas with and without defense structures, respectively, therefore exceeding the observed tidal level (T_{obs}) by 1.63 m. The vertical seawall in the west zone increased total water elevations along this part of the coastline by 0.61 m in comparison to the natural zone. This is at least partly a result of the proximity of the wall to the still water level. The wall faces frequent wave impacts during such tidal conditions and the wall toe is completely submerged by more than 1.7 m of water depth, regardless of the wave component. Focusing on the wave breaking depth (D_b) enables a better understanding of the wave-breaking pattern both near the wall and on the natural beach (Figure 6). According to Equation (1), D_b = 1.96 m, providing a geodetic elevation of 0.09 m. Waves therefore break close to the toe of the wall, whilst in natural areas, breaking occurs on the foreshore slope (Figure 6). Where the seawall is built, no wave runup occurs. Instead, waves reach the defense almost at the breaking point.

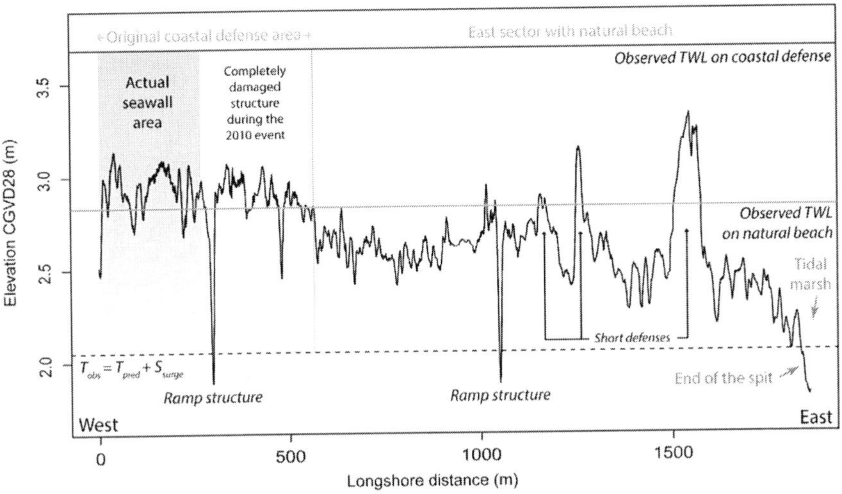

Figure 8. Recorded tide level (dashed black line) and mean observed flood limits along the coast. Longshore coastline crest elevation (D_{high}) variability between the natural zone and areas with defense structures (solid black line). Observed total water levels for the defended zone (dark blue line) exceeded those for the natural areas (light blue line).

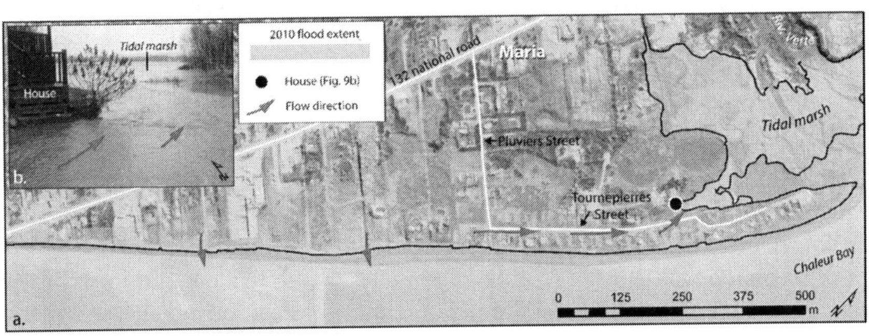

Figure 9. (**a**) Flood extent as identified during the post-storm survey and location of the flow directions in response to land drainage; (**b**) The east zone, the landward side of the spit base acts as a natural outlet for the incoming drained seawater toward the tidal marsh. The picture was taken at 3:00 p.m. (local time) on 6 December, one hour before the flood peak.

During the 6 December 2010 event, unidirectional and landward flow submerged 0.18 km^2 of the coastline at Maria. Overwashed sediments were observed (see Figure 3a) to extend up to 20 m landward in the east and 30 m landward in the central part of the study site. Following north-westward landward propagation, flood waters receded in the opposite direction following land slope, drainage channels and pipes. Much of the floodwater drained along Tournepierres Street and through the tidal marsh (Figure 9b). As a result of this eastwards drainage via the tidal marsh, the landward flow of water in this zone was limited (Figure 9a). Houses located on embankments landwards of short defense structures (Figure 8) were spared from flooding. Results show that, despite the observed tide not reaching the beach or the defense crest on 6 December 2010, heavy flooding occurred along the entire Pointe Verte coastline in Maria due to wave effects. In the western zone, almost half of the original wooden seawall was destroyed during the event, probably due to different material robustness along the defense structure (see Figure 8). Where the wall material was older and thus the protection less sturdy, much of the beach was returned to a natural state in front of many properties.

4.1.2. Wave Runup Implications in Modelling Total Water Level for Flood Mapping

Calculated beach slopes show minimal variability and have a mean value of 0.14, with a mean standard deviation of 0.013 for $\tan\beta_{ufs}$ and 0.017 for $\tan\beta_{1.5Ho}$. We considered this mean value representative of the study area and valid for the cross-shore transect approach assuming an inshore wave direction perpendicular to the shoreline. The linear regression between the upper foreshore beach slope and the slope located between the berm crest and T_{obs}-$1.5H_0$ shows a good correlation ($r^2 = 0.87$, p-value < 0.001). Therefore, the upper foreshore beach slope was used in the RS06 formulation associated with the 2010 event.

The RS06 modelled runup (Figure 7d) and TWL for the natural beach was 1.02 m and 3.07 m, respectively. Observed total water level on the natural beach was 2.83 m (Table 2), thus resulting in a coastal inundation throughout Pointe Verte. The RS06 model predicts TWL to within uncertainty ranges (RMS = 0.38 m) of the RS06. Considering the relatively constant coastline elevation (D_{high}) (2.66 ± 0.28 m standard deviation), observed tidal levels fail to overtop the coastline crest (D_{high}) without the influence of wave effects. Wave breaking effects transfer wave energy into potential energy in the form of wave runup and thus create the potential for flooding to occur. Along the coastal defense, however, results show a higher observed TWL by at least 0.86 m compared to the natural beach (Figure 6 and Table 2), giving a water level of 3.69 m. With a 0.01 m underestimation difference, the estimated TWL on coastal defense gives 3.68 m, suggesting that adding the wave height to the still water level is a good approach to predict TWL in this case.

Table 2. Flood characteristics according to observed and predicted total water levels along natural and defended areas during the 2010 event.

2010 flood characteristics			
Coastline state	Natural beach		Coastal defense
Wave components (m)	Runup		H_0
	1.02		1.63
Time (UTC)		20:00	
Predicted tide		1.42	
Observed tide + Surge		2.05	
Surge only		0.63	
Estimated TWL (m)	3.07		3.68
Mean observed TWL (m) *(post-storm survey)*	2.83		3.69
	Estimated flood area (km^2)		Observed flood area (km^2)
Entire study area	0.267		0.179
West of Pluviers Street only	0.096		0.092
East of Pluviers Street only	0.171		0.087
West area with coastal structure	0.005		0.006

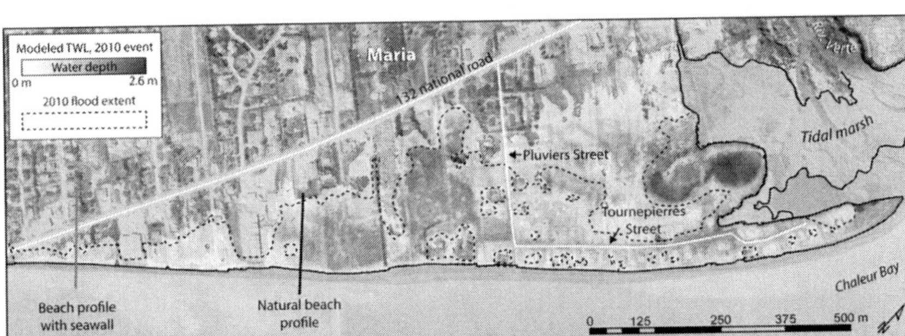

Figure 10. Modelled total water levels for the 2010 flood extent. The dashed line represents the observed flood extent during the post-storm survey. Red and black lines show profiles illustrated in Figure 6.

Estimated total water levels obtained from Equations (4) and (5) for the 2010 event were mapped over the 2007 LiDAR grid elevation whilst the data from the mobile terrestrial LiDAR system were used for defining crest elevations (D_{high}) (Figure 8). For the west and east areas, the predicted TWL estimates well the flooded surface. The static approach, however, does not take into account the flow pattern landward of the spit and therefore overestimates the overall flood extent by 0.088 km^2 (Table 2). The overestimation is mainly attributed to the area eastward of Pluviers Street, where the model overestimates TWL by 0.084 km^2. Considering westward of Pluviers Street, modelled and measured flooded surfaces are strongly correlated with a difference of only 0.004 km^2. The area protected by the seawall also offers good correlation between both surfaces where only 0.001 km^2 is underpredicted by the model. The static approach thus represents well the flood extent when a landward driven flow occurs compared with a drained area. The results also show that Environment Canada's (EC) operational wave forecasting model based on WAM cycle 4 MW3 "PROMISE" produced reliable data for the mapping of coastal flooding at Maria. This model was used for the first time to validate coastal flooding mapping applications.

4.2. Tidal Analysis and Sea-Level Trend

According to the Belledune tide station, the measured relative SLR trend was 4.06 mm/year between 1964 and 2014 (Figure 11). Once modelled with the peak over threshold method with a generalized Pareto distribution, the estimated total water level (including tides, surge and wave runup estimations) during the 6 December 2010 event exceeded the 100 years return period (Figure 12a) and corresponds to a level of 3.01 ± 0.78 m in 2010 and 3.09 ± 0.78 m in 2030. The observed water level on the natural beach (2.83 m) did not exceed the modelled water level and corresponds to a return period of 46 years. In the zone with vertical defense structures, the total water level (including tides, surge and deep-water significant wave height) reached 3.69 m during the 6 December 2010 event. This value has an annual probability of 0.0005 (Figure 12b).

Return periods (Figure 12) are particularly high considering only eight years of data are available. The confidence intervals suggest that, for the natural beach, the level reached during the December 2010 event could have a return period from 30 to over 200 years. An analysis of the extreme water levels return periods for the December 2010 event was conducted in the St. Lawrence maritime estuary from the longest available time series (1897–2010) at Pointe-au-Père-Rimouski station and indicates a return period of 150 years [75], which is closer to our upper interval.

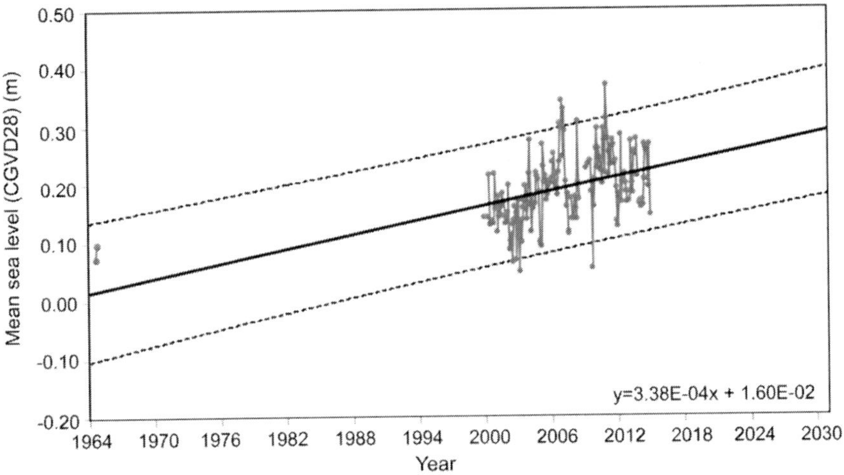

Figure 11. Linear model (solid line) fitted to 1964–2014 monthly mean sea levels at Belledune with 95% confidence bands. Linear term is statistically significant (99%).

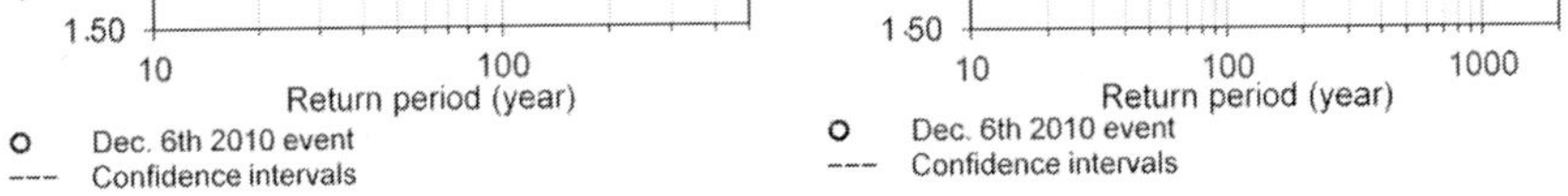

Figure 12. Estimated total water levels return periods at Belledune tide station according to the peak over threshold method including (**a**) wave runup for the natural beach on the eastern part of the study site and (**b**) significant wave height for the defended area on the west zone.

Table 3. Estimated extreme total sea levels for (**a**) (high water tide + sea-level rise + storm surge + wave runup) and (**b**) (high water tide + sea-level rise + storm surge + significant wave height) for different return periods for years 2005, 2010, 2020 and 2030 at Belledune tide station.

Estimated total water levels return periods				
		(a)		
Return Period	**Level 2005**	**Level 2010**	**Level 2020**	**Level 2030**
1 year	1.78 ± 0.11	1.80 ± 0.11	1.84 ± 0.11	1.89 ± 0.11
2 years	1.95 ± 0.14	1.97 ± 0.14	2.02 ± 0.14	2.06 ± 0.15
10 years	2.36 ± 0.30	2.38 ± 0.30	2.42 ± 0.30	2.46 ± 0.30
30 years	2.63 ± 0.47	2.65 ± 0.47	2.69 ± 0.47	2.74 ± 0.47
50 years	2.76 ± 0.57	2.79 ± 0.57	2.83 ± 0.57	2.87 ± 0.57
100 years	2.94 ± 0.73	2.96 ± 0.73	3.00 ± 0.73	3.05 ± 0.73
		(b)		
Return Period	**Level 2005**	**Level 2010**	**Level 2020**	**Level 2030**
1 year	1.95 ± 0.11	1.97 ± 0.11	2.01 ± 0.11	2.05 ± 0.11
2 years	2.14 ± 0.15	2.16 ± 0.15	2.20 ± 0.15	2.24 ± 0.15
10 years	2.56 ± 0.31	2.58 ± 0.31	2.62 ± 0.31	2.66 ± 0.31
30 years	2.82 ± 0.47	2.84 ± 0.47	2.88 ± 0.48	2.92 ± 0.48
50 years	2.93 ± 0.57	2.95 ± 0.57	2.99 ± 0.57	3.03 ± 0.57
100 years	3.08 ± 0.71	3.10 ± 0.71	3.14 ± 0.71	3.18 ± 0.71

Moreover, rising sea levels are likely to cause an increase in flood events and a shortening of return periods. A hypothetical 1 in 100 year event at Maria may become a 1 in 70 year event by 2030 (Table 3). This represents a 14% increase in frequency.

5. DISCUSSION

Based on flood limits from post-storm survey, crest elevation and beach slope determination, wave climate modelling, wave runup estimations and sea-level rise projections, the proposed static flood approach presented here has emphasized the importance of integrating the wave component into flood assessments for low-lying coasts (Figure 13). Flood levels were identified with *in situ* measurements assuming that wave runup superimposes on tidal and surge components to increase water levels. Although runup values are mainly acquired by modelling, *in situ* measurements, or video monitoring systems [111,112,113,114], they can be estimated using measurements of field deposits [32,52,55,79]. The approach proposed here is based on deep-water wave modelling, as no wave buoy data were available for the study site during the 2010 event. The modelled results show good accordance with *in situ* measurements from the field. This is despite the empirical formula failing to account for physical processes in the surf zone [111] or the time-varying morphology of the beach in response to continuously changing wave characteristics [28].

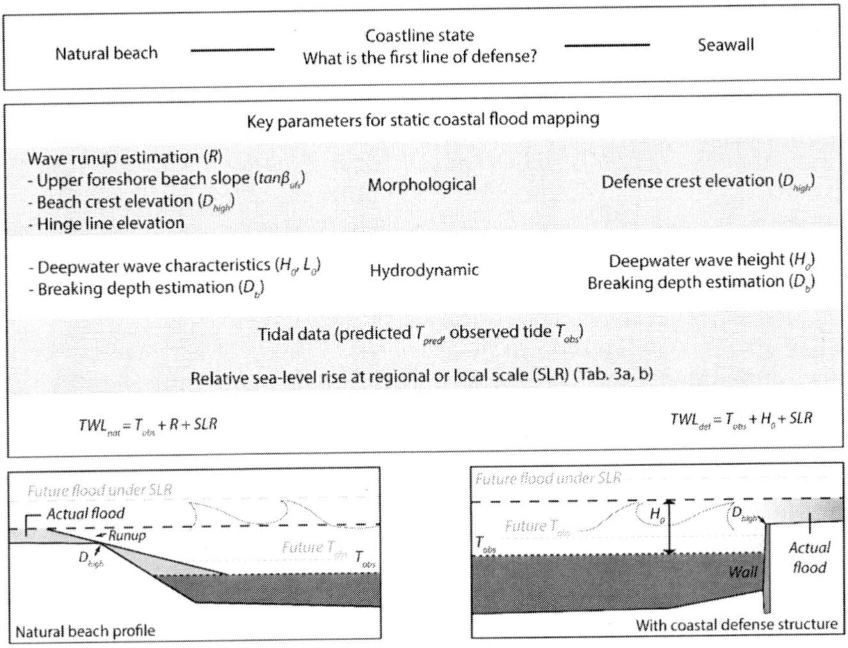

Figure 13. Conceptual and methodological approach for the static coastal flood assessment.

5.1. The Need for a Wave-Tide-Surge Integrated Approach

On 6 December 2010, no flood would have occurred in the area of Pointe Verte without a wave effect on the beach. In regards to the alongshore crest height variability and the water level observed at the tide gauge, only the concrete ramps, the end of the spit and the tidal marsh were affected by the tidal and surge components during the 2010 event. The crest elevation of both the natural beach and coastal structures was generally too high to enable a situation where the backshore would be entirely flooded by the storm surge alone [24]. Our results clearly show that the use of extreme water levels (tide + storm surge) alone, as often recommended and used for coastal flood mapping [59,61,63,64], is not necessarily sufficient for predicting coastal flooding, even in a fetch limited environment like the Chaleur Bay. The same site in 2005 experienced similar storm conditions, *i.e.* with upper foreshore limited tidal levels combined with high waves and overtopping of defense structures and overwashing of natural berms [52]. The flood extent elevation in 2005 was lower in comparison to the 2010 event (2.44 m on natural beach and 3.34 m on coastal structures), possibly due to a deep-water wave height of only 1.16 m in 2005, as obtained by the forecasting wave model in this study. Similar results have been observed on low-lying and rocky coasts of the St. Lawrence maritime estuary, where only 1% of the flooded zone between Rimouski and Sainte-Flavie would have been inundated by the tidal and surge components combined [77]. In all cases, total water levels, which caused major flooding, were increased by the wave effect. The results show the importance of integrating the wave component into the analysis of coastal flooding even for sheltered coastlines where their effects are often ignored or underestimated. Similar findings were observed in California where infrastructures at risk of flood are often concentrated around sheltered embayments. Originally, wave effects were ignored in flood analyses [58] yet subsequently have been integrated into more complete hydrodynamic flood models [67]. Other results were recently observed on the other side of the Atlantic, where in February 2010, Xynthia generated a large storm surge contributing to coastal flood in France. Bertin *et al.* [115] underlined the exceptional surface stress originating from young and steep waves propagating toward the shore, increasing water levels at coastline and contributing to flooding. For more than a decade, considering the wave component is also part of the FEMA's procedure for coastal flood mapping in the US [116,117,118]. Using computer programs such as Runup 2.0 and ACES (Automated Coastal Engineering System), numerical models like CSHORE, or empirical formulas, wave runup and wave overtopping are included in the flood map methodology for the National Flood Insurance Program [94].

For the natural beach of Pointe Verte, we modelled deep-water wave characteristics and integrated these into a static flood model under a wave runup empirical formulation [29]. The estimated wave implication for the 2010 event increased the still water level from 2.05 m to 3.07 m, resulting in a theoretical wave runup of 1.02 m in natural areas. This increase in water levels was also recorded by Bernatchez *et al.* [52], where an amplitude of 0.4 m over the still water level was recorded on the natural beach during the storm in December 2005.

Along the natural beach, wave runup occurs on the upper foreshore beach slope. Using this beach slope in the Stockdon runup formulation and adding this value to the observed tide and surge showed good correspondence with field debris elevation measurements. Although the foreshore slope has been used in previous works [29,32,34,67,96,119], the mean upper foreshore slope defined here was located between the beach crest and the hinge line, corresponding to the "slope of the active section of the beach", as suggested by Cariolet and Suanez [32] for macrotidal environments. Using this beach slope, the Stockdon model showed good results in estimating wave runup and mean flood limit elevations were within model errors (±0.38 m). This is in accordance with authors who suggest using the Stockdon formula in order to estimate wave runup as a first step in coastal flood assessments [94,95]. Some studies show that the Stockdon formula tends to over-estimate the flooding extent [55,67] or occasionally underestimate it [111].

Other runup empirical formulas based on beach slope and deep-water wave characteristics exist [30,34,57,97,120] and demonstrated different results [32,77]. In our case, without precise field recordings of high frequency wave runup and near-shore hydrodynamics, we find the Stockdon model based on deep-water wave characteristics and upper foreshore beach slope suitable to our static flood model on the natural beach. The upper foreshore slope offers a practical advantage to coastal managers because it can be readily defined using GIS [51]. By digitizing the coastline and the hinge line from a series of aerial photographs and LiDAR products, one is able to estimate wave runup at high tide if wave climate data is available. This is a measureable morphological aspect that can be used in historic flood assessments and for predicting future extreme events.

As demonstrated along the seawall at Maria, wave crest elevation in relation to the height of the defense crest is also an important parameter in coastal flood management [121]. Upper foreshore sediment volumes are diminished in front of coastal structures in response to wall reflection and scouring at the toe [52,122]. Therefore, waves do not break on the foreshore until reaching the vicinity of the wall front. As a result, almost no dissipation occurs on the sloping beach in front of the wall and flood limits become higher landward of seawalls in comparison to behind natural beaches. No wave runup occurs in such cases [43] and flooding is due primarily to wave overtopping. Chini and Stansby [53] showed that the overtopping discharge was higher when beach level was lower at the toe of the seawall. This may explain the much higher TWL measured in front of defense structures compared to the natural beach in Maria. Since no wave overtopping discharge calculations were made in the present study, we propose adding significant deep-water wave heights to the still water level in order to estimate flood levels landward of the seawall. This approximation is a useful approach for coastal managers and showed good results in this study. Since coastal flood assessment and mapping for large areas can easily require time consuming and expensive studies [20], especially where complex coastal defense structures are present and require computationally demanding effort for bulk application [95], adding deep-water wave height to the observed tide (with surge) as a simple rule-of-thumb in order to obtain total water levels facing seawall can be suitable as a first approach to flood mapping in the present case where waves do not break before reaching the vicinity wall.

Another element vital to flood model robustness is the accuracy with which we can identify and map the first line of defense [25] both in natural environments and along defense structures. Recent studies show that airborne LiDAR digital elevation models, usually with a spatial resolution of 1 m and a vertical accuracy of ±15 cm [123,124] and sometimes up to 30 cm [125,126,127], are inadequate to determine this limit with sufficient precision as micro-scale topographic features relevant to hydraulic connectivity are overlooked [58]. The result is an over- or under-estimation of the flood extent. The mobile terrestrial LiDAR system (MTLS) that we used in this study produced a point cloud with a spatial resolution of less than 5 cm and vertical errors of ± 3 cm (RMS), which effectively solves the problem of having to determine defense crest elevation manually [25]. The MTLS also allows for much faster field surveys in comparison to fixed terrestrial laser scanning (TLS) or DGPS, which are sometimes used in lieu of airborne LiDAR data [128]. This is an important advantage in the coastal environment as the time available to map beach morphology can be short at low tide. Surveys with the MTLS also provide potential in urban flood modelling where the representation of micro-scale topographic effects and complex features (curbs, ditches roads, defense structure crests, walls, and/or buildings) or drainage systems have significant impact on flood propagation [67,69,70]. Airborne LiDAR data covering larges areas in a rapid survey can be combined with MTLS surveys along coastal borders to maximize the advantage of both platform to generate high-resolution digital elevation models for accurate coastal flood mapping.

5.2. Inland Drainage and Sea-Level Rise

During the 6 December 2010 event, water levels in the tidal marsh at Maria were lower than the predicted flood levels due to the protection offered by the coastal spit. Marshes attenuate wave energy and provide natural resilience to flooding exposure [129,130]. Due to a land gradient landward of the coastline, however, the Stockdon model did not predict well the flood limits. This model overestimation has been reported previously when applied to other flooded peninsula environments [67] and a decelerating landward flow with distance is normally seen in this particular situation [113]. Flood limits at the tidal marsh in east Maria were below the 2.83 m average for the region. Due to the morphology of the marsh, only the astronomical tide and storm surge components affected this flooding surface during the event, demonstrating the natural protection offered by this type of environment.

The relatively low 2.05 m still water level was too low to produce an inundation regime during the 2010 event. An inundation regime occurs when a constant inland flow of water under storm surge effects (tide + surge) is realized, causing continuous submersion of the coastline [24,25]. Under future sea-level rise, conditions where the still water line is raised, an inundation regime could occur over the Pointe Verte area, which has the potential to flood the entire spit and tidal marsh. In such a case, the land drainage effect would diminish and the static flood model presented here would likely give an accurate account of flooding for the area eastward of Pluviers Street.

According to our linear SLR prediction of 4.06 mm/year (1964–2014), the return period for an equivalent storm to generate a still water level (tide + surge) reaching the beach crest of 2.66 m causing an inundation situation is 150 years. This modelled linear trend is in accordance with the IPCC's RCP8.5 sea level rise scenario for at least the next two decades.

As an example, 20 years of SLR at this constant rate produces a net rise of 0.082 m from 2010 to 2030. Integrating the IPCC's RCP8.5 sea-level scenario [131] with current isostatic land motion (−1.78 mm/year; [100]) results in a net SLR of 0.122 m by 2030, relative to 2010 levels. By integrating the IPCC's projection and Koohzare's isostatic model, we can estimate the year when an inundation regime event (at least equal to the mean coastline elevation of 2.66 m) would happen. A hypothetical 2.66 m still water level could be reached by 2081, after 71 years. Moreover, in the context of climate change, the significant reduction of sea-ice cover will promote an increase in winter storm waves hitting the coast [132].

The trend of relative SLR has been 6.96 mm/y between 2000 and 2014 at the Belledune tide gauge. This evidence of acceleration in SLR is also observed at tide gauge stations in Eastern Canada and Northeastern United States [104]. The SLR acceleration since 1990 is a global response to the increased radiative forcing of the climate system [133]. Due to the short time series of tide gauge data, we opted for a conservative linear trend compared to the quadratic approach used by other authors to project sea-level rise [8,104].

The use of tide gauge data has the advantage of integrating both isostatic and eustatic components to project relative SLR at the regional scale [104]. However, due to the short time series and discontinuity of the tide gauge data from the Belledune station, the projection of relative SLR was made only for a short period. The recent acceleration in global SLR averages [3,4] and the large differences between scenarios of eustatic rise by 2100 [7] demonstrate the need to develop and use proxies to better establish regional relative sea-level trends [134,135,136]. These data will be needed to establish a more robust coastal flood risk mapping.

6. CONCLUSIONS

Wave runup estimation and modelling of water levels and deep-water waves have reproduced observed flood limit elevations during the storm event of 6 December 2010, as experienced along a natural and armoured coastline. Results clearly show the importance of integrating wave data and runup in assessing and mapping the risk of coastal flooding. They also demonstrate that the lowering of beach levels at a seawall toe increases total water level elevations in the backshore and consequently the risk of flooding due to a wave breaking occurring at the vicinity of the seawall. Our approach integrates runup, which takes into account beach slope (natural areas), or deep-water wave height (at seawalls), the return period of extreme water levels and future relative sea-level rise in order to establish, at the municipality scale, coastal flood risk maps. This approach, which takes into account several variables, is easy to use for coastal managers, which is an important factor in the process of adaptation. Such static approaches do not, however, account for land drainage. In the context of accelerating sea-level rise, it becomes necessary to establish, at the regional scale, a longer series of relative sea-level trends, a validation of wave, drainage, and numerical runup models in different geomorphological settings and hydrodynamic conditions. The next step will be to integrate wave overtopping discharge models to better represent overland flow patterns. In this perspective, the use of a mobile terrestrial laser system offers excellent potential to model and map more accurately coastal flooding. Integration of validated modelled data into a warning system offers a promising avenue for managing and mitigating coastal risks.

ACKNOWLEDGMENTS

We thank the Quebec government for funding this project as part of its program for preventing the principal types of natural risks. Thanks as well to Tarik Toubal for the storm debris line mapping with DGPS and Susan Drejza, Patrick Bouchard and David Lacombe for the mobile terrestrial LiDAR survey. We thank Denis Jacob and Viateur Turcotte from Environment Canada for the wave data provided by the wave forecasting model, and to Phillip MacAulay (CHS) for the Belledune station data information. We also thank Azadeh Koohzare for sharing vertical crustal movement data. Thanks for the two anonymous reviewers and editors for their very pertinent and helpful comments.

AUTHOR CONTRIBUTIONS

Field surveys (DGPS, LiDAR, flood limit): C.F., D.D., P.B., S.V.W. Hydrodynamic data analysis: D.D., A.L. Sea-level analysis: G.B.B., P.B., D.D. Flood mapping: D.D., C.F. Discussion on the results: D.D., P.B., A.L., G.B.B., C.F., R.L.B. Major contribution to the writing of the entire paper: D.D., P.B. with a main contribution of A.L. for introduction, discussion and wave analysis sections, G.B.B. for sea-level and return period for extreme water levels and S.V.W. for the methodology section of the mobile terrestrial LiDAR data. Edited English syntax: R.L.B.

CONFLICTS OF INTEREST

The authors declare no conflict of interest.

REFERENCES

1. Nicholls, R.J.; Tol, R.S.J. Impacts and responses to sea-level rise: A global analysis of the SRES scenarios over the twenty-first century. Philos. Trans. A 2006, 364, 1073–1095.
2. Nicholls, R.J.; Cazenave, A. Sea-level rise and its impact on coastal zones. Science 2010, 328, 1517–1520.
3. Church, J.A.; White, N.J. Sea-Level Rise from the Late 19th to the Early 21st Century. Surv. Geophys. 2011, 32, 585–602.
4. Rahmstorf, S.; Foster, G.; Cazenave, A. Comparing climate projections to observations up to 2011. Environ. Res. Lett.2012, 7, 044035.
5. Pfeffer, W.T.; Harper, J.T.; O'Neel, S. Kinematic Constraints on Glacier Contributions to 21st-Century Sea-Level Rise. Science 2008, 321, 1340–1343.
6. Church, J.A.; Clark, P.U.; Cazenave, A.; Gregory, J.M.; Jevrejava, S.; Levermann, A.; Merrifield, M.A.; Milne, G.A.; Nerem, R.S.; Nunn, P.D.; et al. Sea Level Change. In IPCC. Climate Change 2013: The Physical Science Basis; Stocker, T.F., Qin, D., Plattner, G.K., Tignor, M., Allen, S.K., Boschung, J., Nauels, A., Xia, Y., Bex, V., Midgley, P.M., Eds.; Cambridge University Press: Cambridge, UK; New York, NY, USA, 2013.
7. Horton, B.P.; Rahmstorf, S.; Engelhart, S.E.; Kemp, A.C. Expert assessment of sea-level rise by AD 2100 and AD 2300. Quat. Sci. Rev. 2014, 84, 1–6.
8. Parris, A.; Bromirski, P.; Burkett, V.; Cayan, D.; Culver, M.; Hall, J.; Horton, R.; Knuuti, K.; Moss, R.; Obeysekera, J.; et al. Global Sea Level Rise Scenarios for the US National Climate Assessment; NOAA: Silver Spring, MD, USA, 2012.

9. Yin, J.; Schlesinger, M.E.; Stouffer, R.J. Model projections of rapid sea-level rise on the northeast coast of the United States. Nat. Geosci. 2009, 462, 262–266.
10. Howard, T.; Pardaens, A.K.; Lowe, J.A.; Ridley, J.; Hurkmans, R.T.W.L.; Bamber, J.L.; Spada, G.; Vaughan, D. Sources of 21st century regional sea level rise along the coast of North-West Europe. Ocean Sci. Discuss. 2013, 10, 2433–2459.
11. Obeysekera, J.; Park, J. Scenario-based projection of extreme sea levels. J. Coast. Res. 2012, 29, 1–7.
12. Vafeidis, A.T.; Nicholls, R.J.; Mcfadden, L.; Tol, R.S.J.; Hinkel, J.; Spencer, T.; Grashoff, P.S.; Boot, G.; Klein, R.J.T. A new global coastal database for impact and vulnerability analysis to sea-level rise. J. Coast. Res. 2008, 24, 917–924.
13. Gornitz, V.M.; Daniels, R.C.; White, T.W.; Birdwell, K.R. The Development of a Coastal Risk Assessment Database: Vulnerability to Sea-Level Rise in the U.S. Southeast. J. Coast. Res. 1994, 371, 327–338.
14. Jallow, B.P.; Toure, S.; Barrow, M.M.K.; Mathieu, A.A. Coastal zone of The Gambia and the Abidjan region in Cote d'Ivoire: Sea level rise vulnerability, response strategies, and adaptation options. Clim. Res. 1999, 12, 129–136.
15. Titus, J.G.; Richman, C. Maps of lands vulnerble to sea level rise: Modeled elevations along the US Atlantic and Gulf coasts. Clim. Res. 2001, 18, 205–228.
16. Solomon, S.; Qin, D.; Manning, M. IPCC. Climate Change 2007. The Physical Science Basis; Cambridge University Press: Cambridge, UK; New York, NY, USA, 2007.
17. Stocker, T.F.; Qin, D.; Plattner, G.K.; Tignor, M.; Allen, S.K.; Boschung, J.; Nauels, A.; Xia, Y.; Bex, V.; Midgley, P.M.; et al. IPCC. Climate Change 2013: The Physical Science Basis; Cambridge University Press: Cambridge, UK; New York, NY, USA, 2013.
18. Rao, K.N.; Subraelu, P.; Rao, T.V.; Malini, B.H.; Ratheesh, R.; Bhattacharya, S.; Rajawat, A.S. Sea-level rise and coastal vulnerability: An assessment of Andhra Pradesh coast, India through remote sensing and GIS. J. Coast. Conserv. 2008, 12, 195–207.
19. Nicholls, R.J.; Hoozemans, F.M.J.; Marchand, M. Increasing flood risk and wetland losses due to global sea-level rise: Regional and global analyses. Glob. Environ. Chang. 1999, 9, S69–S87.
20. FEMA. Atlantic Ocean and Gulf of Mexico coastal guidelines update. Final draft; Federal Emergency Management Agency: Washington, DC, USA, 2007.
21. Walsh, K.J.E.; Betts, H.; Church, J.; Pittock, A.B.; McInnes, K.L.; Jackett, D.R.; McDougall, T.J. Using sea level rise projections for urban planning in Australia. J. Coast. Res. 2004, 20, 586–598.
22. Gehrels, W.R.; Milne, G.A.; Kirby, J.R.; Patterson, R.T.; Belknap, D.F. Late Holocene sea-level changes and isostatic crustal movements in Atlantic Canada. Quat. Int. 2004, 120, 79–89.
23. Koohzare, A.; Vaníček, P.; Santos, M. Pattern of recent vertical crustal movements in Canada. J. Geodyn. 2008, 45, 133–145.
24. Sallenger, A.H. Storm impact scale for barrier islands. J. Coast. Res. 2000, 16, 890–895.
25. Stockdon, H.F.; Sallenger, A.H.; Holman, R.A.; Howd, P.A. A simple model for the spatially-variable coastal response to hurricanes. Mar. Geol. 2007, 238, 1–20.
26. Benavente, J.; Del Río, L.; Gracia, F.J.; Martínez-del-Pozo, J.A. Coastal flooding hazard related to storms and coastal evolution in Valdelagrana spit (Cadiz Bay Natural Park, SW Spain). Cont. Shelf Res. 2006, 2, 1061–1076.
27. Soldini, L.; Antuono, M.; Brocchini, M. Numerical Modeling of the Influence of the Beach Profile on Wave Run-Up. J. Waterw. Port Coast. Ocean Eng. 2012, 139, 115.
28. Bauer, B.O.; Greenwood, B. Surf zone similarity. Geogr. Rev. 1988, 78, 137–147.
29. Stockdon, H.; Holman, R.; Howd, P.; Sallenger, A. Empirical parameterization of setup, swash, and runup. Coast. Eng.2006, 53, 573–588.
30. Komar, P.D. Beach processes and sedimentation, 2nd ed.; Prentice Hall, Inglewood Cliffs: Upper Saddle River, NJ, USA, 1998.
31. Cariolet, J.M.; Suanez, S.; Meur-Férec, C.; Postec, A. Cartographie de l'aléa de submersion marine et PPR: Éléments de réflexion à partir de l'analyse de la commune de Guissény (Finistère, France). Cybergeo 2012, 2012, 1–21.

32. Cariolet, J.M.; Suanez, S. Runup estimations on a macrotidal sandy beach. Coast. Eng. 2013, 74, 11–18.
33. Ruggiero, P. Is the intensifying wave climate of the U.S. Pacific Northwest increasing flooding and erosion risk faster than sea level rise? J. Waterw. Port Coast. Ocean Eng. 2012, 139, 88–97.
34. Ruggiero, P.; Komar, P.D.; McDougal, W.G.; Marra, J.J.; Reggie, A. Wave runup, extreme water levels and the erosion of properties backing beaches. J. Coast. Res. 2001, 17, 407–419.
35. Seabloom, E.W.; Ruggiero, P.; Hacker, S.D.; Mull, J.; Zarnetske, P. Invasive grasses, climate change, and exposure to storm-wave overtopping in coastal dune ecosystems. Glob. Chang. Biol. 2013, 19, 824–832.
36. Battjes, J.A. Surf similarity. In Proceedings of the 14th International Conference on Coastal Engineering ASCE, Copenhagen, Denmark, 24–28 June 1974; pp. 466–480.
37. Cariolet, J.M. Quantification du runup sur une plage macrotidale à partir des conditions morphologiques et hydrodynamiques. Geomorphol. Reli. Process. Environ. 2011, 1, 95–108.
38. Stockdon, H.F.; Thompson, D.M.; Plant, N.G.; Long, J.W. Evaluation of wave runup predictions from numerical and parametric models. Coast. Eng. 2014, 92, 1–11.
39. Holman, R.A.; Sallenger, A.H. Setup and swash on a natural beach. J. Geophys. Res. 1985, 90, 945–953.
40. Hughes, S.A. Wave momentum flux parameter: a descriptor for nearshore waves. Coast. Eng. 2004, 51, 1067–1084.
41. Mase, H.; Asce, M.; Tamada, T.; Yasuda, T.; Hedges, T.S.; Reis, M.T. Wave runup and overtopping at seawalls built on land and in very shallow water. J. Waterw. Port Coast. Ocean Eng. 2013, 139, 346–357.
42. Na, S.J.; Do, K.D.; Suh, K.-D. Forecast of wave run-up on coastal structure using offshore wave forecast data. Coast. Eng. 2011, 58, 739–748.
43. Pullen, T.; Allsop, N.W.H.; Bruce, T.; Kirtenhaus, A.; Schüttrumpf, H.; van der Meer, J.W. Eurotop: Wave overtopping of sea defences and related structures: assessment manual; Environment Agency: Rotherham, UK; Expertise Netwerk Waterkeren: Utrecht, The Netherlands, Kuratorium fur Forschung im Kusteningenicurwesen: Hamburg, Germany, 2007.
44. Carbone, F.; Dutykh, D.; Dudley, J.M.; Dias, F. Extreme wave runup on a vertical cliff. Geophys. Res. Lett. 2013, 40, 3138–3143.
45. Chen, X.; Hofand, B.; Altomare, C.; Suzuki, T.; Uijttewaal, W. Forces on a vertical wall on a dike crest due to overtopping flow. Coast. Eng. 2015, 95, 94–104.
46. Peng, Z.; Zou, Q.-P. Spatial distribution of wave overtopping water behind coastal structures. Coast. Eng. 2011, 58, 489–498.
47. Van der Meer, J.W.; Verhaeghe, H.; Steendam, G.J. The new wave overtopping database for coastal structures. Coast. Eng. 2009, 56, 108–120.
48. Molines, J.; Medina, J.R. Calibration of overtopping roughness factors for concrete armor units in non-breaking conditions using the CLASH database. Coast. Eng. 2015, 96, 62–70.
49. Van Gent, M.R.A.; Van den Boogaard, H.F.P.; Pozueta, B.; Medina, J.R. Neural network modelling of wave overtopping at coastal structures. Coast. Eng. 2007, 54, 586–593.
50. Sabino, A.; Rodrigues, A.; Araujo, J.; Poseiro, P.; Reis, M.T.; Fortes, C.J. Wave overtopping analysis and early warning forecast system. In Proceedings of the ICCSA 2014, Le Havre, France, 23–26 June 2014; Murgante, B., Misra, S., Rocha, A.M.A.C., Torre, C., Rocha, J.G., Falcão, M.I., Taniar, D., Apduhan, B.O., Gervasi, O., Eds.; Springer: Cham, Switzerland, 2014.
51. Bernatchez, P.; Fraser, C. Evolution of coastal defence structures and consequences for beach width trends, Québec, Canada. J. Coast. Res. 2012, 285, 1550–1566.
52. Bernatchez, P.; Fraser, C.; Lefaivre, D.; Dugas, S. Integrating anthropogenic factors, geomorphological indicators and local knowledge in the analysis of coastal flooding and erosion hazards. Ocean Coast. Manag. 2011, 54, 621–632.

53. Chini, N.; Stansby, P.K. Coupling TOMAWAC and Eurotop for uncertainty estimation in wave overtopping predictions. In Advances in Hydroinformatics, Springer Hydrogeology; Gourbesville, P., Cunge, J., Caignaert, G., Eds.; Springer: Singapore, Singapore, 2014.
54. Dawson, R.J.; Dickson, M.E.; Nicholls, R.J.; Hall, J.W.; Walkden, M.J.A.; Stansby, P.K.; Mokrech, M.; Richards, J.; Zhou, J.; Milligan, J.; et al. Integrated analysis of risks of coastal flooding and cliff erosion under scenarios of long term change. Clim. Chang. 2009, 95, 249–288.
55. Guimarães, P.V.; Farina, L.; Toldo, E.; Diaz-Hernandez, G.; Akhmatskaya, E. Numerical simulation of extreme wave runup during storm events in Tramandaí Beach, Rio Grande do Sul, Brazil. Coast. Eng. 2015, 95, 171–180.
56. Ruggiero, P.; Holman, R.A.; Beach, R.A. Wave run-up on a high-energy dissipative beach. J. Geophys. Res. 2004, 109, 1–12.
57. Holman, R. Extreme value statistics for wave run-up on a natural beach. Coast. Eng. 1986, 9, 527–544.
58. Gallien, T.W.; Schubert, J.E.; Sanders, B.F. Predicting tidal flooding of urbanized embayments: A modeling framework and data requirements. Coast. Eng. 2011, 58, 567–577.
59. Daigle, R.J. Sea level rise estimates for New Brunswick municipalities: Saint John, Sackville, Richibucto, Shippagan, Caraquet, Le Goulet; The Atlantic Climate Adaptation Solutions Association: New-Brunswick, QC, Canada, 2011.
60. Perherin, C.; Roche, A. Évolution des méthodes de caractérisation des aléas littoraux. In Proceedings of the XIèmes Journées Natl. Génie Côtier-Génie Civil, Les Sables d'Olonne, France, 22–25 June 2010; pp. 609–616.
61. Richards, W.; Daigle, R. Scenarios and Guidance for Adaptation to Climate Change and Sea Level Rise- NS and PEI Municipalities; Atlantic Climate Adaptation Solutions Association: Halifax, NS, Canada, 2011; p. 87.
62. Prime, T.; Brown, J.M.; Plater, A.J. Physical and economic impacts of sea-level rise and low probability flooding events on coastal communities. PLoS ONE 2015, 10, e0117030.
63. Daigle, R. Sea-Level Rise and Flooding Estimates for New Brunswick Coastal Sections; The Atlantic Climate Adaption Solutions Association: Halifax, NS, Canada, 2012.
64. Webster, T.; McGuigan, K.; Collins, K.; MacDonald, C. Integrated river and coastal hydrodynamic flood risk mapping of the lahave river estuary and town of Bridgewater, Nova Scotia, Canada. Integrated river and coastal hydrodynamic flood risk mapping of the lahave river estuary and town of Bridgewater, Nova Scotia, Canada. Water2014, 6, 517–546.
65. Pirazzoli, P.A.; Tomasin, A. Estimation of return periods for extreme sea levels: A simplified empirical correction of the joint probabilities method with examples from the French Atlantic coast and three ports in the southwest of the UK. Ocean Dyn. 2007, 57, 91–107.
66. Thompson, K.R.; Bernier, N.B.; Chan, P. Extreme sea levels, coastal flooding and climate change with a focus on Atlantic Canada. Nat. Hazards 2009, 51, 139–150.
67. Gallien, T.W.; Sanders, B.F.; Flick, R.E. Urban coastal flood prediction: Integrating wave overtopping, flood defenses and drainage. Coast. Eng. 2014, 91, 18–28.
68. Webster, T.L.; Forbes, D.L.; Dickie, S.; Shreenan, R. Using topographic lidar to map flood risk from storm-surge events for Charlottetown, Prince Edward Island, Canada. Can. J. Remote Sens. 2004, 30, 64–76.
69. Néelz, S.; Pender, G.; Villanueva, I.; Wilson, M.; Wright, N.G.; Bates, P.; Mason, D.; Whitlow, C. Using remotely sensed data to support flood modelling. Water Management 2006, 159, 35–43.
70. Ozdemir, H.; Sampson, C.C.; de Almeida, G.A.M.; Bates, P.D. Evaluating scale and roughness effects in urban flood modelling using terrestrial LIDAR data. Hydrol. Earth Syst. Sci. 2013, 17, 4015–4030.

71. Forbes, D.L.; Parkes, G.S.; Manson, G.K.; Ketch, L.A. Storms and shoreline retreat in the southern Gulf of St. Lawrence. Mar. Geol. 2004, 210, 169–204.
72. Masson, A.; Catto, N. Extratropical Transitions in Atlantic Canada: Impacts and Adaptive Responses. Geophy. Res. Abstr. 2013, 15, 3149.
73. Boyer-Villemaire, U.; Bernatchez, P.; Benavente, J.; Cooper, J.A.G. Quantifying community's functional awareness of coastal changes and hazards from citizen perception analysis in Canada, UK and Spain. Ocean Coast. Manag. 2014, 93, 106–120.
74. Drejza, S.; Frieseinger, S.; Bernatchez, P. Exposition des infrastructures routières de l'Est du Québec (Canada) à l'érosion et à la submersion. In Proceedings of the Actes du Colloque International Connaissance et Compréhension des Risques Côtiers: Aléas, Enjeux, Représentations, Gestion, Brest, France, 3–4 July 2014.
75. Bernatchez, P.; Boucher-Brossard, G.; Sigouin-Cantin, M. Contribution des archives à l'étude des événements météorologiques et géomorphologiques causant des dommages aux côtes du Québec maritime et analyse des tendances, des fréquences et des temps de retour des conditions météo-marines extrêmes; Chaire de recherche en géoscience côtière, Laboratoire de dynamique et de gestion intégrée des zones côtières, Université du Québec à Rimouski; Rapport remis au ministère de la Sécurité publique du Québec: Rimouski, QC, Canada, 2012; p. 140.
76. MacCaulay, P.; Canadian Hydrographic Service, Dartmouth, N.S., Canada. Personal communication, 2015.
77. Didier, D.; Bernatchez, P.; Marie, G. Évaluation de la submersion côtière grâce à l'estimation in situ du wave runup sur les côtes basses du Bas-Saint-Laurent, Canada (Québec). In Proceedings of the Actes du Colloque International Connaissance et Compréhension des Risques Côtiers: Aléas, Enjeux, Représentations, Gestion, Brest, France, 3–4 July 2014.
78. Ramana Murthy, M.V.; Reddy, N.T.; Pari, Y.; Usha, T.; Mishra, P. Mapping of seawater inundation along Nagapattinam based on field observations. Nat. Hazards 2012, 60, 161–179.
79. Cariolet, J.-M. Use of high water marks and eyewitness accounts to delineate flooded coastal areas: The case of Storm Johanna (10 March 2008) in Brittany, France. Ocean Coast. Management 2010, 53, 679–690.
80. Van-Wierts, S.; Bernatchez, P. Relevé LiDAR terrestre à Sainte-Luce dans le secteur de l'Anse aux Coques dans le cadre d'une étude de recharge de plage en zone d'affouillement; Chaire de recherche en géoscience côtière, Laboratoire de dynamique et de gestion intégrée des zones côtières, Université du Québec à Rimouski; Rapport remis au ministère de la Sécurité publique du Québec: Rimouski, QC, Canada, 2012; p. 29.
81. WAMDI Group. The WAM Model: A Third Generation Ocean Wave Prediction Model. J. Phys. Oceanogr. 1988, 18, 1775–1810.
82. Komen, G.J.; Cavaleri, L.; Donelan, M.; Hasselmann, K.; Hasselmann, S.; Janssen, P.A.E.M. Dynamic and Modelling of Ocean Waves; Cambridge University Press: New York, NY, USA; p. 556.
83. Monbaliu, J.; Hargreaves, J.C.; Carretero, J.C.; Gerritsen, H.; Flather, R. Wave modelling in the PROMISE project. Coast. Eng. 1999, 37, 379–407.
84. Jacob, D.; Perrie, W.; Toulany, B.; Saucier, F.; Lefaivre, D.; Turcotte, V. Wave model validation in the St. Lawrence river eastuary. In Proceedings of the 7th International Workshop on Wave Hindcasting and Forecasting, Banff, AB, Canada, 21–25 October 2004.
85. Janssen, P.A.E.M. Quasi-linear theory of wind-wave generation applied to wave forecasting. J. Phys. Oceanogr. 1991, 21, 1631–1642.
86. Cavaleri, L.; Rizzoli, P.M. Wind wave prediction in shallow water: Theory and applications. J. Geophys. Res. 1981, 86, 10961–10973.
87. Hasselmann, K.; Barnett, T.P.; Bouws, E.; Carlson, H.; Cartwright, D.E.; Enke, K.; Ewing, J.A.; Gienapp, H.; Hasselmann, D.E.; Kruseman, P.; et al. Measurements of

Wind-Wave Growth and Swell Decay During the Joint North Sea Wave Project (JONSWAP); Deutches Hydrographisches Institute: Delft, Netherlands, 1973.

88. Battjes, J.A.; Janssen, J.P.F.M. Energy loss and set-up due to breaking of random waves. In Proceedings of the 16th International Conference Coastal Engineering, Hamburg, Germany, 27 August–3 September 1978; pp. 569–587.

89. Lambert, A.; Neumeier, U.; Jacob, D. Évaluation du modèle WAM opéré par Environnement Canada dans le Golfe du Saint-Laurent; résultats préliminaires pour les tempêtes de l'automne 2010; Institut des sciences de la mer de Rimouski, Université du Québec à Rimouski; Rapport technique remis au Ministère des Transports du Québec: Rimouski, QC, Canada, 2012.

90. Lambert, A.P.; Neumeier, U.; Jacob, D.; Savard, J.-P. Are regional operational wind-waves models usable to predict coastal and nearshore wave climate? In Proceedings of the 2012 AGU Fall Meeting; San Francisco, CA, USA: 3–7 December 2012.

91. Lambert, A.; Neumeier, U.; Jacob, D.; Savard, J.-P. Évaluation du modèle WAM opéré par Environnement Canada dans le Golfe du Saint-Laurent; résultats intermédiaires pour les années 2010-2011; Institut des sciences de la mer de Rimouski, Université du Québec à Rimouski; Rapport technique remis au Ministère des Transports du Québec: Rimouski, QC, Canada, 2013.

92. Dalrymple, R.A.; Eubanks, R.A.; Birkemeier, W.A. Wave-induced circulation in shallow basins. J. Waterw. Ports Coast. Ocean Div. 1977, 103, 117–135.

93. Matias, A.; Williams, J.J.; Masselink, G.; Ferreira, Ó. Overwash threshold for gravel barriers. Coast. Eng. 2012, 63, 48–61.

94. Melby, J.A.; Nadal-Caraballo, N.C.; Kobayashi, N. Wave runup prediction for flood mapping. Coast. Eng. Proc. 2012, 33, 1–15.

95. Kergadallan, X. Analyse statistique des niveaux d'eau extrêmes. Environnement maritime et estuarien; CETMEF: Margny Lès Compiègne, France, 2013; p. 179.

96. Vousdoukas, M.I.; Wziatek, D.; Almeida, L.P. Coastal vulnerability assessment based on video wave run-up observations at a mesotidal, steep-sloped beach. Ocean Dyn. 2011, 62, 123–137.

97. Nielsen, P.; Hanslow, D.J. Wave runup distributions on natural beaches. J. Coast. Res. 1991, 7, 1139–1152.

98. Dean, R.G.; Dalrymple, R.A. Water waves mechanics for engineers and scientists. Advances Series on Ocean Engineering; World Scientific: Singapore, Singapore, 1991; p. 368.

99. Shaw, J.; Taylor, R.B.; Solomon, S.; Christian, H.A.; Forbes, D. Potential impacts of sea-level rise on Canadian coasts. Can. Geogr. 1998, 42, 365–379.

100. Koohzare, A.; University of New-Brunswick, Fredericton, NB, Canada. Personal communication, 2015.

101. Fisheries and Oceans Canada. Canadian Tides and Water Levels data Archives. Available online: http://www.isdm-gdsi.gc.ca/isdm-gdsi/twl-mne/index-eng.htm (accessed on 20 March 2015).

102. CHS–Canadian Hydrographic Service. Predicted Water Levels, numerical dataset. 2012. Available online: http://www.tides.gc.ca/eng/info/WebServicesWLD (accessed on 24 October 2012).

103. Garner, K.L.; Chang, M.Y.; Fulda, M.T.; Berlin, J.A.; Freed, R.E.; Soo-Hoo, M.M.; Revell, D.L.; Ikegami, M.; Flint, L.E.; Flint, A.L.; et al. Impacts of sea level rise and climate change on coastal plant species in the central California coast. PeerJ 2015, 3, e958.

104. Boon, J.D. Evidence of sea level acceleration at U.S. and Canadian tide stations, Atlantic Coast, North America. J. Coast. Res. 2012, 285, 1437–1445.

105. Gumbel, E.J. Statistics of Extremes; Columbia University Press: New York, NY, USA, 1958; p. 375.

106. Coles, S. An Introduction to Statistical Modeling of Extreme Values; Springer: London, UK, 2001.

107. Pugh, D.T.; Vassie, J.M. Extreme sea-levels from tide and surge probability. In Proceedings of the 16th on Coastal Engineering Conference, Hamburg, Germany, 27 August–3 September 1978; pp. 911–930.

108. Arns, A.; Wahl, T.; Haigh, I.D.; Jensen, J.; Pattiaratchi, C. Estimating extreme water level probabilities: A comparison of the direct methods and recommendations for best practice. Coast. Eng. 2013, 81, 51–66.
109. Petrov, V.; Guedes Soares, C.; Gotovac, H. Prediction of extreme significant wave heights using maximum entropy. Coast. Eng. 2013, 74, 1–10.
110. Tebaldi, C.; Strauss, B.H.; Zervas, C.E. Modelling sea level rise impacts on storm surges along US coasts. Environ. Res. Lett. 2012, 7, 014032.
111. Laudier, N.A.; Thornton, E.B.; MacMahan, J. Measured and modeled wave overtopping on a natural beach. Coast. Eng. 2011, 58, 815–825.
112. Holman, R.A.; Guza, R.T. Measuring run-up on a natural beach. Coast. Eng. 1984, 8, 129–140.
113. Holland, K.; Holman, R.; Sallenger, A. Estimation of overwash bore velocities using video techniques. In Proceedings of 1991 coastal sediments conference, Seattle, WA, USA, 25–27 June 1991; pp. 489–497.
114. Holman, R.; Stanley, J. The history and technical capabilities of Argus. Coast. Eng. 2007, 54, 477–491.
115. Bertin, X.; Li, K.; Roland, A.; Bidlot, J.-R. The contribution of short-waves in storm surges: Two case studies in the Bay of Biscay. Cont. Shelf Res. 2015, 96, 1–15.
116. FEMA. Guidelines Specifications for Flood Mapping Partners, Appendix D. Guidance for Coastal Flooding Analysis and Mapping; Federal emergency management agency: Wichita, KS, USA, 2012.
117. FEMA. Wave Runup and Overtopping-FEMA Coastal Flood Hazard Analysis and Mapping Guidelines; Focused Study Report for FEMA: Seattle, WA, USA; February; 2005.
118. FEMA. Great Lakes Coastal Guidelines, Appendix D.3. http://greatlakescoast .org/pubs/reports/Great_Lakes_Coastal_Guidelines_Update_Jan2014.pdf (accessed on 2 June 2015).
119. Holland, K.T.; Holman, R.A. The statistical distribution of swash maxima on natural beaches. J. Geophys. Res. 1993, 98, 10271.
120. Mase, H. Random wave runup height on gentle slope. J. Waterw. Port Coast. Ocean Eng. 1989, 115, 649–661.
121. Saitoh, T.; Kobayashi, N. Wave transformation and cross-shore sediment transport on sloping beach in front of vertical wall. J. Coast. Res. 2012, 280, 354–359.
122. Sabatier, F.; Anthony, E.J.; Héquette, A.; Suanez, S.; Musereau, J.; Ruz, M.H.; Regnauld, H. Morphodynamics of beach/dune systems: Examples from the coast of France. Géomorphol. Reli. Processes Environ. 2009, 1, 3–22.
123. Jones, A.F.; Brewer, P.A.; Johnstone, E.; Macklin, M.G. High-resolution interpretative geomorphological mapping of river valley environments using airborne LiDAR data. Earth Surf. Processes Landforms 2007, 32, 1574–1592.
124. Leon, J.X.; Heuvelink, G.B.M.; Phinn, S.R. Incorporating DEM Uncertainty in Coastal Inundation Mapping. PLoS ONE 2014, 9, e108727.
125. Hodgson, M.E.; Jensen, J.R.; Schmidt, L.; Schill, S.; Davis, B. An evaluation of LIDAR- and IFSAR-derived digital elevation models in leaf-on conditions with USGS Level 1 and Level 2 DEMs. Remote Sens. Environ. 2003, 84, 295–308.
126. Aguilar, F.; Mills, J.P.; Delagado, J.; Aguilar, M.A.; Negreiros, J.G.; Pérez, J.L. Modelling vertical error in LiDAR-derived digital elevation models. ISPRS J. Photogramm. Remote Sens. 2010, 65, 103–110.
127. Han, J.; Kim, S. Spatial zonation of storm surge hazardous area in the Nakdong Estuary of Korea using high precision terrain data acquired with airborne LiDAR system and geospatial analysis. J. Coast. Res. 2013, 65, 1385–1390.
128. Schubert, J.E.; Gallien, T.W.; Majd, M.S.; Sanders, B.F. Terrestrial Laser Scanning of Anthropogenic Beach Berm Erosion and Overtopping. J. Coast. Res. 2015, 299, 47–60.
129. Arkema, K.K.; Guannel, G.; Verutes, G.; Wood, S.A.; Guerry, A.; Ruckelshaus, M.; Kareiva, P.; Lacayo, M.; Silver, J.M. Coastal habitats shield people and property from sea-level rise and storms. Nat. Clim. Chang. 2013, 3, 913–918.

130. Gedan, K.B.; Kirwan, M.L.; Wolanski, E.; Barbier, E.B.; Silliman, B.R. The present and future role of coastal wetland vegetation in protecting shorelines: Answering recent challenges to the paradigm. Clim. Chang. 2011, 106, 7–29.
131. Prather, M.; Flato, G.; Friedlingstein, P.; Jones, C.; Lamarque, J.-F.; Liao, H.; Rasch, P. Annex II: Climate System Scenario Tables. In IPCC. Climate Change 2013: The Physical Science Basis; Stocker, T.F., Qin, D., Plattner, G.-K., Tignor, M., Allen, S.K., Boschung, J., Nauels, A., Xia, Y., Bex, V., Midgley, P.M., Eds.; Cambridge University Press: Cambridge, UK; New York, NY, USA, 2013.
132. Senneville, S.; St-Onge Drouin, S.; Dumont, D.; Bihan-Poudec, A.-C.; Belemaalem, Z.; Corriveau, M.; Bernatchez, P.; Bélanger, S.; Tolszczuk-Leclerc, S.; Villeneuve, R. Rapport final : Modélisation des glaces dans l'estuaire et le golfe du Saint-Laurent dans la perspective des changements climatiques; ISMER-UQAR, Rapport final présenté au ministère des Transports du Québec: Rimouski, QC, Canada, 2014.
133. Church, J.A.; Monselesan, D.; Gregory, J.M.; Marzeion, B. Evaluating the ability of process based models to project sea-level change. Environ. Res. Lett. 2013, 8, 014051.
134. Bittermann, K.; Rahmstorf, S.; Perrette, M.; Vermeer, M. Predictability of twentieth century sea-level rise from past data. Environ. Res. Lett. 2013, 8, 014013.
135. Gehrels, W.R.; Woodworth, P.L. When did modern rates of sea-level rise start? Glob. Planet. Chang. 2013, 100, 263–277.
136. Barnett, R.L.; Gehrels, W.R.; Charman, D.J.; Saher, M.H.; Marshall, W.A. Late Holocene sea-level change in Arctic Norway. Quat. Sci. Rev. 2015, 107, 214–230.

CHAPTER 7

Role of Beach Morphology in Wave Overtopping Hazard Assessment

Benjamin T. Phillips [1,2,*], Jennifer M. Brown [2], Jean-Raymond Bidlot [3] and Andrew J. Plater [1]

[1]*Department of Geography and Planning, School of Environmental Sciences, University of Liverpool, Roxby Building, Chatham St., Liverpool L69 7ZT, UK*

[2]*National Oceanography Centre, Joseph Proudman Building, 6 Brownlow St., Liverpool L3 5DA, UK*

[3]*European Centre for Medium-Range Weather Forecasts, Shinfield Park, Reading RG2 9AX, UK*

ABSTRACT

Understanding the role of beach morphology in controlling wave overtopping volume will further minimise uncertainties in flood risk assessments at coastal locations defended by engineered structures worldwide. XBeach is used to model wave overtopping volume for a 1:200 year joint probability distribution of waves and water levels with measured, pre- and post-storm beach profiles. The simulation with measured bathymetry is repeated with and without morphological evolution enabled during the modelled storm event. This research assesses the role of morphology in controlling wave overtopping volumes for hazardous events that meet the typical design level of coastal defence structures. Results show that disabling storm-driven morphology under-represents modelled wave overtopping volumes by up to 39% under high Hs conditions and has a greater impact on the wave overtopping rate than the variability applied within the boundary conditions due to the range of wave-water level combinations that meet the 1:200 year joint probability criterion. Accounting for morphology in flood modelling is therefore critical for accurately predicting wave overtopping volumes and the resulting flood hazard and to assess economic losses.

Keywords: sea defence breaching; storm morphology; wave overtopping; XBeach; flood hazard; joint probability; beach morphodynamics; wave modelling

1. INTRODUCTION

For storm events at and beyond the design level of modern, engineered flood defences, it is expected that the defence will experience partial wave overtopping or infiltration through the defence materials by high water levels. This may lead to breaching, which is the reduction of a defence's effective crest height, where flood water propagates through gaps in the defences as a result of structural failure. Once overtopping is replaced by significant breaching, inflow volume to the floodplain can increase by orders of magnitude [1].

The consequences of these increases in inflow volume to the floodplain have been demonstrated by Hurricane Katrina, where approximately 84% of the Greater New Orleans flood extent entered through breaches [2], some of which were up to 622 m in length [3]. Due to high flood velocity in the vicinity of breaches, there is an increase in the dangers associated with flooding, such as instability, drowning and impact from debris and building collapse [4]. These additional risks were clear in the U.K. during the 1953 North Sea floods, when defences were breached in 1200 places [5]. Here, the highest densities of casualties were found to be associated with their proximity to a sea defence breach [6]. On Canvey Island, 56 out of the 58 people who died during the 1953 floods were in close proximity to a breach [7]. These increased risks reinforce the need for effective flood risk assessment to account for risks of inundation from sea defence breaching.

The morphological response of a beach to storm events has been shown to be mostly dependent on wave action [8,9], with other parameters including surge duration, pre-existing morphology and still water level also being capable of influencing storm-driven morphodynamics [10,11]. This research aims to determine the role of the initial beach profile and storm-driven changes in beach morphodynamics in controlling wave overtopping volume through a sea defence breach. This study models selected joint probability extreme wave-water level combinations with fixed pre- and post-storm morphologies (see Section 2.2) and time-evolving morphology. Determining the role of morphology in controlling wave overtopping is important for coastal managers to monitor and manage beach profiles to maximize their natural coastal resilience function and to inform model setup for flood risk assessment.

1.1. Controls on Breaching

The majority of breach-triggering mechanisms can be captured under three main headings, with excessive wave overtopping being the most common mechanism for breach initiation [12,13]:

- Hydraulic failure: overtopping damage causing significant erosion or scouring to the landward side.
- Geo-hydraulic failure: porewater-related effects, such as erosion of the internal structure through seeping.
- Global static failure: pressure forces (ice, wind, waves) exceeding the structure's resistance.

Breaching dynamics have been extensively researched and are characterised by an initiation phase followed by a growth phase. Inflow discharge to the floodplain is dependent on both of these phases. Two unique erosional processes, hydraulic erosion and slumping, are large controls on breach growth [14]. Any erosion-resistant defence, such as a concrete sea wall with an in situ stepped concrete revetment (used as the breaching case study here), can restrict breach initiation by hindering erosion and down-cutting. This applies even when a breach has been initiated, as it reduces the severity of slumping on cracks in the defence. A concrete foundation or an underlying more resistant geological substrate can also reduce breach expansion [1]. Morphological response during Hurricane Sandy was shown to be largely dependent on sea defence condition, where wave attack on Bay Head was shown to be a factor of 1.7 less with the seawall present, due to the prevention of 15 m of vertical dune erosion [15].

The hydraulic gradient (the difference between the extreme water level relative to the surrounding floodplain) is the initial driver of flow across the coastal floodplain. The higher the hydraulic gradient, the faster the flow of flood water and the more severe the erosion. Flow velocities gradually decay due to a reduction in the hydraulic gradient as a result of repeated erosion and slumping. The elevation and extent of the coastal floodplain ultimately controls the maximum breach size, as this determines the extent to which the original hydraulic gradient persists during the expansion of the defence breach [1].

1.2. Flood Risk Modelling

In the context of this study, joint probability refers to significant wave height (Hs) and water level (WL), of which astronomical tide and storm surge are components as these occur simultaneously to produce an extreme event and potentially cause coastal inundation [16]. It is necessary to consider the joint probability of the two variables in flood risk management and coastal engineering, as it provides a more realistic understanding of a defence's resilience under various Hs-WL combinations, which together have the same probability of occurrence. Noticeable differences in inundation extent and flood hazard can arise from different Hs-WL combinations, that have the same probability of occurrence [17]. Therefore, a joint probability approach is more suitable for flood risk assessment when compared with independent, univariate consideration of Hs and WL. Considering that excessive wave overtopping is the predominant cause of sea defence breaching [13], joint probability analysis can also reduce uncertainty in developing fragility curves for various types of sea defence.

A key component of flood risk analysis is breach dimension, as this is a key control on flood outcome and inundation extent [18,19]. Studies have attempted to model breaching dynamics that result from wave impacts on the seaward slope [20] and landward side [21]. Further studies have attempted to contextualise sea defence breaching within an integrated flood risk analysis [22,23] and to derive thresholds for storm events based on morphological thresholds [24].

Early wave overtopping models were one-dimensional and empirically derived, where beaches are assumed to be impermeable. Early numerical one-dimensional models to solve cross-shore sediment transport formulation include [25,26]. Two-dimensional models capable of solving along-shore sediment transport, erosion and time-varying wave overtopping on sandy beaches have also been developed [27]. Previous research has focussed on storm-driven hydrodynamic and morphodynamic changes on gravel beaches [28,29], where the applied model (XBeach-G) was shown to accurately predict initial wave overtopping and wave run-up. Similar studies have been applied for sandy beaches [30,31] and barrier beaches, where one-dimensional wave overtopping models were shown to yield reasonable predictions of 0.72–0.87 (R^2 value) when validated against field data of wave overtopping rates [32].

More recent improvements in accounting for storm-driven changes in beach morphodynamics in academic studies have moved beyond assuming a fixed beach profile, for example with SWAB (Shallow Water and Boussinesq, developed by [33]) [34], to accounting for gravel barrier response using XBeach-G [17]. This study quantified uncertainty in terms of flood hazard and inundation extent from various Hs-WL scenarios with the same annual joint probability for a gravel beach morphology and found that the greatest flood hazard is posed by a swell wave regime under high water levels [17]. However, this conclusion is heavily dependent on the prevailing wave climate and cannot be safely extrapolated to areas, such as the Irish Sea, that are wind wave dominant. Our study addresses the need to understand the uncertainties regarding storm-driven morphodynamics under joint probability forcing conditions and sea defence breaching. The approach presented focuses on the use of datasets available to coastal managers to help inform them of the appropriate model setup required when performing flood risk assessments that use coastal impact models.

1.3. Site Selection

Given that estimates of population residing less than 1 m above sea level range from 1%–2.3% of the total global population [35], there is a global need to understand breaching dynamics. Following historical examples of sea defence breaching leading to fatalities in the U.K. (e.g., the North Sea floods of 1953), we use a case study in a location that is vulnerable to wave overtopping and has recently experienced it during the winter 2013/2014 storms [36]. The selected frontage for this study extends from Rhyl to Prestatyn in North Wales, U.K. (Figure 1). The area experiences a macrotidal regime, with a mean spring tidal range of 7.20 m [37]. The wave climate is dominated by wind waves generated in the eastern Irish Sea, where significant wave height is less than 5.5 m and peak period is less than 12 s, and there is an absence of swell waves due to limited fetch conditions [38]. This region of North Wales lies within a low and expansive coastal floodplain and has experienced inundation exacerbated by sea defence breaching. The most notable example of this is the Towyn flood in 1990. During this event, 467 m of sea wall were breached by a 1:500 year extreme water level [39], composed of high tides, a 1.3-m storm surge and 4.5 m-high waves. Previous beach erosion coupled with poor maintenance of the sea defences resulted in a lack of protection and waves having maximum impact on the sea wall. The flood extent reached up to 2 km inland up to a depth of 1.8 m due to the expansive, low lying nature of the floodplain.

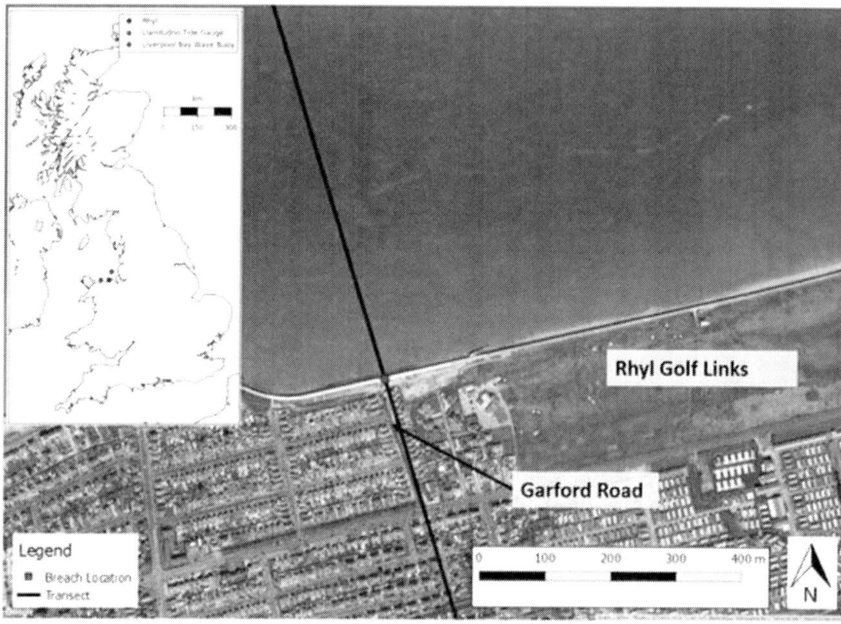

Figure 1. Satellite imagery of the study area, the breach at Garford Road and Rhyl Golf Links [43] and the location of Rhyl, Llandudno Tide Gauge and Liverpool Bay Wave Buoy within the U.K.

A more recent example occurred in December 2013 at Garford Road in Rhyl where significant wave overtopping led to the breaching of a secondary sea defence, allowing flood water to propagate into properties [36]. Garford Road in Rhyl serves as the location for the breach in this study given the flooding in December 2013, its poor condition (Grade 4) and its eroding beach [36]. The defences here comprise an in situ concrete sea wall, with a stepped concrete revetment and rear recurved wall and were constructed in the 1950s. Low beach levels could exacerbate the probability of a sea defence breach, by enabling scouring at the toe of the structure, which is the most likely failure mechanism for a vertical sea wall [40]. Other issues with flood risk management here include poor drainage of Rhyl Golf Links, where flood waters drained towards properties or remained on the golf course for several months [41]. In terms of morphology, the beach exhibits a nearshore swash bar system. Such features are common in wave-dominant regions of large tidal range and influence storm wave impact [42]. The location is therefore suitable to test the impacts of fixed morphology during storm conditions on wave overtopping and breach inflow volumes within flood risk assessment models, typically used to inform shoreline management plans. Given these factors and that flood risk here is driven by storm tides enabling wave overtopping, this section of defence is appropriate for addressing our research aims.

2. MODELLING APPROACH

To simulate the flood hazard, a storm impact model (XBeach, October 2015 "Kingsday" version) is applied in non-hydrostatic mode to the observed beach profile at Garford Road (Figure 1). This study utilises XBeach to model wave overtopping. XBeach solves coupled 2D horizontal equations for wave propagation, flow and sediment transport for wave and tidal boundary conditions [31]. Studies have shown XBeach to effectively hindcast storm impact, overwash and breaching on sandy beaches [30,31]. It has also been used in the operation of coastal impact forecasting systems to forecast when storm impact hazard thresholds are exceeded by hydrodynamic and wave conditions [44,45]. As the beach at Rhyl is composed of sand and a fixed sea wall, XBeach is a suitable model to use for this study as it can represent the variable and fixed evolution of the beach and sea wall, respectively. These sections describe the methods and data used to set up the transect and the boundary conditions used to force XBeach, which is applied here as a 1D cross-shore profile model.

2.1. Joint Probability Extreme Water Level Dataset

Joint probability analysis requires a dataset of Hs at corresponding high waters, and to reduce uncertainty, the length of this dataset should be as long as possible. For this study, multiple datasets of measured and modelled data were combined to generate joint probability boundary conditions. The Environment Agency methodology recommends using the Llandudno tide gauge for the study area [46], the tide gauge record for which extends from 1994–2015. The Liverpool Bay wave buoy, deployed and maintained by Centre for Environment, Fisheries and Aquacultural Science (CEFAS), provides a 13-year record from 2002–2015. Here, we use data from a rerun of the European Centre Wave Model (ECWAM, [47]) Cycle 41R1 (28-km resolution) to provide a longer dataset. The model is forced by a six hourly ERA-interim wind field with no wave data assimilation. The effect of water level change and surface current due to tides and surge is negated. The output used is hourly Hs and Tp from 1979–2014 and was extracted from the offshore limit of the XBeach boundary. This global hindcast was produced in preparation of the ECMWF (European Centre for Medium Range Weather

Forecasts) next reanalysis (ERA5). This allows us to utilise a longer data record, from 1994–2014, in our joint probability analysis. Significant wave heights that exceeded 4.89 m were excluded from the joint probability inputs as these were rare and in some cases over-estimated due to the coarse resolution of the model, causing the joint probability analysis to become skewed. This upper limit ensures that the applied modelled Hs are within the maximum 5.5-m conditions for this region, as determined by [38]. The ECMWF wave hindcast exhibits an RMSE (root mean square error) of 0.31 m, a bias (the difference between the estimator's expected value and the true value of the parameter being estimated) of 0.06 m and a symmetric slope value of 1.11 for Hs when validated against the Liverpool Bay Wave Buoy. For each high water, the closest Hs and Tp were extracted. The JOIN-SEA software uses the generalised Pareto distribution (GPD) model and simultaneous records of Hs, WL and Tp (handled within a wave steepness variable) to calculate joint probability extreme water levels. Hsand WL are transformed to normal scales using the generalised Pareto distribution model to values greater than specified thresholds [16]. An empirical distribution function is applied to values below this threshold. A single bivariate-normal distribution is then used to determine the dependence between Hs and WL. A large range of synthetic records containing the same statistical properties as the original inputs is then simulated, and the exceedance of Hs-WL thresholds for specified return periods is calculated in the form of a distribution curve. For this study, we use a 1:200 year or a 0.5% probability event, as the Natural Resources Wales flood maps demonstrate Garford Rd. as being at risk from this severity of event if flood defences were not present [48].

In order to understand the role of morphology in controlling wave overtopping volume across the joint probability curve, we first identified the Hs-WL combinations that caused wave overtopping. We found wave overtopping to occur between a 2.99 m above Ordnance Datum Newlyn (m ODN) WL and 5.75 m Hs (where the water level is sufficient to allow large waves to impact the defence) and 5.12 m ODN WL and 1.5 m Hs (where Hs is insufficient in causing wave overtopping despite elevated WL). To select suitable scenarios to test the impacts of morphology, we used scenarios with two identical WLs to allow the direct comparison of morphological variability under varying Hs only (Table 1). This section of the joint probability distribution represents the tipping point between water levels allowing waves to overtop the sea wall and elevated water levels allowing smaller waves to cause overtopping. Storm-driven morphology impacts have been shown to be mostly dependent on high intensity wave energy [24] and the morphological variability under different conditions representing the same storm severity. The maximum recommended wave height with which to force XBeach is 5 m and is therefore not exceeded by our boundary conditions.

Table 1. Hs-WL combinations extracted from the 1:200 year joint probability curve. Time-integrated overtopping volumes are provided from scenarios with morphological beach evolution enabled. Scenario 3 is shown in bold, as this was used as the most extreme forcing for the breach event, and for the extraction of pre- and post-storm profiles. MHWST, mean high water spring tide.

Scenario	WL (m ODN)	H_s (m)	Overtopping Volume (m^3 per m)	Skew Surge above MHWST (m)
1	4.56	5.00	8089	0.73
2	4.56	4.75	7701	0.73
3	**4.80**	**4.00**	**9244**	**0.97**
4	4.80	3.75	8755	0.97

Joint probability analysis has been applied to historic high waters using data from Llandudno tide gauge and Liverpool Bay wave buoy. The joint probability analysis created a 1:200 year return period curve, shown in Figure 2. In this study, we present results for conditions that pose the greatest flood hazard, given in Table 1. Their selection is in agreement with other studies or previously-observed flood events that occurred due to defence overtopping and breaching. For the high water of 26 February 1990 (12:00), a water level of 6.13 m ODN (univariate return period of 1:80 year) coincided with a 4.32 m Hs (univariate return period of one in 1.5 years). When considered in joint probability, this event can be classified as a 1:500 year return period [39]. The same analysis is provided for the high water of 5 December 2013, when a 1.09-m surge coincided with a spring tide of 5.13 m ODN and 4.55 m Hs. The univariate probabilities for the water level and wave heights were calculated as 1:44 year and 1:5 year, respectively. This extreme event may have exceeded 1:200 year when considered in joint probability [49]. Given that this event caused significant wave overtopping and the breaching of a secondary sea defence, a 1:200 year WL and the values extracted from our joint probability analysis can be considered appropriate in addressing our research aims.

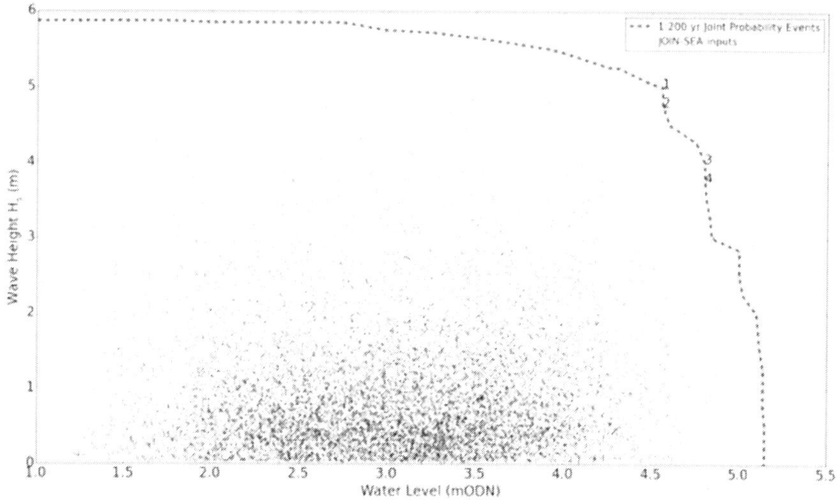

Figure 2. The 1:200 year joint probability distribution calculated from JOIN-SEA. Scenarios 1–4 are annotated, with specific parameters of these events shown in Table 1.

2.2. Model Setup

To generate tidal boundary conditions, we used Proudman Oceanographic Laboratory Tidal Prediction Software (POLTIPS) (provided by National Oceanography Centre Marine Data Products team) to identify a tidal cycle with a high water of 3.83 m ODN, as this is the mean high water spring tide (MHWST) for Llandudno for 2008–2026 [37]. The designed storm surge profiles for each Class A tide gauge across the U.K. are provided by the Environment Agency [46]. The designed surge profile from Llandudno is used to scale the MHWST curve to the required WL at 15-min temporal resolution. To do this, the difference between MHWST and the selected WL is calculated. These values were then multiplied by the normalised surge ratio and added to the tidal predictions to generate a time-varying storm tide. We use the peak tidal cycle of 12.5 h rather than the full 99-h profile, as overtopping is likely to only occur around high water during a tidal cycle. Skew surges up to 1.01 m have been recorded at Llandudno (and up to 1.09 m at Liverpool), with at least 10 skew surges higher than 0.73 m occurring since 1996 [50]. The skew surge values required to obtain the appropriate WLs (see Table 1) do not exceed any skew surges on record and are therefore suitable for this study. Given that our aim is only to test the role of morphology, the peak surge and tidal high water are set to coincide following the methodology recommended by [46].

Along with water level boundary conditions, XBeach was forced with a constant Hs and a 9.08 s Tp using a JONSWAP (Joint North Sea Wave Project) spectrum, which was also kept constant for each of the morphological setups. Tp was derived from the ECMWF hindcast; the chosen value is representative of the maximum Tp associated with Hs with similar values to those used in this study (Table 1). The maximum value is chosen to ensure maximum wave impact for overtopping. A constant Tp is applied in each scenario to limit the overtopping variability between scenarios to only the Hs-WL conditions. Wave direction was not accounted for in the model boundary conditions, again to restrict the range of input parameters being varied to Hs-WL conditions alone. In each simulation, the waves are assumed to propagate directly onshore for maximum impact. A constant wave condition is applied to simulate the maximum hazard, reducing the complexity of having to represent the time-variability in wave conditions relative to the tidal elevations for events at this location.

By exploring a full range of profiles in relation to high water, we can gain a better understanding of total and time-varying wave overtopping volumes across the joint probability distribution. Post-storm profiles were selected to determine whether changes in morphology could lead to greater wave overtopping under storm conditions in subsequent tidal cycles. Pre-storm profiles were selected, as high, healthy beach levels damp approaching waves and promote breaking prior to impact. We selected the highest pre-storm beach level (Scenario 1) and the lowest post-storm beach level (Scenario 4) at the toe of the sea defence from all four Hs-WL scenarios as input bathymetries, to fully explore morphological variability. The variability between pre-storm and post-storm profiles is demonstrated in Figure 3 and shows reasonable agreement between profiles. The upper beach profiles of the smoothed pre- and post-storm profiles are representative of a more reflective and a more dissipative profile [51], respectively. The consideration of event-driven morphological evolution was also investigated to account for the potential impact of models that may not be capable of resolving morphological evolution, in order to understand the implications for flood modelling. XBeach was therefore used to test the following morphological settings and input profiles, which represent the possible combinations used by local authorities for coastal flood risk assessments:

1. A realistic (measured seasonal profile, Figure 3) nearshore swash bar system, extracted from the original 1D transect, with morphology disabled.
2. A realistic (measured seasonal profile, Figure 3) nearshore swash bar system, extracted from the original 1D transect, with morphology enabled.

3. Pre high water (transect extracted 33% of the way through the XBeach simulation of Scenario 1, pre-storm, Figure 3), where the water level had begun to smooth the nearshore swash bar system to represent the pre-storm profile, with morphology disabled.

4. Post high water (transect extracted at the end of the XBeach simulation of Scenario 4, post-storm, Figure 3), to be representative of a post-storm beach profile, with morphology disabled.

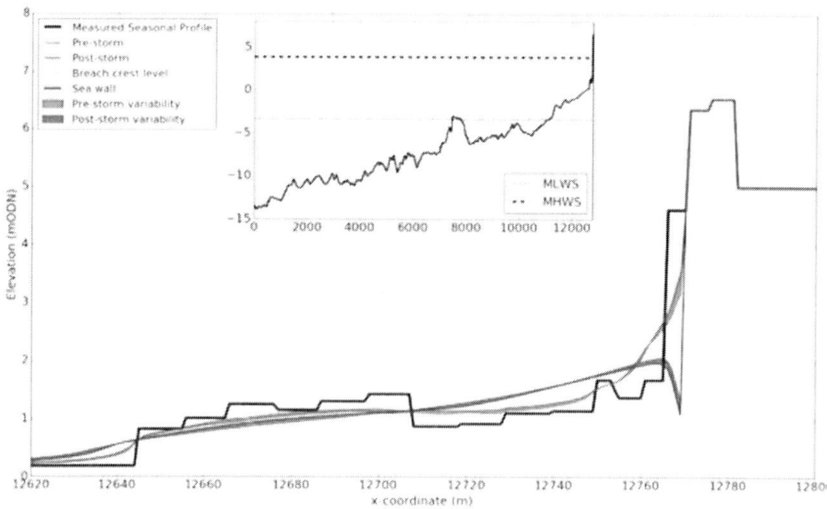

Figure 3. Applied 1D transect in XBeach, with all considered initial bathymetries. The post-storm variability (blue shading) shows the difference between the lowest (Scenario 4, Table 1, selected profile) and highest (Scenario 1, Table 1) elevation at the sea defence toe. The pre-storm variability (red shading) shows the difference between the lowest (Scenario 4) and highest (Scenario 1) elevation at the sea defence crest toe. Mean high water spring (MHWS) and mean low water spring (MLWS) tidal elevations are provided by [37].

XBeach contains several options for morphological evolution. To completely disable morphological evolution for the pre-storm, post-storm and actual, but fixed profiles, we entered "morphology = 0" into the parameters file. XBeach also contains an option for allowing certain grid coordinates to remain fixed whilst others morphologically evolve. This approach is appropriate for the morphology enabled scenario, where the sea wall is fixed, but the beach evolves through the simulation. Each grid point in the transect was set to a value of zero for non-erodible (i.e., the sea wall) or 10 for the beach. This file is then set under the "ne_layer" keyword in the parameter file. When running the model, the typical simulation time was approximately 8 h, with a time step of 0.02 s. Therefore, time steps were sufficiently small to capture overtopping events, which are in the order of seconds.

The one-dimensional model transect (1 m wide) consisted of a combination of a nearshore topographic survey (carried out on 5 October 2010 using real-time kinematic GNSS in accordance with the then Environment Agency National Standard Contract and Specification for Surveying Services). The transect was extended by EDINA Digimap gridded bathymetry capturing depths up to approximately −13 m ODN at 13 km offshore (see Figure 3), both at 1-m grid resolution. This offshore limit represents the locations at which the offshore wave hindcast for Rhyl was provided. Linear interpolation was used to ensure a smooth transition between the two datasets. Standard cross-shore discharge (Qx in XBeach) from the model output was used to calculate wave overtopping volume. The time-varying discharge for the defence crest was extracted and multiplied by the save interval (5 s) to convert into a volume. Using this approach means that applying a constant, planar value to the floodplain is an acceptable approach, as once the water is landward of the defence crest, it becomes irrelevant. Breaches were integrated by subtracting the loss of crest height from the crest elevation.

2.3. Modelling Sea Defence Breaching

We impose a 1-m loss of crest height for the sea wall at 60, 30 and 15 min before high water, at high water and 15 min after high water. The bed profile, wave spectrum and water level from Scenario 3 were extracted at the time of the breach. The upper section of the pre-breach beach profile was modified to incorporate a lower crest level for the seawall. XBeach was then "warm started" at the time of breaching with the breached defence crest from the extracted pre-breach water level and wave spectrum, and the simulation continued for the remainder of the tidal cycle to simulate post-breach wave overtopping. A 5-s save interval was applied in XBeach, with results showing wave overtopping volume averaged out over 7.5 min (i.e., 1% of a tidal cycle). Wave overtopping did not occur more than 150 min either side of high water, so these sections of the tidal curve are excluded in the analysis of our results.

3. RESULTS

3.1. Beach Morphology under Wave Overtopping Scenarios

Figure 4 clearly shows high variability in time-integrated total wave overtopping volumes between the scenarios, despite all having the same probability of occurrence. Disabling morphology on the measured bathymetry yields the lowest wave overtopping volume in every scenario. The pre-storm profiles exhibit the highest wave overtopping in each scenario. This is likely to be attributable to the smoothing of the nearshore swash bar (Figure 3), and the low elevations of the upper beach, before rapid steepening at the defence toe, allowing waves to break closer to the seawall. There is a negligible difference between the morphology enabled and post-storm profiles in all scenarios, suggesting that the cross-shore sediment transport rapidly redistributes sediment to form the post-storm profile. The variability in overtopping volume between the scenarios is much less when morphology is considered for the measured bathymetry than when it is not (black compared with white bars, Figure 4), with differences of 1542 m^3 and 2363 m^3 between the minimum and maximum volumes for morphology enabled and disabled, respectively. The variability in overtopping across the actual profile with enabled and disabled morphology is greater under Scenarios 1 and 2, which both experience greater wave action compared with Scenarios 3 and 4 (Table 1). Our results clearly demonstrate that, particularly under high Hs conditions, excluding morphology considerably under-estimates wave overtopping volume by up to 2959 m^3 and 3010 m^3 under Scenarios 1 and 2, 36% and 39% lower than the morphology-enabled wave overtopping volumes, respectively. This shows that morphological evolution has the potential to exert greater influence on the uncertainty of overtopping volumes than the choice of Hs-WL combinations representing the most extreme flood hazard conditions for the required event severity.

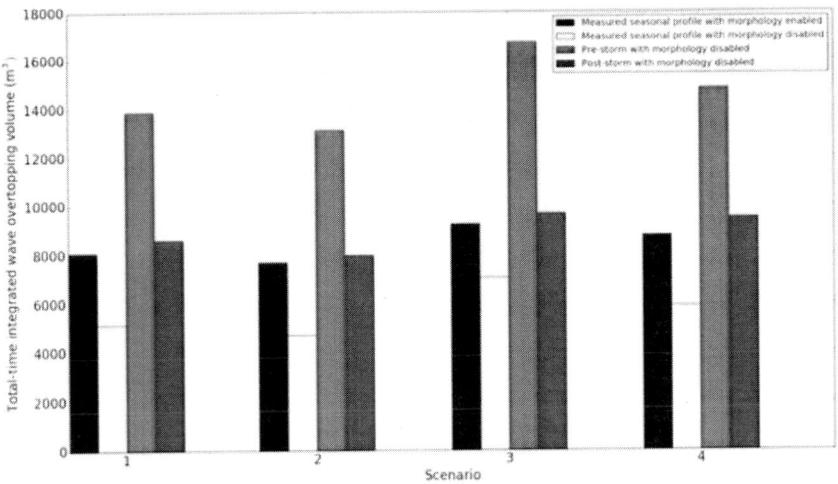

Figure 4. Total time-integrated wave overtopping for each scenario under the applied morphological profiles: measured seasonal profile with morphology disabled, measured seasonal profile with morphology enabled, pre-storm with morphology disabled and post-storm with morphology disabled.

In addition to the total time-integrated wave overtopping volumes, there is important temporal variability in the time-varying wave overtopping across the scenarios (Figure 5). The largest variability occurs between 75 min before and 50 min after high water. The variability in wave overtopping is greatest for the pre-storm profile with morphology disabled. For this model setup, the greater overtopping volumes occur either side of high water in Scenario 2 and more noticeably in Scenarios 3 and 4. This suggests that the increased WLs and lower waves enable an increase in wave reflection at high water, reducing the overtopping for this beach profile. The measured profile with morphology disabled has a continually low overtopping volume. This suggests that the morphological features and their evolution dampen the influence of the wave field variability as it propagates towards the shoreline. Variability in the overtopping duration is negligible across all scenarios, with overtopping occurring just earlier than 100 min before high water. The decay in overtopping volume occurs at a similar rate and time across all four scenarios each model setup. However, overtopping is seen to end just earlier than 100 min after high water, suggesting that the flood and ebb tides cause slight asymmetry in overtopping duration either side of high water. All scenarios display either peaks in wave overtopping either side of high water, typically occurring up to 50 min either side of high water, or little change in the maximum values during the 50 min either side of high water. This suggests there is competition between higher water levels limiting wave dissipation across the beach profile and deeper depths at the structure toe promoting wave reflection.

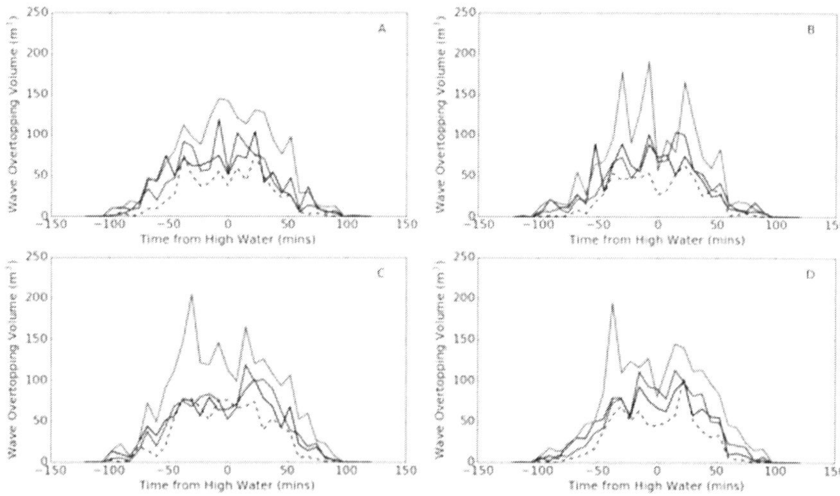

Figure 5. Time-varying wave overtopping for Scenario 1 (**A**), Scenario 2 (**B**), Scenario 3 (**C**) and Scenario 4 (**D**), under the applied morphological profiles: measured seasonal profile with morphology disabled (dashed black line), measured seasonal profiles with morphology enabled (solid black line), pre-storm with morphology disabled (red line) and post-storm with morphology disabled (blue line).

In addition to temporal variability, morphological setting and initial input, bathymetry is shown to be a key control on maximum (5 s averaged) wave overtopping volume. In Scenario 2, the maximum wave overtopping during the pre-storm profile simulation is a factor of three higher than that of the morphology-disabled simulation. The morphological setting and initial input bathymetry have the potential to exercise larger control on wave overtopping volume compared to the applied Hs-WL condition. Noticeably, the morphology-enabled and post-storm profiles display similar temporal variability (particularly under Scenarios 2 and 4) and negligible difference in the total time-varying wave overtopping volumes. The difference in wave overtopping volumes between pre-storm and morphology-disabled profiles exceeds that of the applied Hs-WL conditions.

There is large temporal variability in time-varying scour (Figure 6, calculated as the difference in bed elevation at the grid point adjacent to the structure toe that is enabled to evolve and the adjacent seaward grid cell). Across all scenarios, scour reaches a maximum after high water as the wave action continues to erode the beach after high water. A comparison of Figure 5 and Figure 6 shows that scour is initiated at approximately 80 min before high water just after the onset of wave overtopping. Scenario 3 experiences the most severe rate of scour at its onset, with almost no variability between Scenarios 1 and 2 from 105 min before high water and 90 min before high water. Scenario 4 scours at a more gradual rate. Either side of high water, there is a very variable response in scour level. However, after scouring is initiated, the final depth of the remaining scour hole after the tide has receded seems to be most sensitive to the water level. Higher water elevations enable the wave action to cause scour earlier and the resulting depth to be greater. The results do not suggest that higher water levels enable scour to continue for longer on the falling tide.

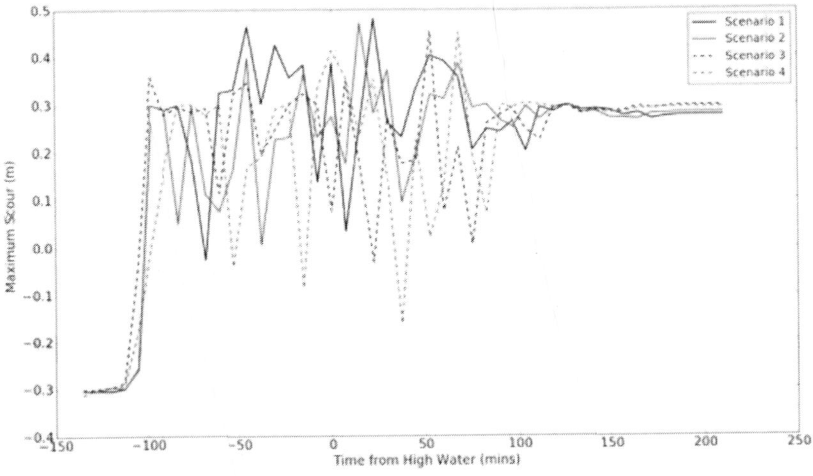

Figure 6. Time-varying scour at the toe of the structure (calculated as time-varying difference in elevation between x-coordinates 12,768 and 12,769 m) for each scenario (see Table 1) with morphology enabled.

The maximum scour under Scenarios 1 and 2 occurs at 15 and 22 min after high water, respectively. The greater extreme water level influence relative to wave action under Scenarios 3 and 4 results in a similar level of maximum scour 50 and 65 min post high water. At approximately 130 min after high water, the scour is within 0.05 m across all four scenarios, until the ebb tide and wave run-up is no longer acting on the beach level at the defence toe and scour plateaus. Here, Scenario 1 (which has the greatest wave action) has resulted in the least resultant scour, with Scenario 3 causing the most resultant scour. Given that there is only a 0.02-m difference in resultant scour across all scenarios, it is not safe to conclude whether tide or wave action controls resultant scour. Accounting for various morphological responses from resultant scour remains important in determining wave overtopping volume on storm clusters occurring on subsequent tidal cycles.

3.2. Beach Morphology under Sea Defence Breaching Scenarios

Figure 7 demonstrates that breach timing and morphological setting have a large control on the time-integrated total wave overtopping volume for Scenario 3 (the Hs-WL combination with the highest overtopping volume, Table 1). The pre-storm profile displays the highest time-integrated total wave overtopping volume for every breach scenario and the morphology-disabled profile the lowest. There is minimal variability in wave overtopping volume between the post-storm and morphology-enabled profiles, particularly at high water. Except for the breach occurring 60 min before high water, wave overtopping for the measured morphology-enabled profile exceeds that of the post-storm profile. The post-storm and morphology-enabled profiles exhibit similar patterns in Figure 4 and Figure 5, when no breach is applied.

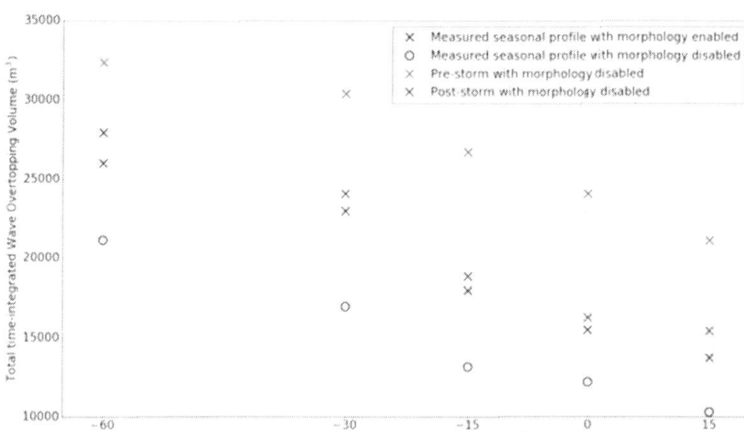

Figure 7. Total time-integrated wave overtopping volume for the duration of the event in Scenario 3 for each morphology setting with a 1-m breach applied at −60, −30, 0 and 15 min from high water. Values represent total overtopping volumes for each breaching simulation, rather than cumulative overtopping through the event.

The breach occurring at 60 min before high water displays the highest cumulative wave overtopping volume; this can be attributed to the water level having more time to overtop the breached defence. There is a reasonably linear relationship between the breach time relative to high water and the total wave overtopping. A “tipping point” seems to exists between 30 min and 15 min before high water, after which the rate of reduction in wave overtopping volume is reduced, for all but the pre-storm profile, which shows an increased rate of reduction. The spread in the results becomes less when the breach occurs after high water. The different morphological settings clearly have an impact on the total time-integrated overtopping volume, despite the large additional discharge due to the breach (Figure 7 volumes compared with Scenario 3 in Figure 4). This further shows that careful selection of an appropriate initial bathymetry within a flood risk assessment model when morphological evolution of the beach is not considered is important. Here, it is again suggested that a post-storm bathymetry is more representative than the other fixed profile simulations at simulating the overtopping volume simulated for the measured profile with morphology enabled. Although, the error is greater than that of the no breach simulation.

4. DISCUSSION

These scenarios and our method cannot be taken as a prediction of a past or future wave overtopping event; rather, they demonstrate the impact of beach morphology on wave overtopping volume under joint probability conditions representative of a 1:200 year event, coupled with sea defence breaching (where applicable). Previous research has primarily been concerned with exploring the impact of a joint probability distribution on inundation extent and flood hazard [17] without comparing the impact of various morphological settings to account for various models that may be applied by coastal managers. Hydrodynamic factors have been applied in a coastal flood forecasting system that parametrised beach levels using canonical correlation analysis (CCA) according to the incident wave conditions [52]. Whilst this may be more computationally efficient over a longer forecasting period, it negates the detailed representation of the time-varying event-driven morphological evolution. This may be an important uncertainty in areas where both tides and waves have a similar impact on morphological evolution. A numerical model, such as XBeach, accounts for the time-varying evolution of the beach profile and the feedbacks on the coupled wave-water level conditions, but is more computationally expensive.

Other studies assessed the overwash regime using multi-decadal time series in a similar coastal setting, but assumed a static morphology [53]. Such an assumption does not capture the potential uncertainty in the flood hazard due to event-driven morphological evolution. Studies using XBeach have been largely limited to studies of morphological evolution during storm conditions [30,42] and have not attempted to quantify the impact of these changes on wave overtopping volumes and the implications for coastal flood modelling.

XBeach has been compared with empirical models (EurOtop and "bathtub" method) in a validation of coastal flooding, where it was found to reasonable predict flood extent [54]. Rigorous validation of wave overtopping models requires inflow volume data rather than just resultant flood extents. It has been shown to yield reasonable predictions of breaching behaviour at Dungeness [55] and to reasonably predict storm morphodynamics on beach systems [44]. Combining appropriate joint probability boundary conditions to create an event of the desired hazard rating, with the understanding of the consequence of the chosen morphological setting and input profile, is therefore important for coastal stakeholders and policy makers to assess the potential range in impact of an extreme event.

4.1. Impact of Morphology on Controlling Wave Overtopping

Our results demonstrate that the influence of morphological forcing creates noticeable variability not only in the time-integrated total wave overtopping volume, but also creates temporal uncertainty in the timing of peak wave overtopping volume. Whilst understanding the temporal uncertainty surrounding the morphological controls on wave overtopping provides a critical insight for managing flood hazard, it is also important to consider the total volume of water, as this will be a key determinant of the total inundation extent and resulting economic cost caused by a flood event. Although we have only applied four scenarios, wind waves dominate the wave climate, and as such, the most severe wave overtopping occurs in the section of the joint probability curve chosen. Given that the variability in wave overtopping due to the representation of the beach morphology (in Scenarios 1 and 2) is considerably greater than the variability due to the considered Hs-WL conditions, this demonstrates that the impact of morphology has the potential to create greater uncertainty in flood hazard for a storm of a given severity than the hydrodynamic boundary conditions that represent the storm conditions.

Sites are typically surveyed bi-annually (summer and winter) to capture the seasonal variability in beach elevation for shoreline management purposes. However, the timing of a survey relative to a storm may cause the profile to represent pre-storm, post-storm or partially-recovered conditions. We therefore indicate the potential uncertainty in a flood risk assessment when using limited available survey data. The variability in the pre- and post-storm profile wave overtopping rates also helps inform managers on profile resilience. Such information is important when considering beach nourishment and pre-winter hazard assessments. Here, a lower beach elevation with a steep gradient at the toe of the defence (as in the pre-storm profile) can allow greater overtopping than a beach face with higher elevations and no increase in elevation at the defence toe (or in this case, scour, as in the post-storm profile).

4.2. Implications for Flood Modelling

The noticeable variability in wave overtopping between the scenarios highlights the importance of accounting for uncertainty in Hs-WL boundary conditions used to force a storm impact model. This reinforces the need for the consideration of joint probability extreme water levels to enable a range of conditions and inform suitable defence design [17]. This approach is being used in commercial studies, e.g., [56], and examples exist where joint probability for historical events is also calculated [57].

Our results clearly show that the applied morphological evolution has a large influence on not only the time-varying wave overtopping, but also the total volume overtopping the defence. A simple modelling approach of taking a nearshore topographic survey and applying a model exclusive of morphological updating is likely to significantly over-estimate or under-estimate the wave overtopping volume, depending on the chosen bathymetry. One with features, such as a nearshore swash bar system, will cause under-prediction, while a smoothed initial profile over-predicts, particularly when the upper beach face is imposed at a low elevation. This is particularly important under joint probability boundary conditions with higher Hs and wave action, as Scenarios 1 and 2 demonstrate greater variability between the morphology enabled and disabled scenarios (up to 39% variability). The importance of accounting for beach morphology in the original transect is shown to be a large control on wave overtopping volume for inundation modelling, whether sea defence breaching occurs or not.

Regardless of the time occurrence of sea defence breaching, the impact of morphological representation during a sea defence breach event is shown to be just as critical as when the sea defence is only overtopped and particularly when the breach occurs more than 15 min before high water. This demonstrates that any flood modelling, even if inclusive of sea defence breaching where wave overtopping volume has the potential to increase by orders of magnitude, should account for the impact of storm-driven changes in morphology. Studies with clusters of storms occurring in close succession should use a morphology model to account for the recovery of features within the beach profile for accurate assessment of the clustered hazard. This has implications for the development of fragility curves, which may benefit from this type of modelling where impacts of beach changes over several tidal cycles on fragility curve inputs (i.e., difference between crest elevation and water level) can be observed. This approach of using multiple morphology forcings is more suitable to locations such as Rhyl, where a nearshore swash bar exists within the beach profile. Here, the flood hazard is due to water levels enabling large wind waves to overtop. Further study would be required to see if morphology is of higher importance than the representative Hs-WL boundary conditions under a bimodal wave regime, in posing a greater secondary flood hazard at higher water levels when the wave regime turns from wind waves to swell waves [17].

4.3. Limitations of This Research

The limitations we discuss are currently widespread due to data availability and are not exclusive to this study. Limitations of this research include the unimodal application of waves due to joint probability analysis providing only a single Hs-WL condition. This may result in the total wave overtopping volume being larger than expected. The relatively short record of Hs and WL (20 years) limits the accuracy of calculating a 1:200 year probability distribution. ECWAM is not a shallow water model and therefore leads to occasional over-prediction of Hs given the coarse resolution of both bathymetry and wind forcing. Considering wave direction would have presented a more realistic representation of wave run-up and beach profile response in the XBeach boundary conditions, but would have further shortened the joint probability dataset and further increased uncertainty. To overcome all of these limitations, a coastal area model could simulate long-term datasets of combined Hs-WL conditions; however, this is computationally demanding [17]. A single deterministic run is used for the four scenario settings, so any variability in overtopping volumes due to the representation of the wave field is not captured. Using an ensemble of simulations would give greater information on the variability in overtopping rates due to the variability in the wave field; however, using the same time-varying wave spectra for each simulation within the four Hs-WL scenarios ensures that the difference in overtopping volumes is due to the input profile and morphological setting.

5. CONCLUSIONS

We have demonstrated a method to use XBeach to determine the role of storm-driven changes in beach morphology under various joint probability Hs-WL combinations. We show that enabling beach morphology increases wave overtopping volumes during high flood hazard events with a one in 200 year severity, particularly under high Hsconditions. This study further shows that the impact of enabling morphological evolution (either enabling morphology in the model or using a profile with smoothed pre-existing features) can have a greater impact on wave overtopping volume than variability in the applied extreme wave height and water level. This applies to the area of a joint probability curve where high water levels enable large wind waves to cause defence overtopping. Morphology is shown to be a key control regardless of sea defence failure, but larger variability exists between forcing when a breach is applied before high water. Accounting for morphology in flood modelling is therefore a critical component for reducing uncertainty under joint probability boundary conditions, particularly under high wave action. As such, storm impact modelling should either be inclusive of morphology or process input topographic datasets to smooth pre-existing features in a wave-dominated, macrotidal area's coastal geomorphology to prevent the impact of geomorphological features (e.g., intertidal swash bar systems or ridge-runnels) that are washed out during the event, dissipating the wave conditions. Uncertainties between seasonal variations in beach levels recorded by bi-annual surveys are important to determine appropriate input bathymetry. Knowledge about the most appropriate input bathymetry to select for typical or maximum overtopping conditions would better inform flood risk management by understanding the uncertainties surrounding the time of peak wave overtopping in relation to high water, the total wave overtopping volume and the resulting flood hazard and economic losses. Here, it is shown that a lower upper beach face elevation (in this case 12,705–12,760 m cross-shore distance) creates a higher overtopping hazard, even when a steep profile is present close to the structure toe, rapidly increasing the beach elevation. Whilst considerable uncertainties in determining the Hs-WL conditions of a given joint probability exist due to only short, simultaneous datasets being available, this study highlights the potential to importantly under-predict wave overtopping volume by excluding morphological evolution, even under sea defence breaching conditions.

ACKNOWLEDGMENTS

Funded by the Engineering and Physical Sciences Research Council as part of the Adaptation and Resilience of Coastal Energy Supply project (EP/I035390/1). B.T.P. was supported by the Natural Environment Research Council's Understanding the Earth, Atmosphere and Ocean Doctoral Training Program (NE/L002469/1) whilst this manuscript was in review. Tide gauge data for Llandudno were provided by the British Oceanographic Data Centre. Liverpool Bay wave buoy data were provided by the Centre for Environment, Fisheries and Aquaculture (CEFAS). The authors would also like to acknowledge Alan Williams of Coastal Engineering UK Ltd. for providing coastal defence inspection reports and the nearshore topographic survey.

AUTHOR CONTRIBUTIONS

The experiment was conceived of and designed by B.T.P., J.M.B. and A.J.P. J.-R.B. modelled the hourly wave hindcast data for joint probability inputs and provided comments on the paper. B.T.P. performed the experiments, analysed the data and wrote the paper along with J.M.B. and A.J.P.

CONFLICTS OF INTEREST

The authors declare no conflict of interest.

REFERENCES

1. Wood, R.M.; Bateman, W. Uncertainties and constraints on breaching and their implications for flood loss estimation. Philos. Trans. R. Soc. Ser. A Math. Phys. Eng. Sci. 2005, 363, 1423–1430.
2. Van Heerden, I.L. The Failure of the New Orleans Levee System Following Hurricane Katrina and the Pathway Forward. Public Adm. Rev. 2007, 67, 24–35.
3. Bernitt, L.; Lynett, P. Breaching of sea dikes. In Proceedings of the 32nd Conference on Coastal Engineering (ICCE No 32), Shanghai, China, 30 June–5 July 2010.
4. Jonkman, S.; Masskant, B.; Boyd, E.; Levitan, M. Loss of life caused by the flooding of New Orleans after Hurricane Katrina: Analysis of the relationship between flood characteristics and mortality. Risk Anal. Off. Publ. Soc. Risk Anal.2009, 29, 676–698.
5. Risk Management Solutions (RMS). 1953 U.K. Floods: A 50-Year Retrospective; Technical Report; Risk Management Solutions, Inc.: Newark, CA, USA, 2003.

6. Department for Environment Food and Rural Affairs (DEFRA). Flood Risks to People Phase 1, R & D Technical Report FD2317; Technical Report; Department for Environment Food and Rural Affairs: London, UK, 2003.
7. Baxter, P.J. The east coast Big Flood, 31 January–1 February 1953: A summary of the human disaster. Philos. Trans. Ser. A Math. Phys. Eng. Sci. 2005, 363, 1293–1312.
8. Wright, L.D.; Short, A.D. Morphodynamic variability of surf zones and beaches: A synthesis. Mar. Geol. 1984, 56, 93–118.
9. He, Z.; Hu, P.; Zhao, L.; Wu, G.; Pähtz, T. Modeling of breaching due to overtopping flow and waves based on coupled flow and sediment transport. Water 2015, 7, 4283–4304.
10. Wright, L.D.; Short, A.D.; Green, M.O. Short-term changes in the morphodynamic states of beaches and surf zones: An empirical predictive model. Mar. Geol. 1985, 62, 339–364.
11. Loureiro, C.; Ferreira, Ó.; Cooper, J. Contrasting morphologic behaviour at embayed beaches in Southern Portugal. J. Coast. Res. 2009, 1, 83–87.
12. Kanning, W.; van Baars, S.; van Gelder, P.H.A.J.M.; Vrijling, J.K. Lessons from New Orleans for the design and maintenance of flood defence systems. In Risk, Reliability and Societal Safety; Aven, T., Vinnem, J.E., Eds.; Taylor and Francis Group: London, UK, 2007.
13. Vorogushyn, S.; Merz, B.; Apel, H. Development of dike fragility curves for piping and micro-instability breach mechanisms. Nat. Hazards Earth Syst. Sci. 2009, 9, 1383–1401.
14. Morris, M.W.; Hassan, M. Breach Formation through Embankment Dams & Flood Defence Embankments: A State of the Art Review, Impact Project Workshop (16–17 May 2002); Technical Report; HR Wallingford: Wallingford, UK, 2002.
15. Smallegan, S.M.; Irish, J.L.; van Dongeren, A.R.; den Bieman, J.P. Morphological response of a sandy barrier island with a buried seawall during Hurricane Sandy. Coast. Eng. 2016, 110, 102–110.
16. Hawkes, P.; Gouldby, B. The Joint Probability of Waves and Water Levels: JOIN-SEA Version 1.0; HR Wallingford: Wallingford, UK, 1998.
17. Prime, T.; Brown, J.M.; Plater, A.J. Flood inundation uncertainty: The case of a 0.5% annual probability flood event. Environ. Sci. Policy 2016, 59, 1–9.
18. Dawson, R.; Hall, J.; Sayers, P.; Bates, P.; Rosu, C. Sampling-based flood risk analysis for fluvial dike systems. Stoch. Environ. Res. Risk Assess. 2005, 19, 388–402.
19. Nicholls, R.; Mokrech, M.; Bates, P.; Dawson, R.; Hall, J. Assessing Coastal Flood Risk at Specific Sites and Regional Scales: Regional Assessment of Coastal Flood Risk; Technical Report 45; Tyndall Centre for Climate Change Research: Norwich, UK, 2005.
20. Stanczak, G.; Oumeraci, H. Model for prediction of sea dike breaching initiated by breaking wave impact. Nat. Hazards 2011, 61, 673–687.
21. D'Eliso, C.; Oumeraci, H.; Kortenhaus, A. Breaching of Coastal Dikes: Reliability Analysis and Validation of the Model System; University of Florence: Florence, Italy, 2007; p. 44.
22. Naulin, M.; Kortenhaus, A.; Oumeraci, H. Failure Probability of Flood Defence Structures/Systems in Risk Analysis for Extreme Storm Surges. Coast. Eng. 2010, 32, 1–15.
23. Wadey, M.P.; Nicholls, R.J.; Hutton, C. Coastal Flooding in the Solent: An Integrated Analysis of Defences and Inundation. Water 2012, 4, 430–459.
24. Gervais, M.; Balouin, Y.; Belon, R. Morphological response and coastal dynamics associated with major storm events along the Gulf of Lions Coastline, France. Geomorphology 2012, 143–144, 69–80.
25. Kobayashi, N.; Tega, Y.; Hancock, M.W. Wave reflection and overwash of dunes. J. Waterw. Port Coast. Ocean Eng.1996, 122, 150–153.
26. Baldock, T.E.; Weir, F.; Hughes, M.G. Morphodynamic evolution of a coastal lagoon entrance during swash overwash. Geomorphology 2008, 95, 398–411.

27. McCall, R.T.; van Thiel de Vries, J.S.M.; Plant, N.G.; van Dongeren, A.R.; Roelvink, J.A.; Thompson, D.M.; Reniers, A.J.H.M. Two-dimensional time dependent hurricane overwash and erosion modeling at Santa Rosa Island. Coast. Eng. 2010, 57, 668–683.
28. McCall, R.T.; Masselink, G.; Poate, T.G.; Roelvink, J.A.; Almeida, L.P.; Davidson, M.; Russell, P.E. Modelling storm hydrodynamics on gravel beaches with XBeach-G. Coast. Eng. 2014, 91, 231–250.
29. Williams, J.J.; de Alegría-Arzaburu, A.R.; McCall, R.T.; van Dongeren, A. Modelling gravel barrier profile response to combined waves and tides using XBeach: Laboratory and field results. Coast. Eng. 2012, 63, 62–80.
30. Dissanayake, P.; Brown, J.; Karunarathna, H. Modelling storm-induced beach/dune evolution: Sefton coast, Liverpool Bay, UK. Mar. Geol. 2014, 357, 225–242.
31. Roelvink, D.; Reniers, A.; van Dongeren, A.; van Thiel de Vries, J.; McCall, R.; Lescinski, J. Modelling storm impacts on beaches, dunes and barrier islands. Coast. Eng. 2009, 56, 1133–1152.
32. Laudier, N.A.; Thornton, E.B.; MacMahan, J. Measured and modeled wave overtopping on a natural beach. Coast. Eng. 2011, 58, 815–825.
33. McCabe, M.; Stansby, P.; Apsley, D. Random wave runup and overtopping a steep sea wall: Shallow-water and Boussinesq modelling with generalised breaking and wall impact algorithms validated against laboratory and field measurements. Coast. Eng. 2013, 74, 33–49.
34. Prime, T.; Brown, J.M.; Plater, A.J. Physical and economic impacts of sea-level rise and low probability flooding events on coastal communities. PLoS ONE 2015, 10, e0117030.
35. Lichter, M.; Vafeidis, A.; Nicholls, R.; Kaiser, G. Exploring Data-Related Uncertainties in Analyses of Land Area and Population in the "Low-Elevation Coastal Zone" (LECZ). J. Coast. Res. 2011, 274, 757–768.
36. Coastal Engineering UK Ltd. Denbighshire County Council Coastal Defence Monitoring Inspection December 2013; Technical Report December; Coastal Engineering UK Ltd.: Thurstaston, UK, 2013.
37. National Tidal and Sea Level Facility. Highest and Lowest Predicted Tides at Llandudno; The National Oceanography Centre: Southampton, UK, 2008.
38. Wolf, J.; Brown, J.M.; Howarth, M.J. The wave climate of Liverpool Bay-observations and modelling. Ocean Dyn. 2011, 61, 639–655.
39. HR Wallingford. Conwy Tidal Flood Risk Assessment, Stage 1- Interim Report, Report EX 4667; Technical Report; HR Wallingford: Wallingford, UK, 2003.
40. Buijs, F.; Simm, J.; Wallis, M.; Sayers, P. Performance and Reliability of Flood and Coastal Defences. R&D Technical Report FD2318/TR1; Technical Report; Department for Environment Food and Rural Affairs: London, UK, 2007.
41. Parker, S. Flooding to Properties in Rhyl (A548 Coast Road, Garford Road and Surrounding Areas) 5th December 2013; Technical Report; Denbighshire County Council: Denbigh, UK, 2014.
42. Dissanayake, P.; Brown, J.; Karunarathna, H. Impacts of storm chronology on the morphological changes of the Formby beach and dune system, UK. Nat. Hazards Earth Syst. Sci. 2015, 15, 1533–1543.
43. DigitalGlobe. Google Earth Satellite Imagery—QGIS OpenLayers Plugin; DigitalGlobe: Longmont, CO, USA, 2016.
44. Souza, A.J.; Brown, J.M.; Williams, J.J.; Lymbery, G. Application of an operational storm coastal impact forecasting system. J. Oper. Oceanogr. 2013, 6, 23–26.
45. Harley, M.D.; Valentini, A.; Armaroli, C.; Perini, L.; Calabrese, L.; Ciavola, P. Can an early-warning system help minimize the impacts of coastal storms? A case study of the 2012 Halloween storm, northern Italy. Nat. Hazards Earth Syst. Sci. 2016, 16, 209–222.
46. McMillan, A.; Batstone, C.; Worth, D.; Tawn, J.; Horsburgh, K.; Lawless, M. Coastal Flood Boundary Conditions for UK Mainland and Islands. Project SC060064/TR2: Design Sea Levels; Environment Agency: Bristol, UK, 2011.

47. European Centre for Medium Range Weather Forecasts. IFS Documentation| ECMWF; European Centre for Medium Range Weather Forecasts: Reading, UK, 2016.
48. Natural Resources Wales. Natural Resources Wales/Flood Risk Map; Natural Resources Wales: Cardiff, UK, 2015.
49. Wadey, M.P.; Brown, J.M.; Haigh, I.D.; Dolphin, T.; Wisse, P. Assessment and comparison of extreme sea levels and waves during the 2013/14 storm season in two UK coastal regions. Nat. Hazards Earth Syst. Sci. 2015, 15, 2209–2225.
50. National Tidal and Sea Level Facility. Storm Surge Climatology: Wales; National Tidal and Sea Level Facility: Liverpool, UK, 2012.
51. Scott, T.; Masselink, G.; Russell, P. Morphodynamic characteristics and classification of beaches in England and Wales. Mar. Geol. 2011, 286, 1–20.
52. Stansby, P.; Chini, N.; Apsley, D.; Borthwick, A.; Bricheno, L.; Horrillo-Caraballo, J.; Mccabe, M.; Reeve, D.; Rogers, B.D.; Saulter, A.; et al. An integrated model system for coastal flood prediction with a case history for Walcott, UK, on 9 November 2007. J. Flood Risk Manag. 2013, 6, 229–252.
53. Silveira, T.M.; Taborda, R.; Carapuço, M.M.; Andrade, C.; Freitas, M.C.; Duarte, J.F.; Psuty, N.P. Assessing the overwash regime along an embayed urban beach. Geomorphology 2016. Accepted.
54. Gallien, T.W. Validated coastal flood modeling at Imperial Beach, California: Comparing total water level, empirical and numerical overtopping methodologies. Coast. Eng. 2016, 111, 95–104.
55. Prime, T.; Brown, J.M.; Knight, P.J.; Plater, A.J. Sandscaping for Mitigating Coastal Flood and Erosion Risk to Energy Infrastructure on Gravel Shorelines: Dungeness Flood Risk Assessment; Technical Report Research & Consultancy Report No. 49; National Oceanography Centre: Southampton, UK, 2015.
56. EDF Energy Nuclear Generation Ltd. (EDF). EU Stress Test—Dungeness B; EDF Energy Nuclear Generation Ltd.: Barnwood, UK, 2012.
57. Wadey, M.P.; Haigh, I.D.; Nicholls, R.J.; Brown, J.M.; Horsburgh, K.; Carroll, B.; Gallop, S.L.; Mason, T.; Bradshaw, E. A comparison of the 31 January–1 February 1953 and 5–6 December 2013 coastal flood events around the UK. Front. Mar. Sci. 2015, 2.

CHAPTER 8

Effect of Coastal Erosion on Storm Surge: A Case Study in the Southern Coast of Rhode Island

Alex Shaw [1], Mohammad Reza Hashemi [1,*], Malcolm Spaulding [1], Bryan Oakley [2] and Chris Baxter [1]

[1]*Department of Ocean Engineering, University of Rhode Island, Narragansett, RI 02882, USA*

[2]*Environmental Earth Science Department, Eastern Connecticut State University, Willimantic, CT 06226, USA*

ABSTRACT

The objective of this study was to assess the effect of shoreline retreat and dune erosion on coastal flooding in a case study located in the southern coast of Rhode Island, USA. Using an extensive dataset collected during 2011, an ADCIRC model was developed to simulate the propagation of storm surge in the coastal areas, including coastal inlets and ponds. A simplified methodology, based on the geological assessment of historical trends of the shoreline retreat and dune erosion in this area, was incorporated in the model to represent coastal erosion. The results showed that for extreme storms (e.g., a 100-year event), where coastal dunes are overtopped and low-lying areas are flooded, the flooding extent is not significantly sensitive to coastal erosion. However, failure of the dunes leads to a significant increase of the flooding extent for smaller storms. Substantial dampening of the storm surge elevation in coastal ponds for moderate and small storms was associated with coastal inlets connecting to coastal ponds which are often not resolved in regional surge models. The shoreline change did not significantly affect the extent of flooding. It was also shown that the accuracy of a storm surge model highly depends on its ability to resolve coastal inlets, which is critical for reliable storm surge predictions in areas with inlet-basin systems.

Keywords: dune erosion; coastal ponds; storm surge; coastal flooding

1. INTRODUCTION

The northeast of the US, including the coastal regions of Rhode Island, have been impacted by hurricanes in the past, most recently Hurricane Sandy in 2012. Climate change is expected to change the strength and frequency of these events, putting more coastal areas at risk [1]. Further, it is estimated that sea level will rise between 0.2 and 2 m by 2100 in the northeast of the US, which also magnifies the impacts of coastal flooding [2]. As coastal flooding is sensitive to changes in bathymetry and topography of a region, coastal erosion can potentially affect the storm surge propagation. Storm surge and coastal erosion interact in two ways: (1) storm surges (and wave forces) lead to coastal erosion; (2) coastal erosion affects the propagation of storm surge and consequently alters the extent of flooding. While it is possible to examine the two-way interaction processes using morphodynamic models (e.g., [3,4]), which incorporate sediment transport and bed level changes, validating morphodynamic models is very challenging, and developing those models is costly. Alternatively, assuming worst case scenarios (complete dune erosion, shoreline retreat at specified rate, etc.) is an alternative method which allows understanding the effect of coastal erosion on flooding for extreme scenarios (e.g., [5]).

Our case study is located in the southern coast of Rhode Island (Figure 1), which consists of several coastal ponds and barriers. The shorelines are retreating at a rapid rate, in some areas up to 1.15 m per year [6]. The coastal dunes are also eroded during major storm events (Figure 2). The failure of dunes can affect the dynamics of the inlet-basin/pond system.

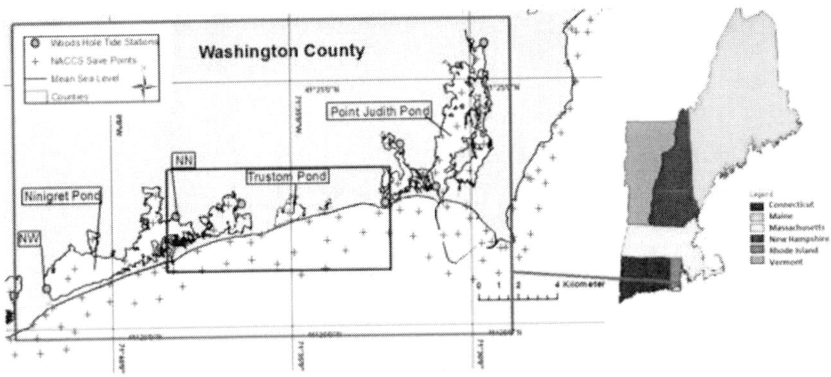

Figure 1. Overview of the the study area in the southern coast of Rhode Island. Other details include save points (blue crosses) from the North Atlantic Coast Comprehensive Study (NACCS) (see Section 2), Woods Hole Group Inc. water level gauge locations (orange dots), and a red box around the study area. Transects in the dark blue area were used to apply erosion scenarios (Figure 8).

Figure 2. Failure of dunes protecting Ninigret Pond after Hurricane Carol (1954); source: Rhode Island (RI) Coastal Resources Management Council.

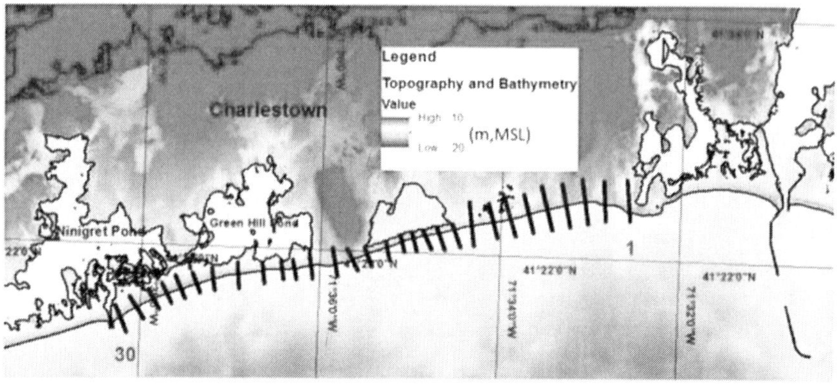

Figure 8. Crossshore transects made to implement coastal erosion between Charlestown Beach and Matunuck Beach (Numbered 1–30).

The objective of this study is to investigate the effects of dune erosion and shoreline retreat (together and separately) on storm surge. The study was carried out using numerical modeling, and analysis of the field data.

Section 2 describes several sources of data (observed/hindcast) which have been used in this study; in particular, the hydrodynamic data, which have been collected during 2011, and other relevant storm surge modeling studies in the region are presented. Section 3explains the simplified methodology which has been used to simulate shoreline retreat and dune erosion. Details of the ADCIRC (ADvanced CIRCulation) model of the study area are provided in Section 4. Several scenarios of coastal erosion and storm surge are discussed in Section 5. Discussions and summary of the results are presented at the end.

2. DATA

From July 2010 to September 2011, Woods Hole Group carried out an extensive data collection program [7], for the US Army Corp of Engineers (USACE) New England District, entitled "Wave, Tide and Current Data Collection, Washington County, Rhode Island". The primary purpose of that work was to collect site-specific data to support a RI Regional Sediment Management Study, and included a collection of water elevation, currents, wave, and meteorological data. Their study included measurement of water elevations inside coastal ponds (Figure 1), as well as waves and currents offshore. This data provided a unique source for understanding the effect of inlet-pond systems on water elevation in this area. Hurricane Irene, which impacted this area during the observation period, was also used for model validation.

For simulation of synthetic storms (i.e., 100-year event), the North Atlantic Coast Comprehensive Study (NACCS; [8,9]) was used. NACCS is based on a system of numerical models including ADCIRC [10], WAve Model (WAM), and STeady state spectral WAVE model (STWAVE) [11]. It has simulated hydrodynamic and wave fields of 1050 synthetic tropical storms as well as 100 extratropical historical storms over the Atlantic Coast. The model was based on a relatively high resolution unstructured mesh (30 m–50 m near the coast). The synthetic storms were generated based on the statistical analysis of past storms. The NACCS provides model results at the save points (Figure 1), including time series of the wind, wave and water levels for the events and return period analyses for the tropical storms. These data were used to force the model at the boundary for a synthetic storm representing a 100-year event. It should be added that some of the save points of the NACCS are located inside the coastal ponds which may be inaccurate, as will be discussed later. For the 100-year event, all synthetic storms simulated in NACCS were examined, and a storm surge event which generated the water levels of around 100-year storm surge at Newport (8452660) and Providence (8454000) National Oceanic and Atmospheric Adminstration (NOAA) water level stations was selected. This storm had a maximum surge of 3.20 m (Mean Sea Level (MSL)) at Newport (Figure 3), which is close to 3.35 m (MSL) or 2.7 m (Mean Higher High Water (MHHW)) for the 100-year event, considering the 100-year event at the upper 95% confidence level.

Since coastal erosion occurs during major storms, several hurricanes were considered in this study. For validation, Hurricane Irene (late August 2011) was selected as observed data was available during this hurricane in several locations inside the model domain. A larger storm event, Hurricane Bob, a strong tropical storm which occurred on 19 August 1991 was chosen. Hurricane Bob provides a good representation of large storms in the area, but it was not large enough to overtop the barriers. Also, two synthetic storms from NACCS including a storm representing the 100-year event (which is important for planning purposes) were simulated.

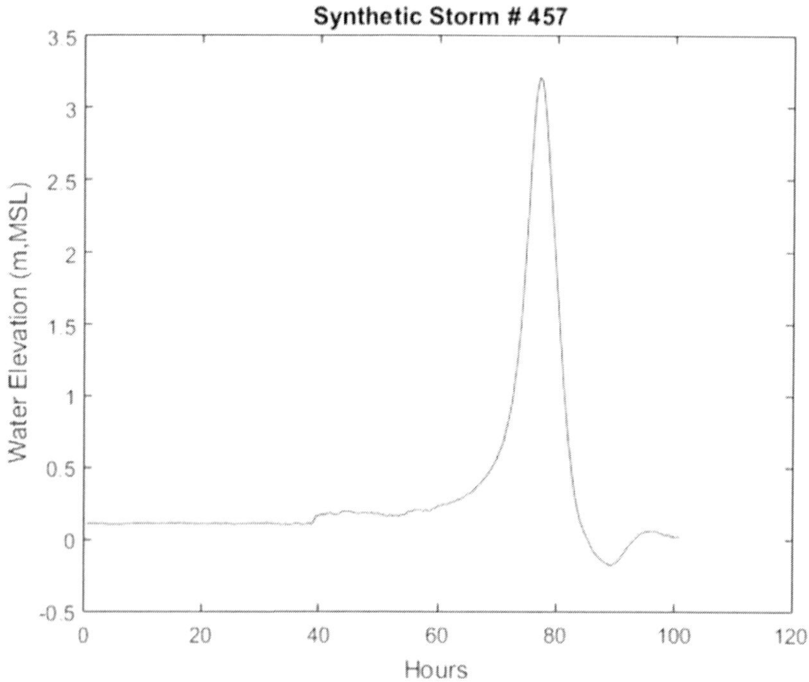

Figure 3. The time series of storm surge for synthetic storm 457—from NACCS—which approximately produces 100-year storm surge (at the upper 95% confidence level) near the Newport NOAA water level station.

For the surge model, both bathymetry and topography of the domain (a digital elevation model: DEM) were necessary due to wetting and drying. A DEM with a resolution of 10 m was used based on the National Geography Data Center (NGDC) Bathymetry Data and the USACE 2010 coastal Light imaging, Detection, And Ranging (LiDAR) survey. The LiDAR survey focused on the south coast and extended about 1 km offshore (Figure 4).

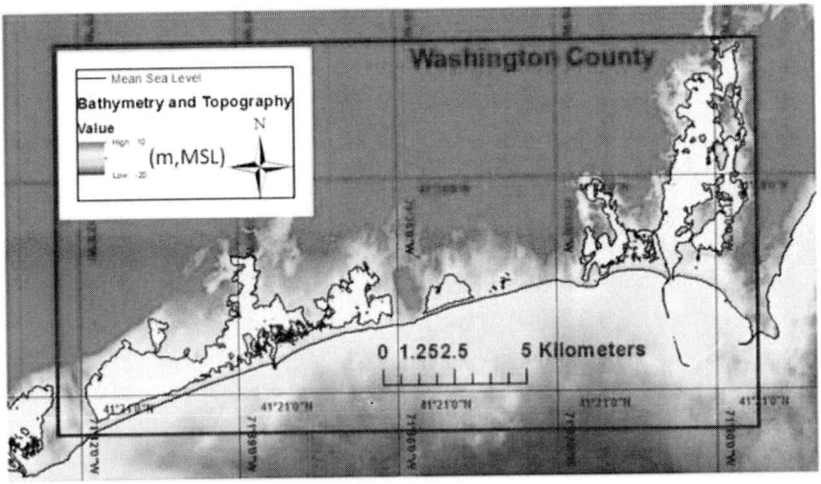

Figure 4. The Digital Elevation Model (DEM) around the study area.

Wind data (for forcing the ADCIRC model) were extracted from the USACE Wave Information Study (WIS) hindcasts near the domain. The WIS data covers a period from 1980 to 2012. For this study, the wind fields from large storm events were of interest. It should be mentioned, as the model domain covered just the southern coast of RI, the spatial variability of wind was considered negligible in this small area. For this 30-year period, Hurricane Bob which made land fall in RI on 19 August 1991, was chosen. Hurricane Bob gives a good representation of large storms in the area [9]. It is the fifth largest storm recorded at the NOAA tide gauge at Newport RI, and approximately corresponds to a 20-year event, according to the extremal analysis for the site (Figure 5). The Newport water elevation station is the closest station to the study area (71.33 W, 41.51 N) and has a long record including major hurricanes. The wind field for Hurricane Bob extracted from WIS is plotted in Figure 6.

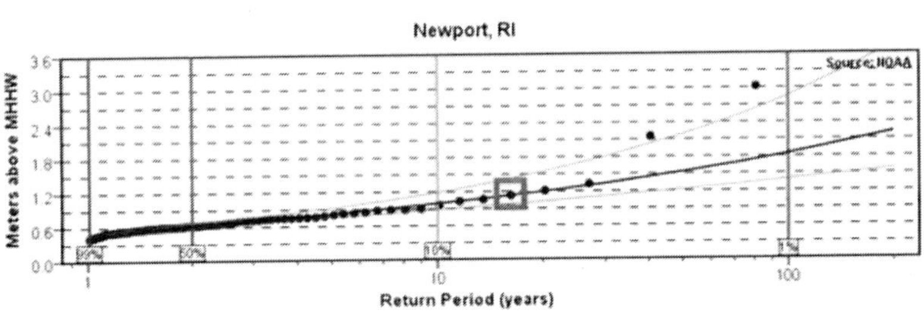

Figure 5. Extremal analysis of water elevation (MHHW = MSL + 0.65) for Newport NOAA station (8452660); the red box shows Hurricane Bob.

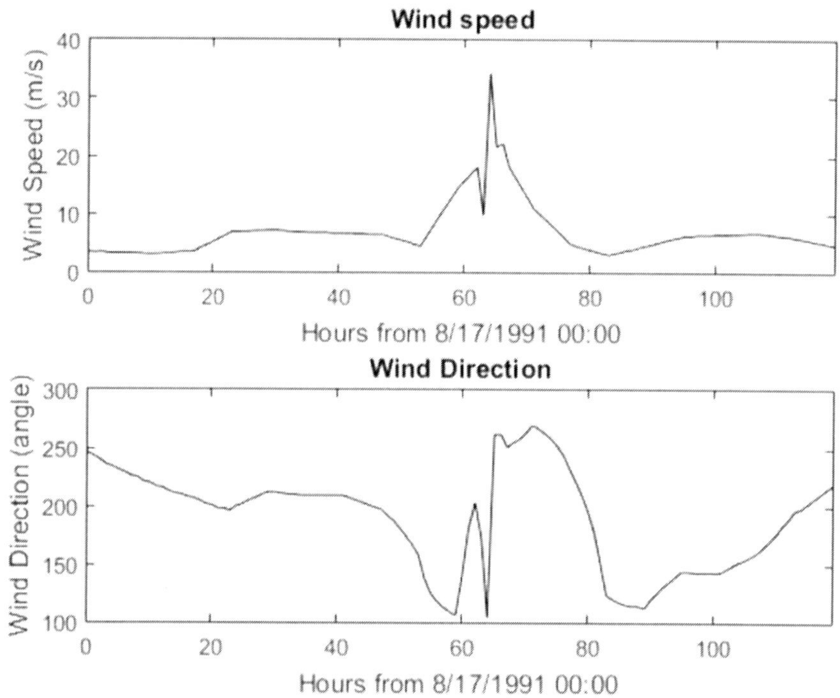

Figure 6. Plots of the wind speed and direction for Hurricane Bob at WIS station number 63079, which is located near the region (71.22 W, 41.25 N).

3. COASTAL EROSION SCENARIOS

Coastal erosion scenarios were based on shoreline retreat and dune erosion during large storm events. The past shoreline retreat rates were used to estimate erosion rates for erosion scenarios, and the DEM was changed according to these rates. It should be mentioned that the rate of erosion is expected to rise due to Sea Level Rise (SLR); nevertheless, this assumption was made to simplify the analysis. Further research is necessary to include the effect of SLR on the rate of erosion. The shoreline retreat rates were calculated using aerial photographs from 1939 to 2014 (Figure 7; [6]). It should be added that shorelines retreat in severe storms and recover during fair weather; however, there is a consistent trend of shoreline retreat over past decades in this region.

Figure 7. A sample shoreline change map for a beach in the study area [6].

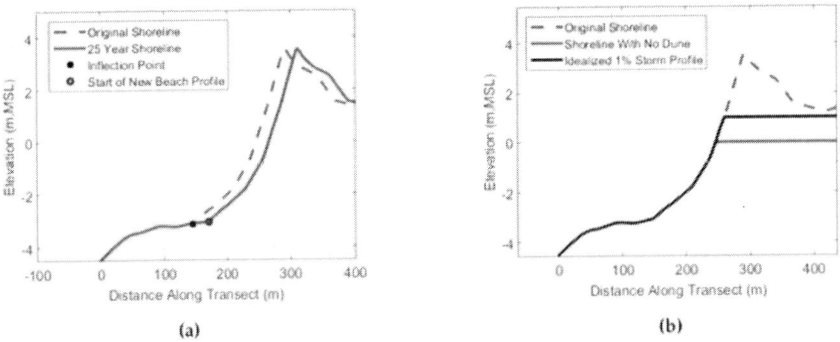

Figure 9. The simplified method which was used to estimate the shoreline geometry after erosion in future. Transect 30 (as an example; Figure 8), with the original shoreline (blue), 25-year shoreline (red), and intersection point (star) are shown. The vertical axis is exaggerated for better clarity. (**a**) shoreline geometry after erosion; (**b**) erosion of the dunes.

The projected shoreline retreat over the next 25 years was considered. The shoreline was divided into the cross-shore profiles shown in Figure 8. In the selected area, the beach profiles consist of an offshore beach slope, a near shore beach slope, and a dune system. The offshore beach slope was extended horizontally to the corresponding 25 years erosion (Figure 9). The same near shore profile and dune system was then assumed at the end of each profile. This method retreats the shoreline while keeping the same beach profile geometry. Once the transects were modified, they were linearly interpolated to modify the DEM of the model.

Coastal erosion during large storms can lead to failure of dunes as well as retreat of the shoreline. During storm events, the combined action of storm surge and waves erode the dune and create surge channels and wash-over fans (Figure 2). To implement dune erosion in the DEM, it was assumed that the dunes were eroded or simply cut off at an elevation Mean High Water (MHW) with a horizontal line (Figure 9). The elevation of the post storm profile was determined by examining the washover fans deposited after past hurricanes in this area, including 1938 Hurricane, the Ash Wednesday storm (1962), and Hurricane Sandy. The elevations of the washover fans were estimated using LiDAR; the slope of 0.003 cm/m was measured for washover fans after Hurricane Sandy, which can be assumed horizontal for the model resolution used in this study.

4. NUMERICAL MODELING

For surge modeling, the ADCIRC model was used. ADCIRC is based on the finite element method and unstructured mesh discretization, allowing areas such as coastal inlets to be resolved with a reasonable computational cost. ADCIRC has been coupled with Simulating WAves Nearshore (SWAN), and can simulate the wave-surge interactions [12]. This model has been extensively used to predict storm surge flooding (e.g., [10,13]).

A mesh was created, resolving coastal inlets, using the Surface water Modeling System Software (SMS) with a resolution of 30 m near the coastline, 150 m farther offshore, and 2 km near the open boundaries. The mesh is plotted in Figure 10. The model was forced along the open boundaries by water elevation, and by wind stress/pressure over the domain. The model was run in the 2-D mode, with a Manning friction coefficient of 0.018 (below MSL), and up to 0.06 in land areas. For the tidal case, the model was forced using five harmonic constituents for tides including M2, N2, K1, S2, and O1 which can be extracted from tidal databases [14]. M4 and other overtide constituents were neglected. M4 is generated in shallower regions by friction, and causes tidal asymmetry. Neglecting the M4 component can change the water level by around 6 cm, which can be neglected during a major storm surge event. These constituents represent the main components of tide for this area (Table 1).

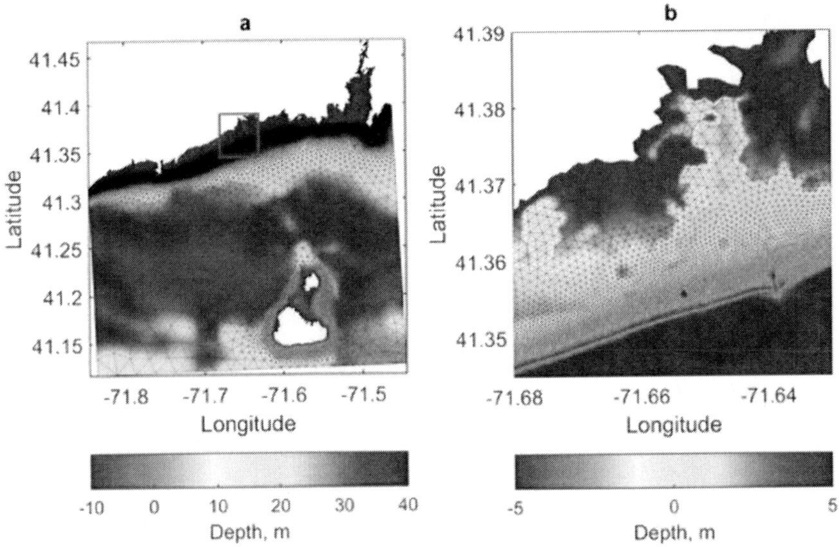

Figure 10. An overview of the mesh used for surge modeling in the southern coast of RI. The model domain is larger than the study area and includes Block Island near the southern boundary. Subfigure (**b**) shows a magnified view of the rectangular area in Subfigure (**a**) around Ninigret Pond.

Table 1. Harmonic constituents at the Newport and Providence NOAA water elevation stations.

Harmonics	Newport Amplitude (m)	Newport Phase (degrees)	Providence Amplitude (m)	Providence Phase (degrees)
M2	0.505	2.3	0.643	9.5
S2	0.108	25.0	0.138	33.6
N2	0.124	345.8	0.152	354.6
K1	0.062	166.1	0.073	169.4
M4	0.057	35.8	0.103	202.2
O1	0.047	202.0	0.027	312.7
M6	0.0005	220.1	0.027	312.7
MK3	0.0008	19.5	0.016	39.3
S4	0.0007	5.1	0.014	23.8
MN4	0.026	347.9	0.014	12.7

5. RESULTS

5.1. Model Skill Assessment

To test the performance of the model for tides and storm surge, the Woods Hole Group Inc. data were used [7]. The observed data were compared with the model results during a spring-neap cycle (for tides), and during Hurricane Irene (for storm surge). For tides, the model was run for 30 days from 14 May 2011 until 13 June 2011 with a one day ramping period. This time period is within the duration of the Woods Hole Group data collection campaign. The model and observed water elevation data were analysed by T_Tidecode to compute tidal constituents [15]. Table 2 shows the comparison between the modeled results and the observed data at the two stations inside Ninigred Pond. As this table shows, the modeled and observed data, in general, are in good agreement. In particular, the performance of the model for the phase and amplitude of the dominant M2 component is good. The model underpredicts the amplitude of S2, but as this component is very small, its effect is not that significant inside the pond. The overall RMSEs (Root Mean Square Error) for amplitude and phase are 0.015 m and 25 , respectively, which are convincing.

Table 2. Skill assessment of the numerical model for tidal predictions inside Ninigret Pond; see Figure 1 for location of the stations. RMSE for amplitude and phase are 0.015 m and 25 , respectively.

Constituents	NW				NN			
	Model		Observation		Model		Observation	
	amp (m)	Phase (deg)	amp (m)	Phase (deg)	amp (m)	Phase (deg)	amp (m)	Phase (deg)
O1	0.022	277	0.018	283	0.022	271	0.017	277
K1	0.024	217	0.019	223	0.024	210	0.019	216
N2	0.021	48	0.017	71	0.020	34	0.017	56
M2	0.083	82	0.081	89	0.075	69	0.077	75
S2	0.008	114	0.021	106	0.007	98	0.021	101

For the storm surge case, Hurricane Irene was simulated, a category 3 storm that occurred in late August 2011. The comparison of the model results and observations are depicted in Figure 11. The performance of the model for both stations is very good with an RMSE of 0.065 m and 0.041 m for NN and NW respectively; however, the model slightly overestimates the surge. Overall, given the magnitude of errors, the performance of the model was considered satisfactory.

5.2. Propagation of Tides/Storm Surge in Coastal Ponds; Effect of Coastal Inlets

As mentioned previously, the southern coast of RI consists of several coastal ponds and barriers, and the failure of dunes can affect the inlet-basin/pond system. At first, a simplified analysis was carried out based on the previous research about the dynamics of inlet-basins, and the collected data in this area. This analysis helped interpret modeling results. Figure 12 shows the comparison of water elevation inside and

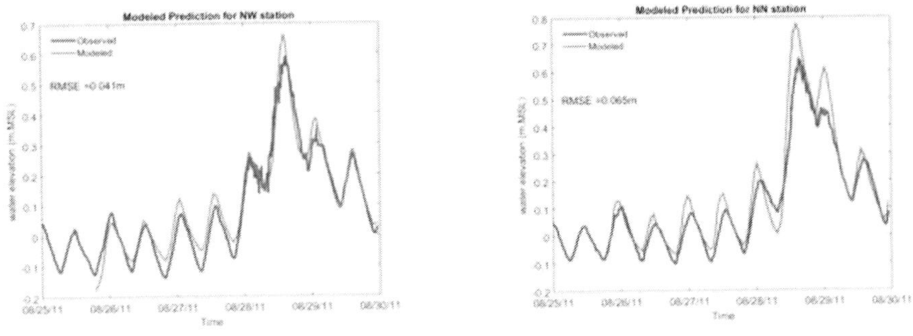

(a) (b)

Figure 11. Comparison between the model predictions and the observed data for Hurricane Irene (see Figure 1 for location of the stations). (**a**) NW station; (**b**) NN station.

outside Ninigret Pond using observed data for a duration of one month. A dramatic reduction of the amplitude can be observed in this figure. A coastal inlet, in general, causes a reduction of water elevation amplitude and a phase-lag or a delay inside coastal ponds relative to offshore. This is mainly associated with the energy dissipation by high velocity currents in an inlet. Simplified analytical methods have been introduced in the literature to compute the reduction of the amplitude, and the phase lag, based on the geometry and physical characteristics of the inlet-basin system. A detailed analytical analysis of inlet-basin hydrodynamics can be found in the Coastal Engineering Manual [16]. Considering a long wave (e.g., tide or surge), with an amplitude of , and a period of *T*, the effect of a coastal inlet on tide/surge signal as it propagates from the ocean to the pond can be written as,

$$[R, \phi_l] = f(A_i/A_b, R_h, L, T, \mathbf{F}); \qquad R = 1 - a_i/a_o \tag{1}$$

where *R* is the reduction in the amplitude, is the amplitude inside the basin/pond, is the phase lag, is the cross sectional area of the inlet, is the area of a basin or a pond, is the hydraulic radius of the inlet, *L* is the length of the inlet, and represents the frictional coefficients for the entrance, exit and channel friction losses. For Ninigret Pond, km , m , km, m. Using these parameters, and assuming entrance, and exit loss coefficients of 0.1 and 1.0, respectively (recommended by the Coastal Engineering Manual [16]), leads to and = 90 . The impact of inlets on tidal signal was also assessed using observed data. By performing a tidal analysis using T_Tide [15] inside (Wood Holes Group

Station) and outside this pond (NOAA, Weekapaug Point 71.76 W, 41.33 N), *R* for the M2 tidal component was found to be 76% for the NW station (Figure 1), with phase lags of 90.5 or about 3 h and 6 min. These values which are based on the observations are very close to the analytical method predictions (i.e., and = 90). Considering that some storm surge events have similar (or longer) periods, if coastal barriers for this pond fail, this reduction of the amplitude no longer exist. Consequently, dune erosion can lead to a significant increase in the flooding area.

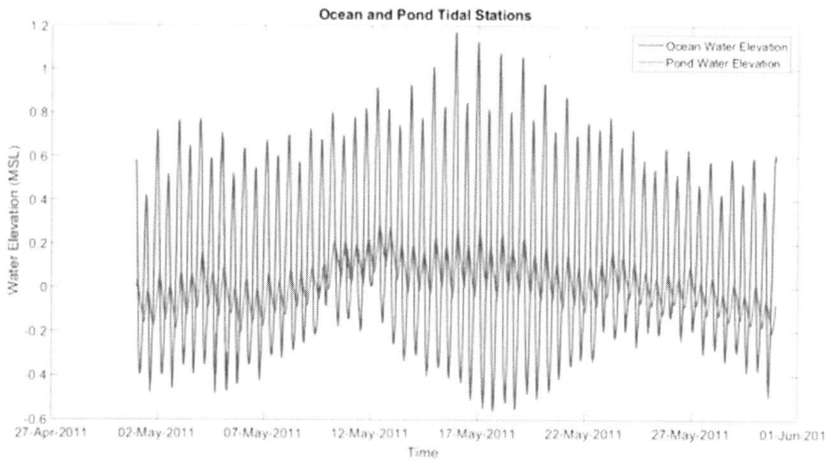

Figure 12. Comparison of observed water elevation data inside and outside Ninigret Pond (NW Gauge, Figure 1).

Further, the geometry of a coastal inlet has a controlling effect on the reduction of the amplitude of water elevation signal. Considering the three coastal ponds in this area (Figure 1; Ninigret Pond, Trustom Pond, and Point Judith Pond), the effect of coastal inlet geometry can be further examined. Point Judith Pond has a wide and deep inlet with a width of 80 m and a depth of around 7 m. Trustom Pond, on the contrary, has no permanent connection to the ocean for tides, but during large storm events, part of its barrier is overtopped or breached (for example in Hurricane Sandy) causing some flooding. Ninigret Pond has a relatively narrow inlet (35 m), protected by hard structures. The water elevations in the three ponds are plotted for a tidal cycle, and for a storm event (Hurricane Bob) in Figure 13 using the ADCIRC model. As this figure shows, the water elevation signal for tide inside and outside of the Point Judith Pond is almost the same due to its wide inlet, but the peak of storm surge slightly attenuates during the storm event. For Trustom Pond, the barrier is not overtopped for tides or the storm surge scenario. Ninigret Pond shows a significant reduction for tides ( ) and for

Hurricane Bob ( ) due to its narrower inlet. Therefore, if a storm surge does not overtop or erode coastal dunes, coastal inlets can significantly decrease the magnitude of a storm surge (inside a coastal pond).

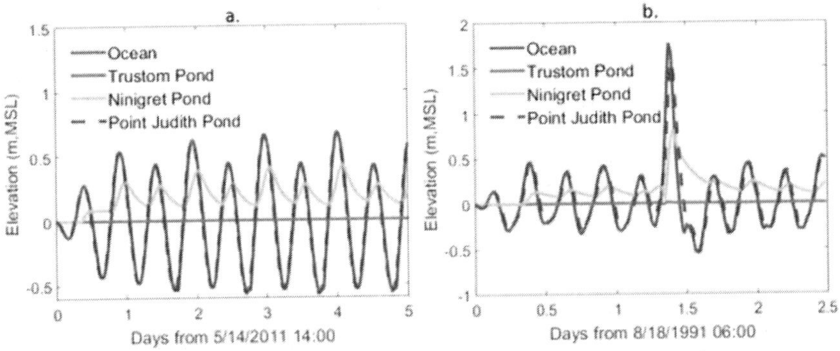

Figure 13. Effect of coastal inlet geometry on surge inside three coastal ponds in the study area; comparison of water elevation in Ninigret Pond, Trustom Pond, and Point Judith Pond for (**a**) Tides; (**b**) Hurricane Bob.

The above analysis shows a significant dampening of tide and surge signal caused by coastal inlets. However, the frequency/period of a water elevation signal and the geometry of an inlet are the two important factors which control this dampening [17,18,19]. Figure 14 shows the reduction of amplitudes of various water elevation signals assuming different periods for Ninigret Pond. The analysis was performed using the simplified analytical method mentioned earlier [16]. As this figure shows, frequencies of 1, 2, and 2.5 days lead to 0.65, 0.38, and 0.25 reductions, respectively. Therefore, a storm surge which has a long period (more than 2 days) will be less effected compared to a tidal signal with a period of 12 h. It should be added that the total water level during a storm surge is due to the combination of tide and surge signals. In addition, the geometry of an inlet, as discussed above, is another important factor, which should be always considered before generalizing these results.

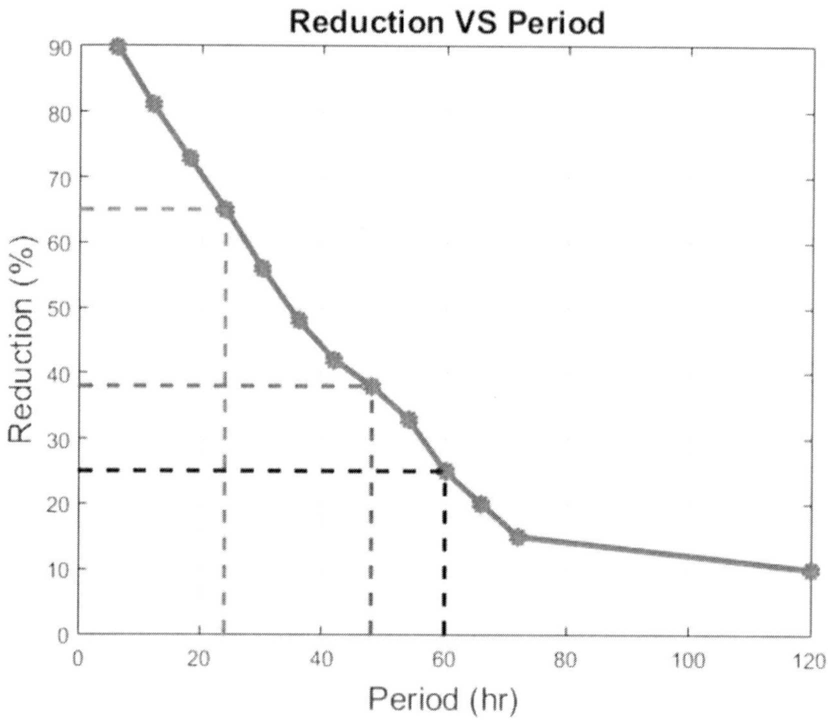

Figure 14. Effect of water elevation signal period on reduction of the amplitude for Ninigret Pond.

5.3. Effect of Erosion on Storm Surge

Two erosion scenarios were considered: shoreline change in 25 years, and dune erosion. As mentioned, the erosion of dunes is a common consequence of large hurricanes in the study area as can be seen in Figure 2, which shows the partial failure of the dune system of Ninigret Pond during Hurricane Carol in 1954. Several scenarios considering the two storm cases (100-year synthetic storm and Hurricane Bob) were considered.

For Hurricane Bob, the flooding areas assuming eroded (retreated) shoreline and the current shoreline were examined. Table 3 shows the summary of results. Considering a retreated shoreline in 25 years, the flooding extent slightly increases by 0.22 km , which is 20% of the original flooded area (1.12 km). This increased flooding area is approximately the advance of the sea (about 30 m) due to coastal erosion; therefore, the shoreline retreat does not significantly increase the extent of flooding. However, when the dunes are eroded, the flooding extent increased by 2.33 km , which is a 207% increase. When dunes erode, the coastal inlets of the ponds can no longer dampen the surge signal, and therefore a much larger area within coastal ponds are flooded. For this scenario, the flooding extent advanced up to 500 m in some areas. Figure 15 shows the flooding extent for existing dunes, and eroded dune profile scenarios.

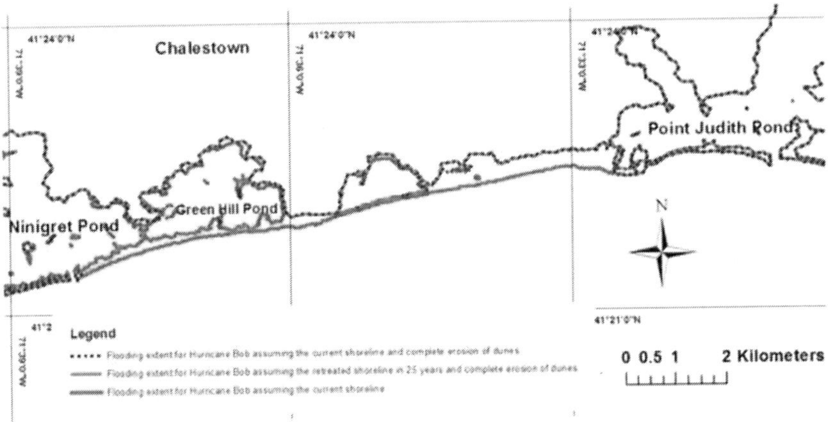

Figure 15. Comparison of Hurricane Bob flooding extent assuming current condition (red) and current shoreline with no dune system (blue), and the 25-year retreated shoreline as well as complete dune failure (black).

Table 3. Differences in flooded areas near the eroded shoreline for Hurricane Bob assuming erosion scenarios.

Erosion Scenario	Current Flooded Area, km^2	Changed Flooded Area, km^2	Difference, km^2	Percentage Increase
Shoreline retreat in 25 years	1.13	1.35	0.22	19.7%
Eroded dunes	1.13	3.5	2.33	207%

For the 100-year event, the erosion scenarios (shoreline retreat and dune erosion) did not lead to a significant change in flooded areas as shown in Figure 16. This is because for this event, the storm surge is large enough to overtop the dunes (the dune top elevation is about 3 m, MSL in this area); therefore, even if the dunes were solid structures and could resist the erosion during storm surge, they could not protect the coastal ponds. It should be noted that the failure of dunes may significantly affect wave propagation for the 100-year event (waves can break over dunes, due to decreased water depth).

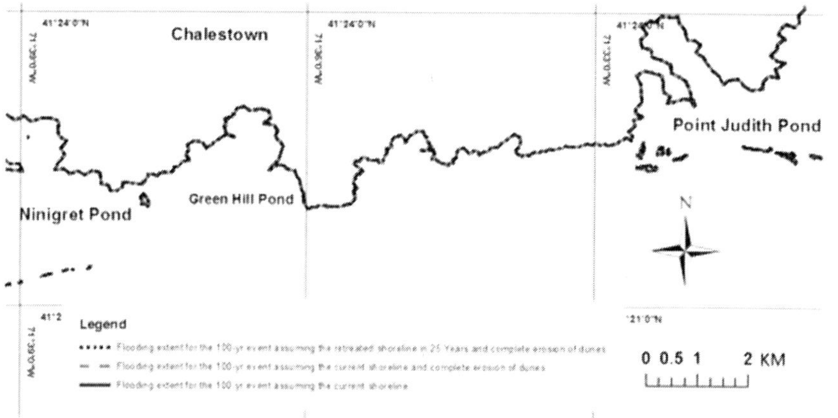

Figure 16. Comparison of 100-year flooding extent assuming the current shoreline (blue), current shoreline with eroded dune system (green), and the 25-year retreated shoreline plus complete dune failure (black). Lines overlap for this scenario.

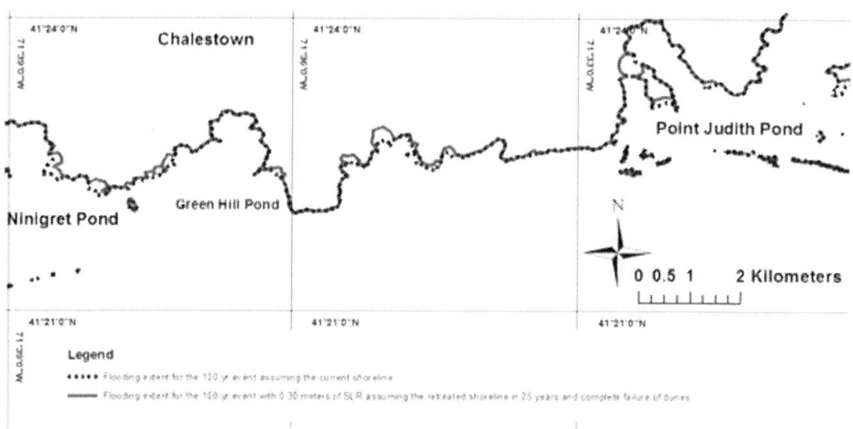

Figure 17. Comparison of flooding extend of the 100-year storm event, assuming 30 cm SLR, and coastal erosion.

SLR, in general, leads to an increase of the flooding extent [20]. A very simple way of investigating the impact of SLR on flooding is the bathtub approach or adding the magnitude of SLR to elevations predicted by a storm surge model; this methods neglects the nonlinearity of the storm surge propagation. A more accurate method includes changing the DEM, and simulating the storm surge assuming a SLR scenario. Consistent to our analysis which assumed a 25 years shoreline retreat, 30 cm or 1 foot SLR was assumed, corresponding to projected values by NOAA (High) for 25 years [21]. Figure 17 shows that the extent of flooding, as expected, increases in some areas. The flooding area increased from 4.72 km to 6.80 km , leading to a 44% increase.

6. DISCUSSION

The geometry of coastal inlets controls the storm propagation for moderate storms in areas with inlet-basin systems. Some regional modeling studies such as NACCS have not resolved these inlets, and their predictions inside coastal ponds may not be reliable beyond barriers and inside ponds. Figure 18 is an example showing the poor resolution of the NACCS mesh around Ninigret Pond inlet. In Figure 19, the prediction of storm surge for a moderate synthetic storm (220, which has a peak elevation of 1.67 m, MSL in Newport [8]), near two save points (see Figure 18b) located inside and outside Ninigret Pond, has been compared with that from our model. The surge event was channeled through the inlet, but given the poor resolution of the NACCS model, water levels are overestimated. It should be noted that NACCS results, unlike the ADCIRC model developed in this study, have not been validated inside coastal ponds and very near shore in RI. Also, waves for this storm are not that significant inside the pond; therefore, wave-surge interaction cannot be associated with higher storm prediction in the NACCS model. The results are identical at the boundary (Point A), as the ADCIRC model was forced by NACCS at the open boundary.

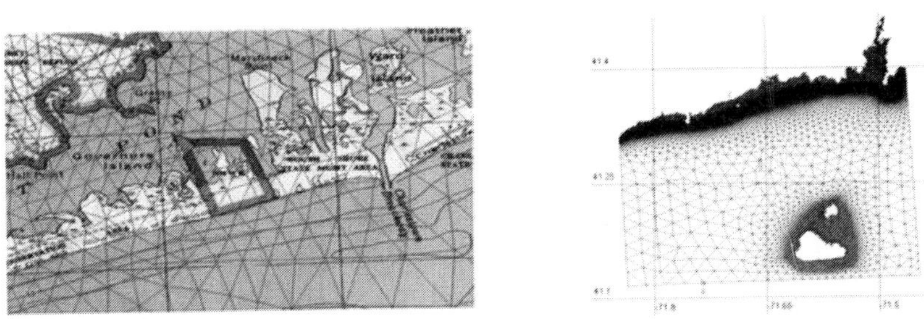

(a) NACCS model resolution around Ninigret inlet

(b) Comparison points: A, B, and C

Figure 18. Effect of model resolution on the results; Subfigure (**a**) shows an example of low resolution NACCS mesh in a coastal inlet; Subfigure (**b**) shows the locations of comparison for NACCS results and those obtained in this study.

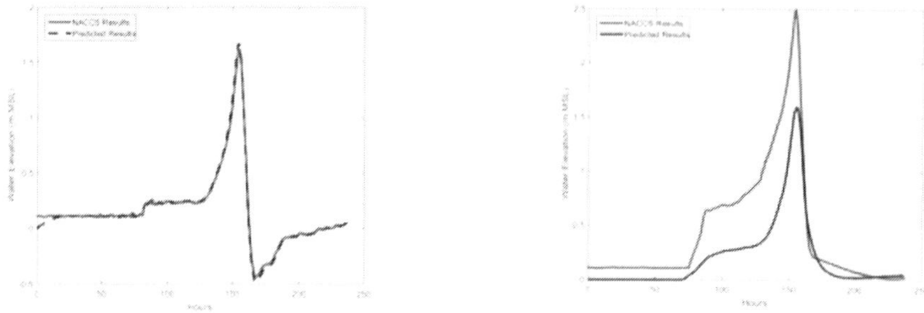

(a) (b)

Figure 19. Comparison of NACCS results and ADCIRC model of this study for synthetic Storm 220. See Figure 18b for locations of comparison. (**a**) Comparison at B: outside Ninigret Pond; (**b**) Comparison at C: inside Ninigret pond.

The dunes along the entire southern coast of RI have an average height of 3.39 m above MSL, but in some areas they are as low as 1.1 m. This means that a storm with a magnitude of 100-year (3.35 m, (MSL) considering the 100-year event at the upper confidence level curve) can potentially overtop all of the dunes. A hurricane such as Carol, which had a surge height of 2.7 m, MSL at Newport RI can breach the dunes (Figure 2), and have a similar but lesser effect on flooding (increasing the flooding extent). Various factors are associated with the erosion of dunes [22], including the geotechnical properties of dunes, the elevation of dunes compared to surge, wave-induced forces, and wave runup/overtopping. Therefore, it is a challenging task to specify a threshold for a storm which leads to dune failure. Morphological modeling (e.g., [4]) along with data collection during and after large storms around coastal dunes can improve our understanding of this process for this area, for future studies.

The analysis carried out in this study was based on two extreme scenarios: complete dune erosion, and no dune erosion. In reality, dunes are partially eroded during major storms (Figure 2), and gradually recover during calmer months. Therefore, a storm cluster can lead to more damage compared with isolated events. The effect of erosion on storm surge is overestimated by assuming complete erosion of dunes. However, the results show the significant impact of dune erosion on flooding, and quantify this impact for this extreme scenario.

7. CONCLUSIONS

We explored the effect of dune erosion and shoreline retreat on coastal flooding in an area which consists of coastal ponds protected by dunes and connected to the ocean by narrow inlets. A storm surge model was developed/validated with a unique dataset, which included water elevation data inside coastal ponds during 2011 and measurements during Hurricane Irene. The conclusions are summarized as follows:

1. The results showed that erosion of dunes has more effect on flooding extent compared with retreat of shorelines.
2. For storms which do not overtop or erode the coastal dunes, the inlets of coastal ponds can significantly decrease the storm surge elevation. This can be explained using the concepts of inlet-basin hydrodynamics. However, for very extreme storms such as a 100-year event where coastal dunes are overtopped, and low-lying areas are flooded, the flooding extent did not significantly change.
3. Assuming complete erosion of the dunes and for the scenario of Hurricane Bob, simulations showed a more than 200% increase in the flooding extent. Several sources of uncertainty can affect these estimations. For instance, in many cases, dunes are partially eroded. Coupled hydrodynamic and morphodynamic models which can simulate dunes erosion more accurately, can lead to more realistic estimations.
4. Numerical surge models which do not fully resolve coastal inlets (e.g., NACCS model in RI) lead to significant errors in the prediction of surge in coastal ponds. Accurate bathymetric and topographic measurement of coastal inlets is essential for storm surge modeling in areas with inlet-basin systems.

ACKNOWLEDGMENTS

This work was under taken with funding support from a Rhode Island Community Development Block Grant (4712) from the U.S. Department of Housing and Urban Development and the State of Rhode Island Division of Planning Office of Housing and Community Development. Thanks to US Army Corps of Engineers (USACE) for sharing the North Atlantic Coastal Comprehensive Studies' dataset. Thanks to National Hurricane Center (NHC) and Center for Operational Oceanographic Products and Services from NOAA for supplying hurricane and water level data. Thanks to Woods Hole Group Inc. for providing the water elevation data. Thanks to Marissa Torres for her comments on the paper.

AUTHOR CONTRIBUTIONS

Alex Shaw developed the numerical models, and analysed the results. Mohammad Reza Hashemi initiated and led the research, and helped in discussion of the results. Malcolm Spaulding advised the research in many aspects, including analysis of the results and development of the models. Bryan Oakley contributed in coastal erosion and general discussions. Chris Baxter helped in the discussion of the results.

CONFLICTS OF INTEREST

The authors declare no conflict of interest.

REFERENCES

1. Wahl, T.; Jain, S.; Bender, J.; Meyers, S.D.; Luther, M.E. Increasing risk of compound flooding from storm surge and rainfall for major US cities. Nat. Clim. Chang. 2015, 5, 1093–1097.
2. Parris, A.; Bromirski, P.; Burkett, V.; Cayan, D.R.; Culver, M.; Hall, J.; Horton, R.; Knuuti, K.; Moss, R.; Obeysekera, J.; et al. Global Sea Level Rise Scenarios for the United States National Climate Assessment; NOAA Technical Report OAR CPO-1; NOAA: Silver Spring, MD, USA, 2012.
3. Roelvink, D.; Reniers, A.; van Dongeren, A.; de Vries, J.V.T.; McCall, R.; Lescinski, J. Modelling storm impacts on beaches, dunes and barrier islands. Coast. Eng. 2009, 56, 1133–1152.
4. McCall, R.T.; de Vries, J.V.T.; Plant, N.; van Dongeren, A.; Roelvink, J.; Thompson, D.; Reniers, A. Two-dimensional time dependent hurricane overwash and erosion modeling at Santa Rosa Island. Coast. Eng. 2010, 57, 668–683.
5. Kurum, M.O.; Edge, B.; Mitasova, H.; Overton, M. Effects of coastal landform changes on storm surge along the Hatteras Island breach area. Coast. Eng. Proc. 2011, 1, 25.
6. Boothroyd, J.; Hollis, R.; Oakley, B.; Henderson, R. Shoreline Change from 1939–2014, Washington County, Rhode Island. 1:2,000 scale, 45 maps; Technical Report; Rhode Island Geological Survey: Kingston, RI, USA, 2016.
7. Woods Hole Group. Wave Tide and Current Data Collection Contract No. W912WJ-09-D-0001-0026; US Army Corps of Engineers: New England District, MA, USA, 2012.
8. Cialone, M.A.; Massey, T.C.; Anderson, M.E.; Grzegorzewski, A.S.; Jensen, R.E.; Cialone, A.; Mark, D.J.; Pevey, K.C.; Gunkel, B.L.; McAlpin, T.O. North Atlantic Coast Comprehensive Study (NACCS) Coastal Storm Model Simulations: Waves and Water Levels; Technical Report, DTIC Document; The U.S. Army Engineer Research and Development Center: Vicksburg, MS, USA, 2015.

9. Hashemi, M.R.; Spaulding, M.L.; Shaw, A.; Farhadi, H.; Lewis, M. An efficient artificial intelligence model for prediction of tropical storm surge. Nat. Hazards 2016, 82, 471–491.
10. Luettich, R., Jr.; Westerink, J.; Scheffner, N.W. ADCIRC: An Advanced Three-Dimensional Circulation Model for Shelves, Coasts, and Estuaries; Report 1. Theory and Methodology of ADCIRC-2DDI and ADCIRC-3DL. Technical Report, DTIC Document; US Army Corps of Engineers: Washington DC, USA, 1992.
11. Smith, J.M.; Sherlock, A.R.; Resio, D.T. STWAVE: Steady-State Spectral Wave Model User's Manual for STWAVE, version 3.0; Technical Report, DTIC Document; US Army Corps of Engineers: Washington DC, USA, 2001.
12. Dietrich, J.C.; Tanaka, S.; Westerink, J.J.; Dawson, C.; Luettich, R., Jr.; Zijlema, M.; Holthuijsen, L.H.; Smith, J.; Westerink, L.; Westerink, H. Performance of the unstructured-mesh, SWAN + ADCIRC model in computing hurricane waves and surge. J. Sci. Comput. 2012, 52, 468–497.
13. Westerink, J.J.; Luettich, R.A.; Feyen, J.C.; Atkinson, J.H.; Dawson, C.; Roberts, H.J.; Powell, M.D.; Dunion, J.P.; Kubatko, E.J.; Pourtaheri, H. A basin-to channel-scale unstructured grid hurricane storm surge model applied to southern Louisiana. Mon. Weather Rev. 2008, 136, 833–864.
14. Mukai, A.Y.; Westerink, J.J.; Luettich, R.A., Jr.; Mark, D. Eastcoast 2001, a Tidal Constituent Database for Western North Atlantic, Gulf of Mexico, and Caribbean Sea; Technical Report, DTIC Document; US Army Corps of Engineers: Washington DC, USA, 2002.
15. Pawlowicz, R.; Beardsley, B.; Lentz, S. Classical tidal harmonic analysis including error estimates in MATLAB using T_TIDE. Comput. Geosci. 2002, 28, 929–937.
16. US Army Corps of Engineers. Hydrodynamics of Tidal Inlets, EM 1110-2-1100 ed.; US Army Corps of Engineers: Washington DC, USA, 2008.
17. Wong, K.C.; DiLorenzo, J. The response of Delaware's inland bays to ocean forcing. J. Geophys. Res. Oceans 1988, 93, 12525–12535.
18. Chuang, W.S.; Swenson, E.M. Subtidal water level variations in Lake Pontchartrain, Louisiana. J. Geophys. Res. Oceans 1981, 86, 4198–4204.
19. Aretxabaleta, A.L.; Butman, B.; Ganju, N.K. Water level response in back-barrier bays unchanged following Hurricane Sandy. Geophys. Res. Lett. 2014, 41, 3163–3171.
20. Woodruff, J.D.; Irish, J.L.; Camargo, S.J. Coastal flooding by tropical cyclones and sea-level rise. Nature 2013, 504, 44–52.
21. Sweet, W.V. Sea Level Rise and Nuisance Flood Frequency Changes around the United States; NOAA Technical Report NOS CO-OPS 073; National National Oceanic and Atmospheric Administration: Silver Spring, MD, USA, 2014.
22. Judge, E.K.; Overton, M.F.; Fisher, J.S. Vulnerability indicators for coastal dunes. J. Waterw. Port Coast. Ocean Eng. 2003, 129, 270–278.

CHAPTER 9

Bias and Efficiency Tradeoffs in the Selection of Storm Suites Used to Estimate Flood Risk

Jordan R. Fischbach [1,2], David R. Johnson [2,3,*,†] and Kenneth Kuhn [2,3,†]

[1]*RAND Corporation, 4570 Fifth Ave., Ste. 600, Pittsburgh, PA 15213, USA*

[2]*Pardee RAND Graduate School, 1776 Main St., Santa Monica, CA 90401, USA*

[3]*RAND Corporation, 1776 Main St., Santa Monica, CA 90401, USA*

ABSTRACT

Modern joint probability methods for estimating storm surge or flood statistics are based on statistical aggregation of many hydrodynamic simulations that can be computationally expensive. Flood risk assessments that consider changing future conditions due to sea level rise or other drivers often require each storm to be run under a range of uncertain scenarios. Evaluating different flood risk mitigation measures, such as levees and floodwalls, in these future scenarios can further increase the computational cost. This study uses the Coastal Louisiana Risk Assessment model (CLARA) to examine tradeoffs between the accuracy of estimated flood depth exceedances and the number and type of storms used to produce the estimates. Inclusion of lower-intensity, higher-frequency storms significantly reduces bias relative to storm suites with a similar number of storms but only containing high-intensity, lower-frequency storms, even when estimating exceedances at very low-frequency return periods.

Keywords: flood risk; joint probability methods; storm selection; computational efficiency

1. INTRODUCTION

Flooding is the most frequently occurring weather-related natural disaster. In the United States, for example, 90 percent of all natural disasters involve flooding [1]. While not all flooding is caused by severe storms, coastal communities are particularly at risk of flooding from storm surge. Sea level rise and, in some areas, land subsidence will compound the problem, making many coastal communities significantly more vulnerable to storm surge hazards in future decades [2,3,4]. In addition, multiple studies have concluded that the average intensity of Atlantic hurricanes is likely to increase in the future due to projected increases in sea surface temperatures and other factors [5,6,7,8].

The Joint Probability Method with Optimal Sampling (JPM-OS) is one approach for estimating the probability distribution function of flooding in a region [9,10,11,12,13]. Storms are characterized using a set of parameters such as their minimum central pressure deficit, forward velocity, and location at landfall. A hydrodynamic model is used to simulate storms whose parameters span the range of plausible values that could occur in nature. A joint probability distribution function (PDF), fit to the historical record of observed storms, is used to estimate the relative likelihood of observing each storm in the suite of simulated, "synthetic" storms. The form of the joint PDF is based on assumptions about conditional relationships between the storm parameters and structural assumptions about how parameters are distributed; for example, the potential intensity of a hurricane is related to its size (radius of maximum windspeed) [14]. The probability masses associated with each synthetic storm are combined with their simulation results to build a cumulative distribution function (CDF) for quantitative metrics of flood risk like surge elevations, flood depths, or direct economic damage.

Long-range planning for flood risk management can involve evaluation of a large number of possible flood control or risk management strategies, such as different combinations of levee and floodwall projects, hazard mitigation, or broader coastal resiliency investments. Benefit-cost analysis entails predicting risk reduction benefits in multiple future time periods over a project's useful lifespan. Uncertainty about climate change, land subsidence, or other deeply uncertain drivers may necessitate that a range of future scenarios should be modeled. All of these factors multiply the number of scenarios, and thus model runs, needed to conduct a thorough and complete comparison of the benefits and costs of different flood risk reduction investments. Computational, budget, and time constraints can thus put downward pressure on the number of storms that can be run per case. We refer to the resulting problem of deciding what storms should be included in the analysis storm suite as the "storm selection" process.

This problem is exacerbated in diverse and complex wetlands ecosystems, such as those present in coastal Louisiana. In these areas, lower-resolution hydrodynamic models may produce biased results, and simplifying assumptions regarding future climate or sea level rise impacts may be invalid [15,16]. A high-resolution hydrodynamic simulation such as the ADvanced CIRCulation (ADCIRC) model is needed to accurately reproduce storm surge behavior from coastal storms [17].

This study has produced estimates of flood depth and damage at different exceedance probabilities associated with storm surge hazard in coastal Louisiana using the Coastal Louisiana Risk Assessment model (CLARA), a quantitative simulation model of storm surge flood risk developed by researchers at the RAND Corporation [18,19,20]. CLARA was developed to better understand how future coastal changes could lead to increased risk from storm surge flooding to residents and assets in Louisiana and assess the degree to which investments in risk mitigation could reduce this risk.

We have expanded upon the original storm selection approach described in Fischbach *et al.* [18] by examining flood depth and damage exceedances rather than surge elevation exceedances. The storm selection analysis was performed as part of a larger model development effort in support of the State of Louisiana's 2017 Coastal Master Plan. This paper draws largely from Fischbach *et al.* [20], a technical report describing the larger project; interested readers may refer to that document and Fischbach *et al.* [18] for additional details about CLARA.

Flood depth exceedance and expected annual damage estimates were produced using a reference set of 446 storms—the complete corpus of storms developed for use in recent JPM-OS studies of the Gulf coast—and with various subsets of the reference set, ranging from 40 to 304 storms. This paper examines the biases in flood depth and damage exceedances introduced by reducing the storm suite relative to a full implementation of JPM-OS using 446 storms. We also discuss the estimated parametric uncertainty associated with results from different storm suites. The results provide guidance for the types and size of storm suites that can be used to produce a good approximation of the results from the reference case when a full implementation of JPM-OS is computationally infeasible or prohibitively expensive for regions like coastal Louisiana.

2. METHODS

The original storm selection problem was to choose a suite of storms to simulate that, when used to fit a surge response surface, produces a probability distribution of surge that closely replicates the true underlying distribution [9]. Prior analyses developed a suite of 304 idealized, synthetic storms that are defined parametrically by their central pressure c_p, radius of maximum windspeeds r_{max}, forward velocity v_f, longitudinal location at landfall x, and angle of incidence at landfall θ_l. For the Louisiana coastline, landfall was defined as the point when the eye of the storm crosses 29.5° N latitude [21,22]. These studies focused on representing storms that produce Category 3 or greater wind speeds on the Saffir-Simpson scale (minimum central pressure of 960 mb or lower); later work increased the number of available storms to 446 by adding intermediate storm tracks and less extreme storms with a minimum central pressure of 975 mb.

The 446 storms were run through a dynamically coupled set of storm surge and wave models: ADCIRC and Simulating WAves Nearshore (SWAN), respectively [17,23,24]. These models were adapted and calibrated to current environmental and landscape conditions by other researchers working with the State of Louisiana [20]. The storms were also run through these models using coastal landscape and sea level conditions in the year 2065 as projected for the "Less Optimistic" scenario used in Louisiana's 2012 Coastal Master Plan [16,25]. An initial screening analysis, described in the Experimental Design section, used results from the current conditions model outputs. The final, detailed analysis focused on the Less Optimistic future scenario, rather than current conditions. The higher level of risk (particularly in areas enclosed by ring levees) associated with the Less Optimistic future allows more information about bias to be drawn from a wider range of return periods, as there are fewer return periods where no flooding occurs.

Using this approach, we estimate flood depth and damage statistics at a large number of spatial grid points. Estimates of flood statistics derived from the full 446-storm suite—referred to as the reference set—are used as a reference standard for comparison. We ran CLARA's modified JPM-OS procedure for a relatively large number of subsets of the reference set to investigate the use of smaller suites of synthetic storms. While we cannot assert that the 446-storm results are themselves unbiased or a true representation of the underlying storm risk, they represent the most complete set currently available for this region and are the best available basis for comparison when evaluating subsets.

In some cases, subsets were formed by eliminating storms from the reference set in ways that we hypothesized would introduce minimal bias. In other cases, subsets were formed by reducing the number of parameters that vary within the storm subset. For example, the storm tracks represented in the reference set are based on analysis of historic storm landfalls, in which a mean angle of incidence θ_l was calculated for each landfall point *x*. The reference set includes storms that follow the mean historic landfall angles, as well as storms that proceed along tracks 45 degrees to either side of the mean angle; some of the subsets we tested were formed by eliminating these "off-angle" tracks from the reference set. Further details on the subsets of storms considered are provided in the experimental design section of this manuscript.

For each subset of the reference set we proceeded as follows.

1. Fit response surfaces for peak storm surge and peak significant wave heights at each location not enclosed by a levee represented in the CLARA model (the "unenclosed grid points") and at special points ("surge and wave points," or SWP) 200 meters in front of levees and floodwalls along the boundary of enclosed protection systems [20]. Separate response surfaces are fit for surge hydrographs (surge elevations over time) and wave periods at the SWPs. The wave heights and periods are calculated at the time of peak surge in the surge hydrograph; the response surfaces for these quantities are fit using these values, and they are assumed to be constant over time (Estimated wave heights are also limited by the depth of the underlying surge. Holding wave heights and periods constant throughout the hydrograph reduces computational complexity and provides a first-order approximation of wave behavior during the points in the surge hydrograph where surge levels are high (*i.e.*, when overtopping may occur).).

2. Estimate peak surge and significant wave heights, using the response surface fits to obtain the predictions; also predict the surge hydrographs and wave periods at SWPs. In unenclosed areas, the effective flood elevations resulting from each synthetic storm are calculated as the sum of peak surge and the free-wave crest height. At SWPs, the predicted surge and wave characteristics are used as inputs to the flood module (as described in Step 4).

The response surface fits are based upon the ADCIRC and SWAN outputs for each storm included in a given subset; these storms are sometimes referred to as the "training set" in the ensuing discussion. In some cases, the response surfaces can be used to predict surge and wave characteristics for a larger number of synthetic storms than those used to fit the response surface. For example, Set D3 includes storms on each track with a central pressure of 960 mb and r_{max} values of 11 and 35.6

nautical miles. The response surface is fit using these storms to estimate the effect of r_{max} on surge and waves. The predictions also include a storm with an intermediate r_{max} of 21 nautical miles. Predictions are made for all storms in the 446-storm set that can be estimated using a response surface based on a given subset. For example, if a subset only includes storms with v_f = 11 knots, the response surface cannot identify the marginal effect of v_f on surge and waves, so synthetic storms with other values for v_f are excluded from the set of synthetic storms used to generate predictions.

A general rule is that for any combination of track and angle represented in a training set, the set of predicted synthetic storms run through the rest of the steps listed below includes all storms from the 446-storm corpus corresponding to that track and angle, excluding storms with v_f = 6 or 17 knots if the training set does not also include storms with variation in the forward velocity.

3. Partition the JPM-OS parameter space by the synthetic storms used for predictions and estimate the probability mass associated with each synthetic storm. The probabilities assigned to each synthetic storm are derived using maximum likelihood methods on a data set of observed historic storms, under certain structural assumptions about the joint probability distribution of the storm parameters [26].

4. Run the predicted synthetic storms through CLARA's flood module using the predicted SWP hydrographs and wave heights to obtain final still-water flood depths on the interior of enclosed protection systems. The flood module includes analysis of overtopping from surge and waves, system fragility and the consequences of breaches, and routing of water between interior polders, also accounting for pumping capacity and rainfall [20,27]. The probability of system failure is modeled as a function of overtopping rates [28].

5. Combine the flood depths and probability masses associated with each predicted synthetic storm to build the distribution functions of flood depths at each CLARA grid point.

6. Extract the flood depth exceedance values from the cumulative distribution functions at various return periods (A flood depth value with a $(1-1/n)$% annual exceedance probability (AEP)—the chance of occurring or being exceeded in a given year—is referred to as having an n-year return period. For this study, we recorded 5-, 8-, 10-, 13-, 15-, 20-, 25-, 33-, 42-, 50-, 75-, 100-, 125-, 150-, 200-, 250-, 300-, 350-, 400-, 500-, 1000-, and 2000-year exceedances.).

7. Run the flood depth exceedances through CLARA's economic module to estimate economic damage exceedances. Also estimate the expected annual damage (EAD) associated with each storm subset.

Full details on how each of these steps are accomplished (for example, the specification of the response surface model) can be found in Fischbach *et al.* [20]. We have found that deriving flood depth exceedance curves from small numbers of storms can be problematic in enclosed areas behind levee systems, with large, sudden jumps in exceedance values; for example, one might estimate a 12-foot difference in flood depths between the 100-year and 125-year exceedances [29]. If one uses a small number of storms, leaving large gaps in the parameter space, each storm may produce very different results from its "nearest neighbor" storms. This is exacerbated by the nonlinear relationship between surge and wave heights and levee overtopping, and also by the impact of system fragility. As a result, using the response surface to generate predicted synthetic storms with intermediate parameter values not represented in the training set is necessary for "smoothing out" the exceedance curve and making it less sensitive to small changes in the probability weights assigned to each storm. For consistency, when running synthetic storms through the CLARA flood module, we use the response surface predictions for all synthetic storms tested as part of a particular test set (as opposed to using the actual hydrodynamic modeling outputs for storms, when available, and response surface predictions only for storms outside of the training set).

The performance of each storm subset was evaluated by summarizing the bias in predicted flood depths at various return periods, relative to the results generated by the reference set. We also analyzed and compared the estimated standard errors associated with each subset.

3. EXPERIMENTAL SECTION

We first conducted a screening-level analysis by evaluating a large number of storm suites—sixteen in total, inclusive of the 446-storm reference set. For the initial screening, flood depths were calculated at a total of 18,273 grid points across the coast. These grid points were chosen by selecting a limited sample of 24 watersheds that varied by (a) size, (b) proximity to the coast, (c) degree of economic development, and (d) whether the watershed is protected by some federally-accredited levee or floodwall system. (The full CLARA v2.0 study region for coastal Louisiana consists of 77,643 grid points comprising 117 watersheds (excluding points consisting of open water). Grid point resolution varies with the concentration of population and assets, with a maximum distance of 1 km between neighboring points.) Watersheds protected by a fully-enclosed ring levee system were excluded from the initial screening sample to avoid running CLARA's flood module (step 4 above) as part of the screening exercise. Damage results were also excluded from the screening analysis.

The ten JPM-OS tracks used in the 2012 Coastal Master Plan analysis are sometimes referred to as primary tracks. They were labeled E1 through E5 and W1 through W5 for tracks in the eastern and western halves of the coast, respectively. Secondary storm tracks correspond to paths in between the primary tracks and were denoted by a B at the end of the track name (e.g., track E1B). Tracks also vary by their angle of incidence made with the coastline upon landfall. Tracks following the mean landfall angle are referred to as central-angle tracks; those making landfall at angles 45 degrees less or greater than the mean angle are referred to as off-angle tracks.

The modeled subsets were chosen to be collections of storms with easily interpretable and describable characteristics, to have variation in the total number of storms, and to have variation in the types of storms represented over the subsets. They were also chosen to avoid groups of storms that could cause identifiability or other performance issues in the response surface model. Louisiana's Coastal Protection and Restoration Authority (CPRA) also identified a need for storm sets with fewer than 154 storms—and preferably fewer than 100 storms—that could be used to evaluate a range of individual structural protection projects during 2017 Coastal Master Plan model production using available computing resources. Table 1 describes what types of storms comprise each of the storm suites evaluated in the screening-level analysis. The number of storms listed in the table represents the number of storms in the training set used to fit the response surface in Step 1 of the procedure described in the Methods section; as noted in the description of Step 2, the set of predicted synthetic storms run through the rest of the procedure may be larger than the training set. However, the size of the training set is the relevant number to discuss when comparing storm suites, as it represents the number of storms that must be run through the more computationally intensive ADCIRC and SWAN models.

Table 1. Characteristics of storm suites selected for screening-level analysis.

Set	Storms	Description
S1	446	Reference set
S2	40	2012 Master Plan (MP) storm suite: 10 primary storm tracks, 4 storms per track varying c_p and r_{max}
S3	90	2012 MP storm suite expanded to 9 storms per track that vary c_p and r_{max}
S4	304	LACPR storm suite: all storms in reference set, except those with 975 mb c_p
S5	194	All storms from LACPR storm set with 11-knot v_f
S6	154	All storms from LACPR storm set with 11-knot v_f on primary storm tracks
S7	216	All storms from reference set with 11-knot v_f on primary storm tracks
S8	144	All storms from reference set on primary storm tracks with 900 mb or 930 mb c_p
S9	120	All LACPR storms on primary, central-angle tracks (includes variation in v_f)
S10	264	All storms from reference set with 11-knot v_f
S11	50	2012 MP storm suite expanded to 5 storms per track varying c_p and r_{max}
S12	60	Set 11, plus storms with 975 mb c_p and central values for r_{max}
S13	120	All central-angle, primary-track storms with 11-knot v_f (includes 975 mb storms)
S14	148	Set 13, plus all 960 mb and 975 mb storms on secondary storm tracks
S15	154	Set 3, plus all 960 mb and 975 mb storms on primary, off-angle storm tracks
S16	182	Set 3, plus all 960 mb and 975 mb storms on secondary tracks or primary, off-angle tracks

Notes: c_p—central pressure; r_{max}—radius of maximum windspeed; v_f—forward velocity; LACPR—Louisiana Coastal Protection and Restoration study [22].

The initial findings led to the development of several new test suites, ranging from 60 to 154 storms, for the detailed analysis. The final sets tested, including number of storms and a description of key characteristics, are shown in Table 2below. Note that the storm set ID numbers for the detailed analysis do not all match up with the IDs used in the screening analysis; although some sets from the screening analysis were included in the detailed analysis, the ID numbers were designated independently. The detailed analysis included estimation of flood depth exceedances at all grid points across the coast, including in areas enclosed by ring levees. Damage exceedances and bias in EAD were also analyzed for these storm suites.

The off-angle tracks for sets D6 and D7 were selected based on their paths passing near to New Orleans, and thus having the potential for significant flooding in urban centers where a large number of grid points and economic assets are densely packed. Our hypothesis was that including off-angle storms affecting the greater New Orleans region would improve accuracy in estimating how likely it is for the city to experience the catastrophic flooding associated with levee failures.

Table 2. Characteristics of storm suites selected for detailed investigation.

Set	Storms	Description
D1	446	Reference set
D2	40	2012 Master Plan (MP) storm suite: 10 primary storm tracks, 4 storms per track varying c_p and r_{max}
D3	60	2012 MP storm suite, expanded to 5 storms per track varying c_p and r_{max}, plus storms with 975 mb central pressure and central values for r_{max}
D4	90	2012 MP storm suite, expanded to 9 storms per track varying c_p and r_{max}
D5	90	2012 MP storm suite, expanded to 7 storms per track (excludes one 930 mb and one 900 mb storm), plus 975 mb storms using extremal (rather than central) r_{max} values
D6	92	Set 3, plus 960 mb and 975 mb storms on off-angle tracks only in E1-E4
D7	92	Set 3, plus 960 mb and 975 mb storms on off-angle tracks only in W3-W4, E1-E2
D8	100	All central-angle, primary-track storms with 11-knot v_f, plus 975 mb storms with central r_{max} values
D9	110	All central-angle, primary-track storms with 11-knot v_f, plus 975 mb storms with extremal r_{max} values
D10	120	All central-angle, primary track storms with 11-knot v_f, including all 975 mb storms
D11	154	Set 4, plus all 960 mb and 975 mb storms on primary, off-angle storm tracks

Notes: c_p—central pressure; r_{max}—radius of maximum windspeed; v_f—forward velocity.

4. RESULTS AND DISCUSSION

An example result from the screening analysis is shown in Figure 1. The depicted storm suite (Set S3 from Table 1) is comprised of nine storms from each of the ten primary tracks. Within a track, storms vary by c_p and r_{max}. The central pressure values represented are 900, 930, and 960 mb; storms with 975 mb pressures were not included. All storms have the central value for v_f of 11 knots. The figure illustrates bias in the estimated 100-year flood depths, relative to the reference set. Note that this particular suite tends to strongly overpredict 100-year depths (blue coloring), relative to the 446-storm reference set, in the far eastern portion of the state, east of New Orleans and the Mississippi River, while it tends to underpredict depths (orange coloring) in the watersheds farthest west.

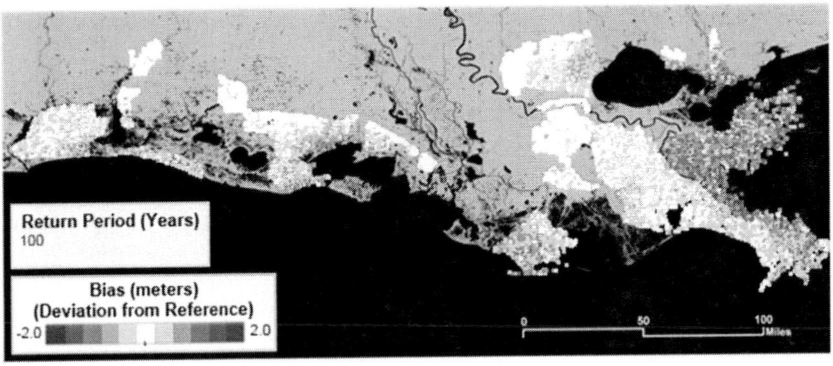

Figure 1. Deviation from reference set in 100-year flood depth exceedances in screening analysis watersheds from a suite consisting of 90 storms (screening set S3).

Figure 2 summarizes the findings of the screening-level analysis. It shows the root mean squared error (relative to the reference set), averaged over all grid points in the screening watersheds, for each modeled storm suite; the figure plots this average bias as a function of the number of storms in each suite. Blue points represent storm suites which contain at least one storm with a 975 mb minimum central pressure, the dot shape represent suites which contain off-angle storm tracks. Sets 8 and 9 (as well as the reference set) from the screening analysis also contain storms with non-central values for forward velocity.

In addition to the dimensions illustrated in the figure, sets S4, S8, and S9 contain storms with non-central values for forward velocity. The screening results suggest that including higher-frequency storms with 975 mb central pressure improved statistical performance. By contrast, secondary or off-angle storm tracks and storms with non-central values for forward velocity did not yield similar improvement and were generally not included in the more detailed final testing and results.

We focused primarily on suites of storms that include 975 mb storms because the screening analysis results suggested that such suites outperform suites that exclude the less intense storms. The relatively poor performance of sets D2 and D4, the suites with no 975 mb storms, suggests that this is a still a valid conclusion when considering the full coastal study region. Improvement is less apparent when off-angle tracks are included, however, especially for sets in which only certain off-angle tracks were included (e.g., Sets D6 and D7).

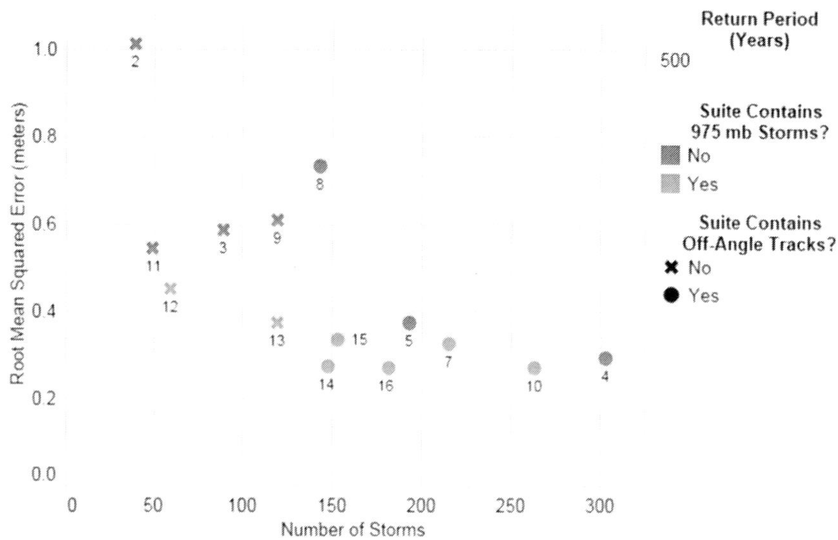

Figure 2. Average coastwide bias and variation by number of storms (screening analysis), 500-year flood depths.

4.1. Flood Depth Bias and Variance Comparisons

Figure 3 summarizes the average coastwide 100-year flood depth bias (root mean squared error, y-axis) and coefficient of variation (point size) for each set in the detailed analysis, plotted against the number of storms in a given storm suite (x-axis). As before, colors indicate whether the set includes 975 mb storms, and shape indicates whether off-angle tracks are included. Bias is estimated relative to the flood depth results from the reference set.

Summary results show that average bias in flood depths at the 100-year interval varies from less than 0.25 m to nearly 1 m, depending on the storm sample. Substantial bias is observed for Set D2, the 2012 Coastal Master Plan suite, compared with the other candidate sets tested. All additional sets with more than 40 storms improve upon the 40-storm results. In fact, increasing the number of storms to 60 (Set D3) leads to substantial improvement, reducing average bias by more than a half a meter. Similar results are observed at other return periods.

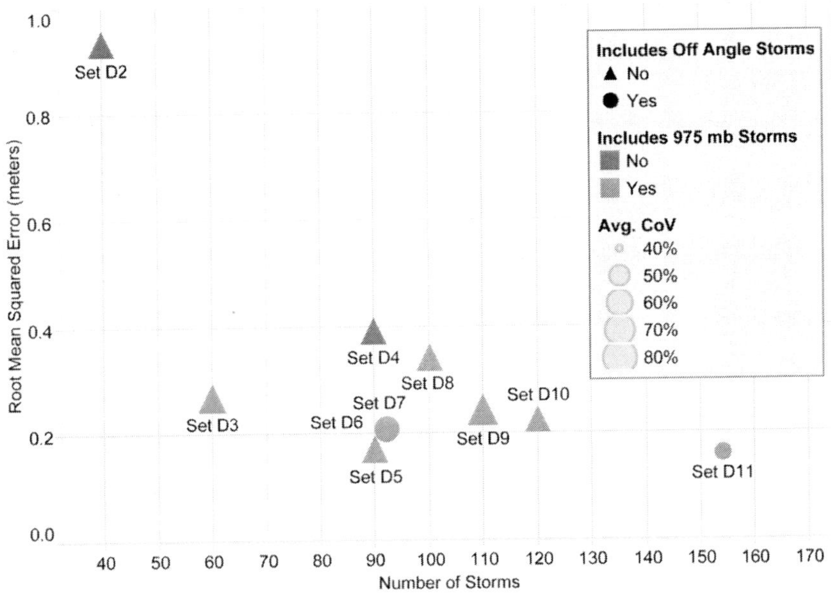

Figure 3. Average coastwide bias and variation by number of storms (detailed analysis), 100-year flood depths.

Figure 4 illustrates how the average bias, again measured by RMSE, changes for each storm suite at different return periods of the flood depth distribution. The x-axis is the return period, on a logarithmic scale. The thickness of the lines shown in the figure reflects the coefficients of variation, a measure of the uncertainty around derived point estimates; thicker lines denote greater uncertainty.

With few exceptions, the relative order of performance across storm subsets is consistent over a wide range of return periods. Sets D2 and D4 are conspicuous in their poor average performance; neither set contains any storms with central pressures of 975 mb, further emphasizing the importance of including higher-frequency events in the training set. Interestingly, this effect is still particularly apparent in the tail of the distribution beyond the 100-year AEP interval, indicating that exclusion of higher-frequency events from the response surface model also skews the model's predictive accuracy for more extreme synthetic storms.

In general, average bias is smaller at 50-year and more frequent return periods. This is due to the large number of points in which no flooding occurs for these exceedances. The absolute error is more likely to be small when the depth estimates themselves are small. Set D11, with 154 storms, performs well across most of the distribution, and better than any smaller sets. However, several of the sets with fewer than 100 storms are not far behind in terms of average bias across the exceedance distribution.

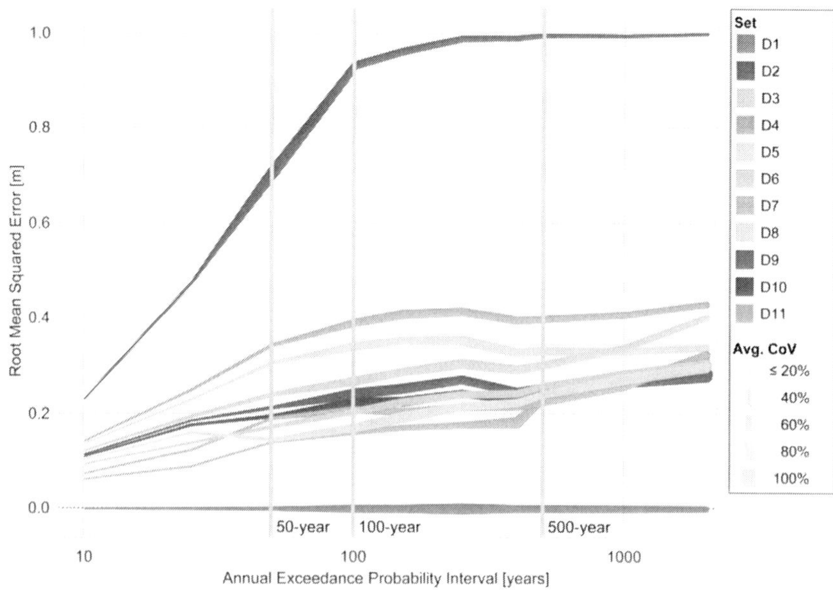

Figure 4. Average bias and variance by exceedance interval for each modeled storm suite.

Figure 5 shows a sequence of maps that display the spatial patterns of bias, relative to the reference set, for three of the storm suites tested. Flood depths at the 100-year return period, as estimated using the reference set, are shown for comparison in Figure 6; when examining the three maps in Figure 5, this gives some indication of the magnitude of the biases relative to the baseline flood depths. When examining the spatial distribution of bias, some interesting patterns emerge. The Set D2 results, representing the 2012 Coastal Master Plan's 40-storm sample, show a substantial overestimate of flood depths across nearly the entire study region. This effect occurs because of the types of storms excluded from Set D2; it only contains extreme central-angle storms on primary tracks, leaving out both 975 mb and off-angle storms. For instance, Set D2 excludes off-angle storms that pass well to the east of New Orleans. These storms would tend to lower the estimated flood depth exceedances in St. Bernard Parish, so as a result Set D2 has positive bias in that region.

Set D3, which includes 60 storms, improves dramatically on the Set D2 results. Some positive bias is still noted in the western portion of the coast and many areas east of the Mississippi River, but the magnitude is notably lower than Set D2. In addition, Set D3 actually underestimates 100-year flood depths compared with the reference in some enclosed areas, including the East Bank of the Greater New Orleans Hurricane Storm Damage Risk Reduction System (HSDRRS) and the Larose to Golden Meadow levee system.

Set D11 yields the lowest overall bias when averaged over all points, and it shows balanced results coastwide when looking at geospatial patterns. Positive bias is still observed in the western parishes, but again with a lower magnitude than Sets D2 or D3. There are some instances where the 100-year flood depth estimates are both positively and negatively biased within the same watershed. When these differences in flood depths are translated into damage, the coastal and parish-level results are more similar to those of the reference set.

The differences between enclosed and unenclosed areas are clearly shown in Figure 7 and Figure 8 (note the different scale on the y-axis between figures). The curves shown in Figure 4 are broken out by showing the same results for points enclosed by (Figure 7) and unenclosed points (Figure 8), respectively.

HSDRRS was accredited in 2014 by the United States Federal Emergency Management Agency as protecting the Greater New Orleans area against at least 100-year surge levels [30]. However, the same standard may not be met in 2065 under the Less Optimistic future scenario assumptions in a future without further mitigating actions. The precise level of protection provided by HSDRRS is uncertain [31]. As such, the average coefficient of variation is relatively large, over 100% for many storm suites over large portions of the distribution.

In unenclosed areas, the performance of all suites but Set D2 is relatively good and largely similar. With the exception of Set D2, all storm suites have an average bias of less than half a meter at all exceedance intervals, and the average uncertainty in the predicted values is also considerably smaller than for points within HSDRRS. This reflects that the exceedance curves in unenclosed areas have lower variance and are more stable, in the sense that few points experience sudden, sharp increases in flood depths at any point of the distribution.

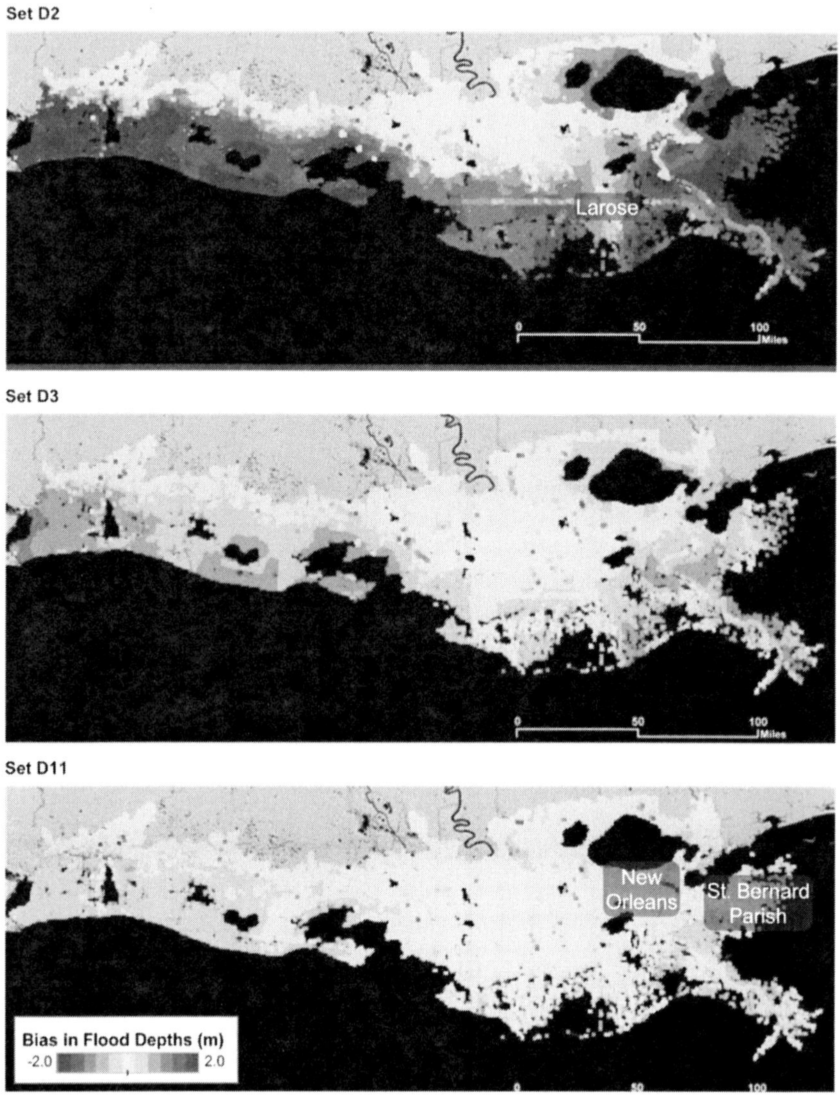

Figure 5. Maps of bias by grid point for detailed analysis sets D2, D3, and D11, 100-year flood depths.

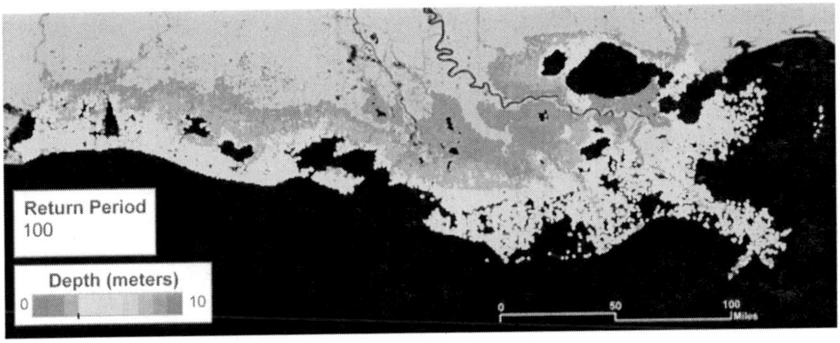

Figure 6. Median 100-year flood depths from reference set D1.

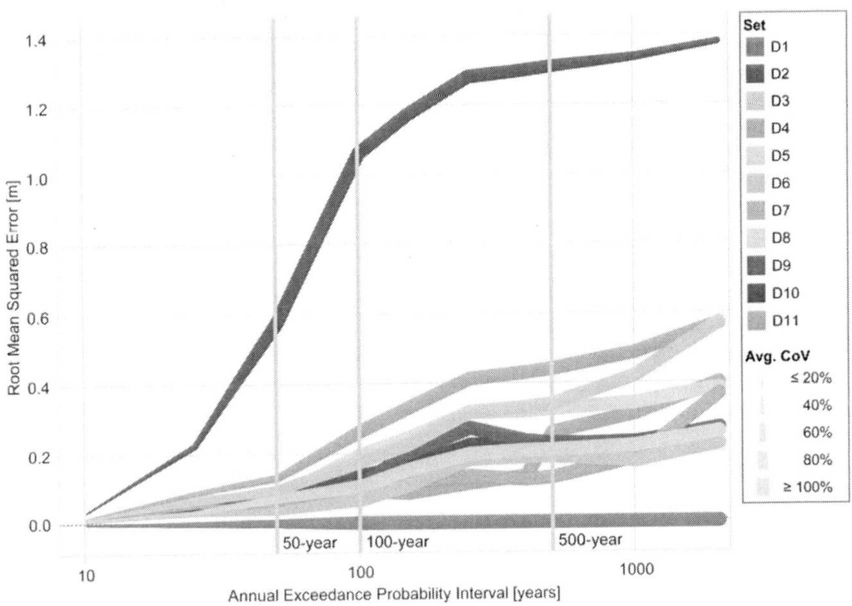

Figure 7. Average bias and variance by exceedance interval (HSDRRS Only).

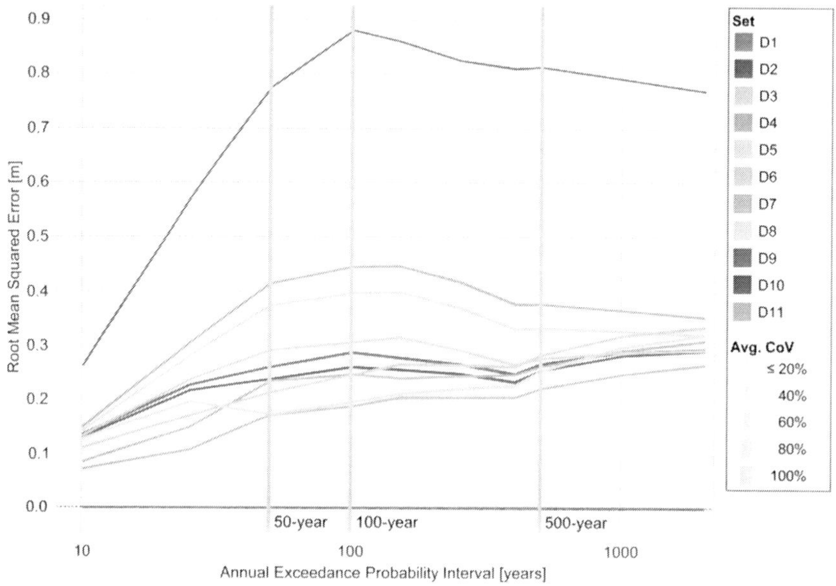

Figure 8. Average bias and variance by exceedance interval (unenclosed points only).

The greater variation in performance in enclosed areas illustrates that flood depth estimates are more sensitive to the choice of storm suite when the accuracy and uncertainty in response surface predictions is further compounded by interactions with engineered flood protection systems. Maintaining accuracy with fewer storms is easier for prediction of storm surge elevation exceedances, as in the original storm selection problem faced by JPM-OS, than for flood depth exceedances when accounting for the levee and floodwall failures, overtopping, and interior drainage dynamics.

Figure 9 focuses in on differences in storm suite performance for enclosed points within HSDRRS, on the east and west sides of the Mississippi River, at the 500-year AEP interval. Consistent with what was shown in Figure 7 and Figure 8, results are most uncertain and show the greatest variance in the East and West Bank areas. Except for Set D2, all sets considered yield an average bias of approximately 0.3 meters (1 foot) or less in enclosed locations on the East Bank. Uncertainty is greater on the West Bank. Similarly, Sets D3 through D11 also result in less than 1 meter of average bias in enclosed areas not within HSDRRS, with many sets also yielding less than 0.3 meters of average bias in these areas (results not shown).

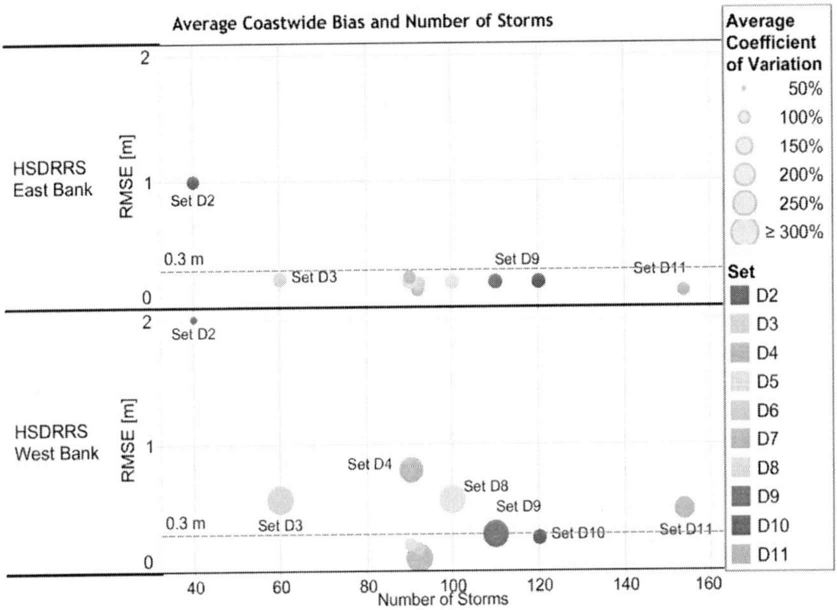

Figure 9. Average bias and variation, 500-year flood depths, east and west banks of the Greater New Orleans Hurricane Storm Damage and Risk Reduction System.

Figure 9 confirms that Set D11 is the best or near-best performer in terms of mean bias across all enclosed areas. For sets with fewer than 100 storms, performance depends on HSDRRS location. Set D3 performs substantially worse, however, on the West Bank, as do many other sets. By contrast, Set D7 was specifically designed to improve performance in the West Bank area of HSDRRS. This suite adds storms to Set D3, including off-angle tracks for the middle of the coast with landfall locations observed to have the greatest effect on this portion of HSDRRS. The results bear this out: Set D7 is the best overall set in terms of bias in the West Bank, even slightly better than the larger Set D11. However, Set D7 produces greater bias in the East Bank HSDRRS, and is among the worst performers when looking at enclosed areas other than the West Bank. It also has a significantly larger coefficient of variation on the West Bank than Sets D5 and D6, which both have nearly identical average RMSE and number of storms to Set D7. This illustrates that while it may be possible to carefully tailor a storm suite to perform well in a particular region, results may be significantly worse in other, even nearby, areas. It can be difficult to predict the overall performance of a storm set, which emphasizes the importance of conducting a storm selection analysis like this on at least one landscape scenario before embarking on a major flood risk assessment encompassing many future scenarios, time periods, and system configurations.

4.2. Damage Bias Comparisons

Next, we sought to understand how the potential bias from different storm suites translates to bias in damage estimates. Somewhat surprisingly, we found that the total bias associated with a suite of storms was typically driven by bias in unenclosed areas, although such regions are more sparsely populated and contain fewer assets than in protected areas like New Orleans. The greater protection in enclosed areas provided by federal levee systems instead meant that there was a lower projected baseline risk in those regions; most storm suites did not predict extensive flooding in enclosed areas except at lower-frequency AEPs, so this resulted in less bias.

Figure 10 below shows a similar spatial breakdown as Figure 9, for example, but instead displays bias in terms of damage (EAD, median results) estimated by the CLARA v2.0 economic damage model. Set D2, which yields very high EAD bias overall, is omitted from the figures below for clarity. Figure 10 confirms the performance noted above, with roughly the same overall ranking of storm suites by EAD bias as with flood depths. Sets D5, D6, D7, and D11 all have a cumulative bias of less than $0.5 billion across all points. D3 shows a total bias over all points of about $1.5 billion; D8 has bias of $2 billion, while D4 has $2.5 billion.

Increasing the number of storms from 60 up to 90–110 does not substantially improve EAD performance for enclosed areas; Sets D4 and D8 consistently produce worse error than Set D3, despite having 90 and 100 storms, respectively. Instead, it is the addition of 975-mb storms that appears to drive improvement. Set D4 has no such storms, while Set D8 contains half as many 975-mb storms than the other suites with 90–110 storms.

Similarly, Figure 11 shows a summary by coastal Louisiana parish (county) of bias in EAD compared to the reference set. Some storm suites, regardless of coastwide performance, produce large bias in specific parishes. For example, Set D4 produces approximately 50% more damage in Jefferson and St. Bernard Parishes than the 446-storm reference set. In turn, this leads to a substantial upwards bias in coastwide EAD because these parishes contain large concentrations of assets.

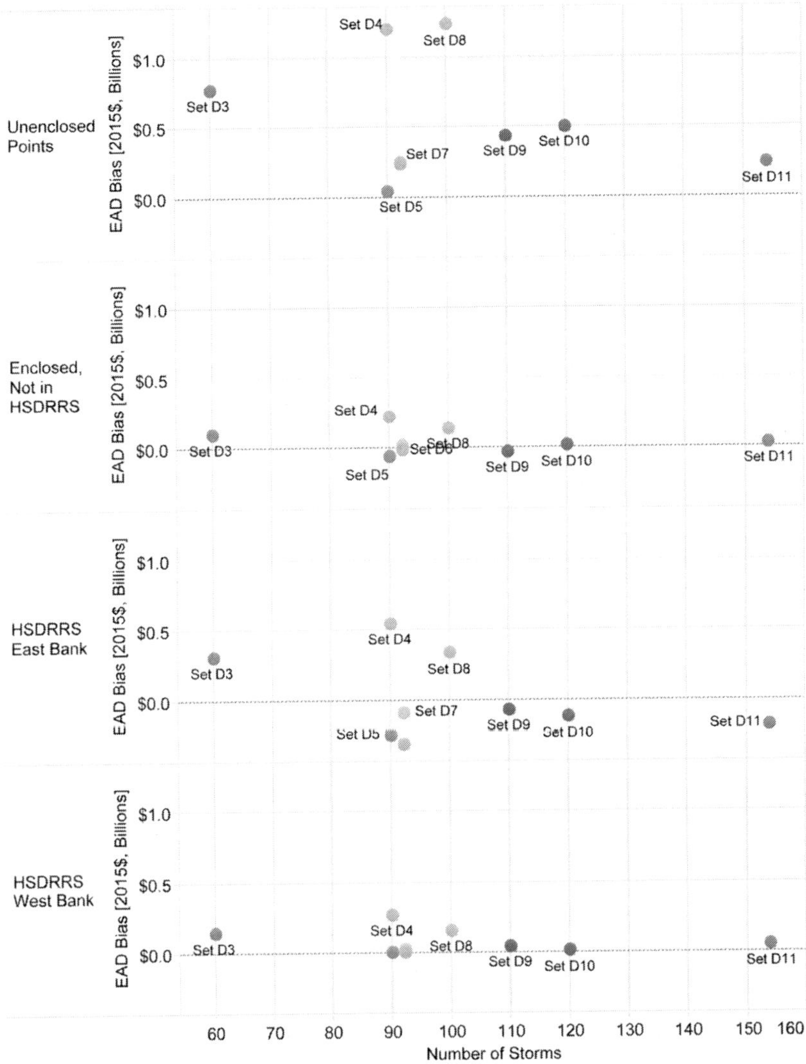

Figure 10. Bias in expected annual damage (median estimates), by location.

Sets D5, D6 and D11 are the highest performers when disaggregating results over the parishes; they are the only storm sets with only one parish's bias over 20% and a maximum bias of magnitude 25% or less. The maximum bias for any parish from Set D11 is 22% in Calcasieu Parish, which is in the far west of the state and includes relatively few assets. All storm suites overestimated EAD in Calcasieu, with errors ranging from 14 to 41%, indicating a general difficulty in reproducing estimates of risk so far west with a small number of storms. Each of the other storm suites produces biases as large or larger in several parishes. Few storm sets performed well, with errors of 10% or less, in both Orleans and Jefferson parishes, the two parishes with the largest value of assets at risk. EAD performance in non-HSDRRS parishes, by contrast, is typically more similar across all sets tested in this round of analysis.

The results discussed so far have portrayed the median (50th percentile) outputs over the sampling design, with the coefficients of variation giving some information about the parametric variation across different suites. Figure 12 illustrates the distribution of EAD in another way. The y-axis indicates the bias in coastwide EAD relative to the full 446-storm reference set; the three points for each subset, from bottom to top, represent the bias associated with the 10th, 50th, and 90th percentile values, respectively, of EAD.

	Set							
Parish Name	D3	D4	D5	D6	D7	D9	D10	D11
Acadia	2%	3%	-2%	0%	8%	-4%	-1%	9%
Ascension	27%	34%	14%	19%	8%	20%	21%	10%
Assumption	2%	3%	0%	3%	0%	1%	1%	-3%
Calcasieu	24%	24%	14%	23%	41%	17%	19%	22%
Cameron	0%	-3%	2%	-2%	2%	-1%	0%	-1%
Iberia	0%	2%	-5%	0%	5%	-3%	-2%	6%
Jefferson	32%	44%	-4%	-17%	-4%	2%	-1%	-5%
Jefferson Davis	9%	13%	1%	9%	18%	8%	8%	15%
Lafourche	5%	11%	0%	2%	2%	1%	4%	2%
Livingston	15%	17%	13%	10%	5%	14%	14%	3%
Orleans	4%	22%	-21%	-15%	0%	-8%	-13%	-7%
Plaquemines	12%	22%	2%	1%	2%	8%	7%	4%
St. Bernard	27%	57%	7%	-8%	0%	15%	9%	7%
St. Charles	5%	8%	2%	3%	0%	5%	5%	1%
St. James	-4%	9%	-12%	-5%	-11%	-12%	-12%	-16%
St. John the Baptist	24%	30%	13%	13%	7%	21%	21%	13%
St. Martin	1%	0%	0%	1%	0%	0%	0%	-1%
St. Mary	0%	3%	-10%	-2%	-3%	-5%	-6%	0%
St. Tammany	12%	14%	9%	7%	7%	10%	10%	8%
Tangipahoa	0%	0%	-1%	-2%	-7%	-1%	1%	-7%
Terrebonne	4%	8%	-4%	2%	-2%	1%	1%	0%
Vermilion	5%	7%	3%	6%	7%	3%	4%	7%

EAD Bias (percent): -50% to 50%

Figure 11. Percentage bias in expected annual damage by parish (median).

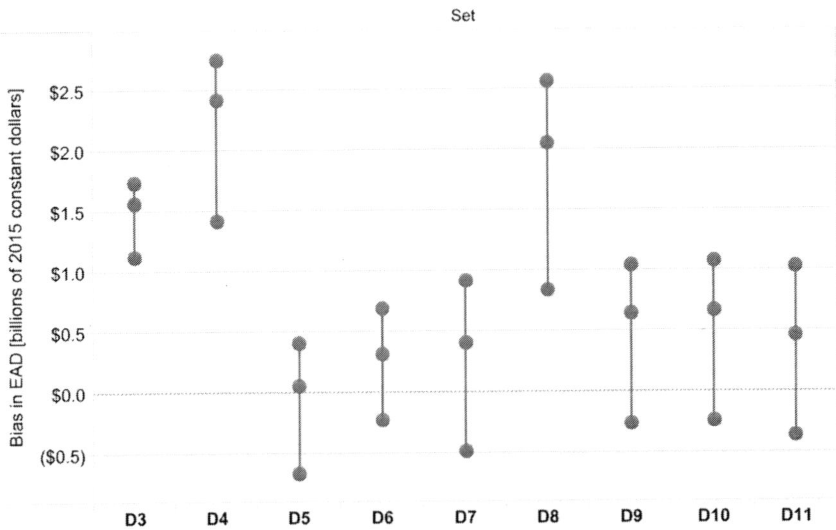

Figure 12. Coastwide bias in terms of expected annual damage (billions of 2010 US dollars).

The primary takeaway here is that many storm suites end up with similar ranges of uncertainty in terms of EAD bias relative to the reference set, when aggregating damage coastwide. Considering only suites with fewer than 100 storms, Set D5 is the best performer at the median, but includes a wider range of results across the parametric distribution than Set D6, and the 10th percentile estimate is that Set D5 may underestimate EAD more strongly than any other set. Set D3 has the smallest range of variation in bias between the 10th and 90th percentile estimates, but it has a larger median bias than most of the larger sets; Set D6 has the second smallest range of uncertainty. Six sets underestimate EAD at the 10th percentile but overestimate it at the median and 90th percentiles.

5. CONCLUSIONS

The storm selection analysis shows a tradeoff between the number of storms and the resulting bias when compared with the reference set of 446 storms. Results show that nearly all storm sets tested produce lower bias when compared with the 2012 Coastal Master Plan 40-storm suite (Set D2). Aside from the fact that Set D2 was the smallest storm suite tested, this should not be surprising, given that the 2012 Coastal Master Plan was intended to model damage associated only with Category 3 or greater hurricanes. Substantial improvement is noted when storms with 975 mb central pressure were included, as well as with the addition of off-angle storms in some cases. The inclusion of storms with variation in forward velocity was found to have little impact on performance.

Of the storm suites tested, Set D6 (92 storms) appears to yield the best balance of results. It shows relatively low bias compared with the reference set in terms of both flood depth and damage, no concerning spatial patterns of bias, and reasonable performance in enclosed areas (particularly Greater New Orleans). It also has the second smallest range of uncertainty in damage among the suites analyzed. Set D3 is notable for producing less average error in flood depths than expected given that it contains only 60 storms.

The results of this storm selection analysis suggest some insights for future planning studies using a JPM-OS approach with high-resolution storm surge and wave modeling. First, the number of storms needed to estimate unbiased flood depths or damages can vary greatly depending on the size of the geographic area of focus, landscape type and configuration, presence of hurricane protection structures, and range of exceedance probabilities targeted. In some cases—for instance, in areas without protection structures or with relatively similar landscape characteristics over a broader area—a smaller storm suite may yield acceptably low bias, with results suitable for screening or planning-level decisions.

In some areas, however, large numbers of storms are needed to avoid bias and assess a wide range of plausible storm impacts from storms with different characteristics. Larger storm suites are also needed to support flood protection engineering and design studies, where decision makers need confidence that the widest range of plausible outcomes are simulated and adequate factors of safety have been developed accordingly.

For the 2017 Coastal Master Plan, a hybrid approach will be adopted: a smaller suite of 60 storms (Set D3) will be used for preliminary evaluation and testing of individual risk reduction projects, but coastwide alternatives—including hurricane protection structures, ecosystem restoration, and nonstructural risk reduction (hazard mitigation for buildings)—will be evaluated and compared using a larger set of 92 storms (Set D6). This two-step approach will help to confirm or validate preliminary project-level results and ensure that potential risk and damage reduction benefits from the plan as a whole are relatively unbiased.

ACKNOWLEDGMENTS

Other RAND researchers and programming staff who have contributed to development of the CLARA model include Ricardo Sanchez, Chuck Stelzner, Rachel Costello, Edmundo Molina-Perez, and James Syme. This work was funded by the Louisiana Coastal Protection and Restoration Authority under the 2017 Coastal Master Plan's Master Services Agreement. Coordination of the process has been managed by The Water Institute of the Gulf. In particular, the authors wish to thank collaborators Hugh Roberts and Zach Cobell at ARCADIS; Mandy Green, Melanie Saucier, Mark Leadon, and Karim Belhadjali at CPRA; Denise Reed at The Water Institute; and Gary Cecchine at RAND Corporation for their intimate involvement in and support of this project.

AUTHOR CONTRIBUTIONS

Fischbach is serving as principal investigator for the 2017 Coastal Master Plan's Risk Assessment team. He led the development of the storm selection analysis methodology and results visualizations, with Johnson's assistance. Johnson is the lead technical developer of the CLARA model; he executed the model runs to support the storm selection analysis and developed the storm suites included in the detailed analysis based on the initial screening results. Kuhn assisted in carrying out the analysis and quality assurance process. All authors contributed to the technical report on which this journal article is based; Johnson led its adaptation for journal submission, with assistance from the other authors.

CONFLICTS OF INTEREST

The authors declare no conflict of interest.

REFERENCES

1. Federal Emergency Management Agency. Fact sheet: Flooding—Our nation's most frequent and costly natural disaster. Federal Emergency Management Agency: Washington, DC, USA, 2010; p. 2.
2. Karl, T.R.; Meehl, G.A.; Peterson, T.C.; Kunkel, K.E.; Gutowski, W.J., Jr.; Easterling, D.R. Weather and Climate Extremes in a Changing Climate, Regions of Focus: North America, Hawaii, Caribbean, and U.S. Pacific Islands; NOAA: Silver Spring, MD, USA, 2008.

3. Burkett, V.R.; Davidson, M.A. Coastal Impacts, Adaptation and Vulnerability: A Technical Input to the 2012 National Climate Assessment; Burkett, V.R., Davidson, M.A., Eds.; Island Press: Washington, DC, USA, 2012.
4. Neumann, J.E.; Emanuel, K.; Ravela, S.; Ludwig, L.; Kirshen, P.; Bosma, K.; Martinich, J. Joint effects of storm surge and sea-level rise on US coasts: New economic estimates of impacts, adaptation, and benefits of mitigation policy. Clim. Change 2014, 1–13.
5. Knutson, T.R.; McBride, J.L.; Chan, J.; Emanuel, K.; Holland, G.; Landsea, C.W.; Held, I.M.; Kossin, J.P.; Srivastava, A.K.; Sugi, M. Tropical cyclones and climate change. Nat. Geosci. 2010, 3, 157–163.
6. Emanuel, K. Downscaling CMIP5 climate models shows increased tropical cyclone activity over the 21st century. Proc. Natl. Acad. Sci. USA 2013, 100, 12219–12224.
7. Villarini, G.; Vecchi, G.A. Twenty-first century projections of north atlantic tropical storms from CMIP5 models. Nat. Clim. Change 2012, 604–607.
8. Webster, P.J.; Holland, G.J.; Curry, J.A.; Chang, H.R. Changes in tropical cyclone number, duration, and intensity in a warming environment. Science 2005, 309, 1844–1846.
9. Resio, D.T.; Irish, J.L.; Cialone, M.A. A surge response function approach to coastal hazard assessment. Part 1: Basic concepts. Nat. Hazards 2009, 51, 163–182.
10. Irish, J.L.; Resio, D.T.; Cialone, M.A. A surge response function approach to coastal hazard assessment. Part 2: Quantification of spatial attributes of response functions. Nat. Hazards 2009, 51, 183–205.
11. Toro, G.R.; Resio, D.T.; Divoky, D.; Niedoroda, A.W.; Reed, C. Efficient joint-probability methods for hurricane surge frequency analysis. Ocean Eng. 2010, 37, 125–134.
12. Ho, F.P.; Myers, V. Joint Probability Method of Tide Frequency Analysis Applied to Apalachicola Bay and St. George Sound, Florida; U.S. Department of Commerce: Silver Spring, MD, USA, 1975.
13. Myers, V. Joint probability method of tide frequency analysis applied to Atlantic City and Long Beach Island, New Jersey; U.S. Department of Commerce: Silver Spring, MD, USA, 1970.
14. Shen, W. Does the size of hurricane eye matter with its intensity? Geophys. Res. Lett. 2006, 33, L18813.
15. Westerink, J.J.; Luettich, R.A.; Feyen, J.C.; Atkinson, J.H.; Dawson, C.; Roberts, H.J.; Powell, M.D.; Dunion, J.P.; Kubatko, E.J.; Pourtaheri, H. A basin- to channel-scale unstructured grid hurricane storm surge model applied to southern Louisiana. Mon. Weather Rev. 2008, 136, 833–864.
16. Cobell, Z.; Zhao, H.; Roberts, H.J.; Clark, F.R.; Zou, S. Surge and wave modeling for the Louisiana 2012 Coastal Master Plan. J. Coast. Res. 2013, 88–108.
17. Luettich, R.A.; Westerink, J.J. Formulation and Numerical Implementation of the 2d/3d Adcirc Finite Element Model Version 44.Xx; R. Luettich: Chapel Hill, NC, USA, 2004.
18. Fischbach, J.R.; Johnson, D.R.; Ortiz, D.S.; Bryant, B.P.; Hoover, M.; Ostwald, J. Coastal Louisiana Risk Assessment Model: Technical Description and 2012 Coastal Master Plan Analysis Results; RAND Corporation: Santa Monica, CA, USA, 2012.
19. Johnson, D.R.; Fischbach, J.R. Using cost-effective and robust strategies to assess the potential for nonstructural risk reduction in coastal Louisiana. In Improving Flood Risk Estimates and Mitigation Policies in Coastal Louisiana under Deep Uncertainty; RAND Corporation: Santa Monica, CA, USA, 2013; pp. 67–101.
20. Fischbach, J.R.; Johnson, D.R.; Kuhn, K.; Pollard, M.; Stelzner, C.; Costello, R.; Molina-Perez, E.; Sanchez, R.; Roberts, H.J.; Cobell, Z. 2017 Coastal Master Plan: Model Improvement Plan, Storm Surge and Risk Assessment Improvements (Subtask 4.9); Coastal Protection and Restoration Authority: Baton Rouge, LA, USA, 2015.
21. Performance Evaluation of the New Orleans and Southeast Louisiana Hurricane Protection System; US Army Corps of Engineers: New Orleans, LA, USA, 2009.
22. Louisiana Coastal Protection and Restoration Technical Report; US Army Corps of Engineers: New Orleans, LA, USA, 2009.

23. Dietrich, J.C.; Tanaka, S.; Westerink, J.J.; Dawson, C.N.; Luettich, R.A., Jr.; Zijlema, M.; Holthuijsen, L.H.; Smith, J.M.; Westerink, L.G.; Westerink, H.J. Performance of the unstructured-mesh, SWAN+ADCIRC model in computing hurricane waves and surge. J. Sci. Comput. 2012, 52, 468–497.
24. Booij, N.; Holthuijsen, L.H.; Ris, R.C. The "swan" wave model for shallow water. Coast. Eng. Proc. 1996, 25.
25. Peyronnin, N.; Green, M.; Richards, C.P.; Owens, A.; Reed, D.; Chamberlain, J.; Groves, D.G.; Rhinehart, K.; Belhadjali, K. Louisiana's 2012 Coastal Master Plan: Overview of a science-based and publicly-informed decision making process. J. Coast. Res. 2013, 67, 1–15.
26. Resio, D.T. White Paper on Estimating Hurricane Inundation Probabilities; US Army Corps of Engineers: New Orleans, LA, USA, 2007.
27. Johnson, D.R.; Fischbach, J.R.; Ortiz, D.S. Estimating surge-based flood risk with the Coastal Louisiana Risk Assessment Model. J. Coast. Res. 2013, 67, 109–126.
28. Final Post Authorization Change Report: Morganza to the Gulf of Mexico, Louisiana; US Army Corps of Engineers: New Orleans, LA, USA, 2013.
29. Johnson, D.R. Improving Flood Risk Estimates and Mitigation Policies in Coastal Louisiana under Deep Uncertainty; Pardee RAND Graduate School: Santa Monica, CA, USA, 2013.
30. Holder, K. FEMA Accredits Hurricane and Storm Damage Risk Reduction System (HSDRRS); US Army Corps of Engineers: New Orleans, LA, USA, 2014.
31. Johnson, D.R.; Fischbach, J.R.; Kuhn, K. Current and Future Flood Risk in Greater New Orleans; The Data Center: New Orleans, LA, USA, 2015.

CHAPTER 10

Influence of Rising Sea Level on Tidal Dynamics in the Bohai Sea

Yanfang Li, Hua Zhang, Cheng Tang, Tao Zou, and Dejuan Jiang

Key Laboratory of Coastal Environmental Processes and Ecological Remediation, Yantai Institute of Coastal Zone Research, Chinese Academy of Sciences, Yantai, Shandong 264003, P. R. China

ABSTRACT

Li, Y. F.; Zhang, H.; Tang, C.; Zou, T., and Jiang, D. J., 2016. Influence of rising sea level on tidal dynamics in the Bohai Sea.

Recent studies highlight that tidal propagation in shallow waters tends to be modified with mean sea level rise, intensifying coastal threats. This study aims to obtain a comprehensive insight into the changes in tidal regimes in the Bohai Sea of China in response to local mean sea level rise. To achieve this goal the hydrodynamic model ROMS, previously calibrated and validated for Bohai Sea, was applied, considering local sea level rise scenario of 0.5 m, 1 m, 2 m and 3 m in the future, respectively. The model results show that under sea level rise conditions the amplitude of the main semidiurnal (M_2) constituent decreases in most areas except two areas where are around two amphidromic points, and changes in amplitude are not proportional to the level of sea level rise. However, compared results implied that it is necessary to consider the spatial-distribution of sea level-rise when predicting the effects of sea level rise in the future. The tidal wave also was distorted in the coastal zone, the changed wavelength made the amphidromic points move to deep water. Tidal current magnitude decreases with mean sea level rise, which could influence transport of sediment nearby coast of Bohai and Laizhou Bay.

Keywords: Sea level rise, amplitude, amphidromic point, tidal current, spatial distribution.

INTRODUCTION

The Bohai Sea is a semi-enclosed sea with shallow water (Figure 1a), which average water depth is only 18 m, surrounded by cities in the north Chinese Sea. In the Bohai Sea barotropic tides are the dominant circulation processes, having considerable amplitudes as well as momentum and energy fluxes. Because of the effects of enhanced bottom friction associated with shallow shelf topography, tides and tidal residual currents play an important role in affecting many other physical and biogeochemical processes in this region (Wei *et al.*, 2003).

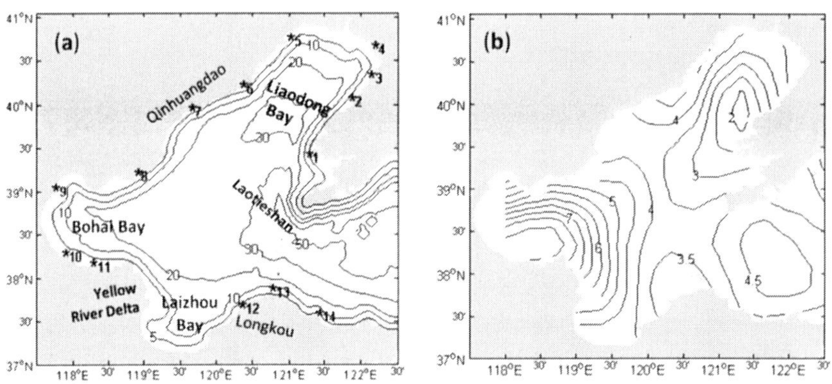

Figure 1.(a) The bathymetry (depth in m) and positions of 14 tide-gauges in the Bohai Sea. Black stars show the location of tidal gauges. (b) The distribution of sea level rising trend (mm/yr) during 1993–2012 in the Bohai Sea

Accelerated economic development and large land reclamation have changed the coastline along the Bohai Sea. In the western Bohai Bay the large land reclamation of about 129.7 km^2 (SOAPRC, 2009), and to the northeast of Laizou Bay an artificial island of 35.2 km^2 was built, which both reconstructed the coastline. Since the shift of the Yellow River mouth has influenced the amphidromic points of M_2 to move (Wang *et al.*, 2014), Huang *et al.* (2015) evaluated the influence of coastline modification in the Bohai Sea based on four different coastlines of 2000, 2008, 2010, and 2012, and found that the changed coastlines could move the amphidromic point and modify tide amplitudes, especially making the amphidromic point near the Yellow River estuary move towards south-east. Therefore it is important to use the newest coastline and accurate topography for tide models in the Bohai Sea.

It is an irrefutable fact that global warming causes sea level to rise. Accelerated sea level rise (SLR) can drastically affect coastal regions, such as submerging and increased flooding of lower-land. Except for these direct effect of increased high water level, on longer term scale SLR could influence tidal system (Pelling *et al.*, 2013a; 2013b; Pickering *et al.*, 2012; Zhang *et al.*, 2013). The basic physical consequence to such changes is the variation in tidal dynamics (Chu-Agor *et al.*, 2011; Muller *et al.*, 2011; Pickering *et al.*, 2012; Poulos *et al.*, 2009; Snoussi *et al.*, 2009; Ward *et al.*, 2012). Some studies have shown that SLR has a direct effect on coastal tides. On the European shelf Pickering *et al.* (2012) found that tidal response of M_2 is nonlinear to future SLR with spatial variation in amplitude. Ward *et al.* (2012) studied the different effects of the way of SLR implemented in the model, flooding could intensify the response caused by SLR (Pelling and Green, 2013), the main difference of tidal response was found in the Bohai Bay and Liaodong Bay between flood and no-flood, but very small difference between flood and partial-flood (Pelling *et al.*, 2013b). Gong *et al.* (2012) studied the effects of sea level rise on Yangtze estuary and found that tidal wave changed slightly in nearshore areas. Zhang *et al.* (2013) studied the tidal system changes due to sea level rise in China, and found that the enhanced amplitude was in the north to amphidromic point, while phase lags were increased to the eastern. The residual currents could be reduced by 20% caused by SLR change in narrow and shallow channels in Portuguese (Valentim *et al.*, 2013). As a major forcing agent SLR could present a substantial influence on coastal erosion worldwide (Cooper and Pilkey, 2004; FitzGerald *et al.*, 2008; Leatherman *et al.*, 2000). To the Yellow River Delta tidal current was one of main factors causing coastal erosion (Ji and Jiang, 1994). All the results showed that future SLR has significant impacts on the tidal amplitudes, currents and associated tidal dissipation.

There have been many results about tide modeling in Chinese Sea (Bao *et al.*, 2001; Fang *et al.*, 2004; Kang *et al.*, 2002; Lefèvre *et al.*, 2000; Song *et al.*, 2013; Wang *et al.*, 2014; Yao *et al.*, 2012; Yu *et al.*, 2003). Some Chinese scientists have studied the effects of SLR on Chinese Sea (Kuang *et al.*, 2014; Yu *et al.*, 2003, 2007), but the conclusions were not consistent, especially the changes of tidal range and phase lags in local Bohai Sea. The influences of SLR on tides depend on local sea level rise and particular coast morphology. In the Bohai Sea from 1993 to 2012 the average satellite altimetry data show that mean sea level had a rising rate of about 3.8 mm/yr in the Bohai Sea, much faster than global average (Church and White, 2006) in Figure 1b. All these conclusions about effects of SLR on coastal tides are most based on a uniform value of SLR, however, there are few studies considering the spatial distribution difference of SLR. To evaluate the potential influence accurately caused by future sea level rise, in this study we use a hydrodynamic model to simulate the tide in Bohai Sea with newest coastline, and according the present sea level rising rate to assess the influence of SLR on tides and residual currents in the Bohai Sea in the future of one hundred years.

This paper is organized as follows: model configuration is described in detail in Section 2, and model results are shown in Section 3, and then followed a discussion and conclusion.

MODEL AND METHODS

Numerical model and set-up

Tides were predicted using the Regional Ocean Modeling System (ROMS) which is a three-dimensional, free-surface, finite difference nonlinear hydrodynamic model, and formulated in a vertical terrain-following sigma coordinate, which was developed by Rutgers and UCLA Universities, USA (Haidvogel *et al.*, 2000; Warner *et al.*, 2008). The model encompassed the whole Bohai Sea from 37°N to 41.05°N meridionally and from 117.5°E to 122.5°E zonally with a grid resolution of 2′ (about 3 km). Bathymetric data for the model was obtained from a combination of 4 digital nautical charts of different scale around this area, which was provided by Navigation Guarantee Department of the Chinese Navy in 2012 (Figure 1a). The depths were raised by an average tidal range of 1.5 m to convert the low-tide datum into a mean sea level, based on the reference mean sea level value provided in each marine chart. The model has one east-open boundary along which the phases and amplitudes of elevation of main astronomical tidal constituents (M_2, S_2, K_1, O_1) were prescribed. The four major tide constituents forcing were derived from the regional ocean model NAO.99b with 5′ resolution (Matsumoto *et al.*, 2000). In the Bohai Sea, with shallow water tidal models are sensitive to bottom friction coefficient, and according to former simulations (Lu and Zhang, 2006; Pelling *et al.*, 2013a; Song *et al.*, 2013; Yao *et al.*, 2012), the drag coefficient in the model was tuned to 1.2×10^{-3} m/s to minimize the simulation errors.

The simulations were set up to enable inter-comparison and therefore all parameters were kept constant with the exception of sea level rise. The first standard experiment represents the present state (case 0), then we run four cases with different value of uniform SLR in the Bohai Sea to access the sensitivity and linearity of tidal response with respect to SLR, the runs indexed with the number 0.5, 1, 2, 3 to represent the increment of SLR scenarios in meter, respectively, and another experiment indexed with '100y' means that spatial-distribution values after the current non-uniform SLR rate lasting for one hundred years.

Since 2011 the altimeter processing upgraded with the improved corrections, there is no adjustment to agree with tide gauge data (Church and White, 2011; Fu and Cazenave, 2001), the processes of sea level variation obtained with TOPEX/Poseidon altimeter and tide gauge data had a remarkably good agreement (Feng *et al.*, 2013; Hu *et al.*, 2014), therefore we used the values calculated from satellite altimeter data (The altimeter products were produced by Ssalto/*Duacs* and distributed by *Aviso*, with support from *Cnes*, http://www.aviso.altimetry.fr/duacs/) to represent spatial-distribution of local sea level rise.

All the simulations use the same start date of the 21th of April 2012, to ensure consistent astronomic forcing. This date was chosen because of the quality of the validation data available for this period. The model was spun up over 10 days and the last 30 days was used for harmonic analysis and to coincide with the period of tidal gauge. The co-tidal charts produced were used both for validation and to evaluate the change in tidal constituents by plotting the difference between those co-amplitude fields.

Tidal energy flux and dissipation

The tidal energy flux was calculated according to (Greenberg 1979):

$$F = \frac{\rho_0}{T}\int_0^T (h + \eta) \cdot [1/2 \times (u^2 + v^2) + g\eta] \cdot (u + iv) \times dt \qquad (1)$$

where ρ_0 is the density of seawater, taken as 1,026 kg/m^3, h is the local depth in meters, η is the sea level fluctuations due to tides, u, v are the depth-averaged tidal currents components from the ROMS solutions and g is the gravitational acceleration, 9.8m/s^2. The integration is averaged over M_2 tidal period T (i.e., 12.42 h).

Mathematically, the tidal energy dissipation can be calculated following the scheme proposed by (Munk, 1997; Taylor, 1919):

$$D = \frac{1}{T}\int_0^T C_d \rho_0 (u^2 + v^2)^{\frac{3}{2}} \, dt \qquad (2)$$

Where C_d is the bottom friction coefficient taken as 0.0012 in our case, the other variables are the same as the above mentioned formula (1) for tidal energy flux. This formula is based on the quadratic bottom stress scheme, which is consistent with our ROMS model configurations.

Model validation

The basic validation was performed to ensure the model was correctly reproducing the observed tide. There were 14 sets of tidal harmonics for validation derived from tidal gauge and published scientific journals (Chen *et al.*, 2011; Li *et al.*, 2011; Wang *et al.*, 2014; Yao *et al.*, 2012; Figure 1a). Because semi-diurnal signal predominated in most parts of the Bohai Sea, we only showed the validation of M_2. Figure 2 showed the accuracy of our simulation. There were two amphidromic points located in Bohai Sea (Figure 2a): one at the offshore area of Qinhuangdao and one near the mouth of the Yellow River, which agreed well with previous publications (Bao *et al.*, 2001; Fang *et al.*, 2004; Yu *et al.*, 2003), and with the new coastline, the changes of Yellow River Delta influenced its surrounding tidal wave and obstructed tidal energy here (Pelling *et al.*, 2013a), the amphidromic point near Old Yellow River Mouth has moved to southeast (Hao *et al.*, 2010). Figure 2b and 2c showed the linear-fit between simulations and observed amplitude and phase lag, respectively, and compared to previous studies (Bao *et al.*, 2001; Song *et al.*, 2013; Zhu and Liu, 2012) the average absolute error of amplitude in this paper is 3.6 cm, and phase error is 7.2°, we also used the variance $VC = 1 - (\frac{R_m}{R_o})^2$ to quantify the model results, where R_o is the root mean squared of observed values and R_m is the root mean squared difference between model and observations (Pelling *et al.*, 2013b). The modeled M_2 amplitudes and phases both achieved a variance captured of 99.4%, which indicated that the errors are well controlled in the whole region, in general the established model of Bohai Sea could capture the tide correctly.

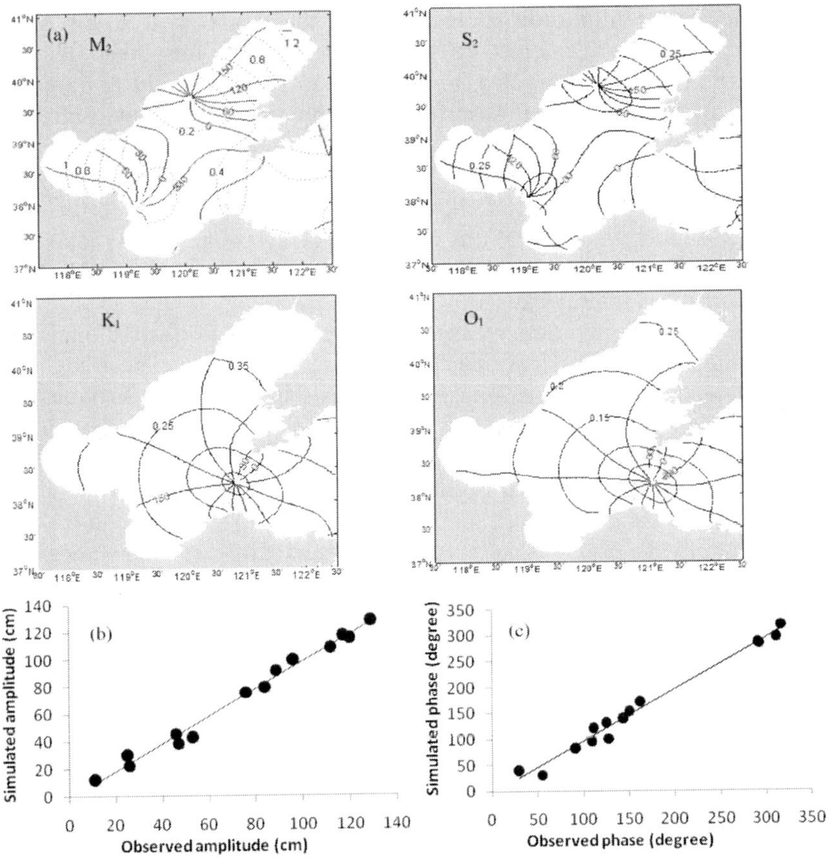

Figure 2. Simulation verification. (a): Cotidal-chart of four main constituents (M_2, S_2, K_1,O_1), dashed line is for amplitude in meter, solid line is phase lag in degree; (b): Comparison between simulated and observed amplitude of M_2; (c): Comparison of phase of M_2

RESULTS

Changes of amplitude and phase

Figure 3 showed the responses of M_2 amplitude and phase to 1 m SLR. When the depth of Bohai Sea increased 1 meter, there exits certain distribution patterns in the differences between the amplitude and phase with different increment of sea level. Figure 3a displays obvious distribution in amplitude change of M_2, there are two areas where amplitude increases (shaded area), one is near the Yellow River Delta and the other in the Liaodong Bay, and the maximum increment of amplitude reached up to 5.5 cm. In the central Bohai basin, Laizhou Bay, Bohai Bay, and northeast of Liaodong Bay, SLR caused amplitude of M_2 to decrease, and the maximum decrement of 12cm appeared in the Laizhou Bay. The spatial distribution pattern of changed amplitude was similar to the result of Pelling *et al.* (2013b)with 2m SLR. However, compared to maximum decrement of 10 cm based on 2m SLR (Pelling *et al.*, 2013b), the discrepancy was mainly caused by flooding. Flooding along the coast could increase energy dissipation, weaken the propagation of reflected tidal wave. The changes of M_2 phase showed a notable distribution. There are some regulations in changes of co-phases, it seems that on the western part near Qinhuangdao (site 7, marked in Figure 1a), co-tidal phase decreased anti-clockwise, while on the right phase lag increased clockwise (Figure 3b); and the same situation appeared in the Yellow River Delta, which was in most agreement with the theoretical model results (Yu *et al.*, 2003), but inconsistent with previous conclusions (Yu *et al.*, 2007), which showed that co-tidal phase changed in a same anti-clockwise direction.

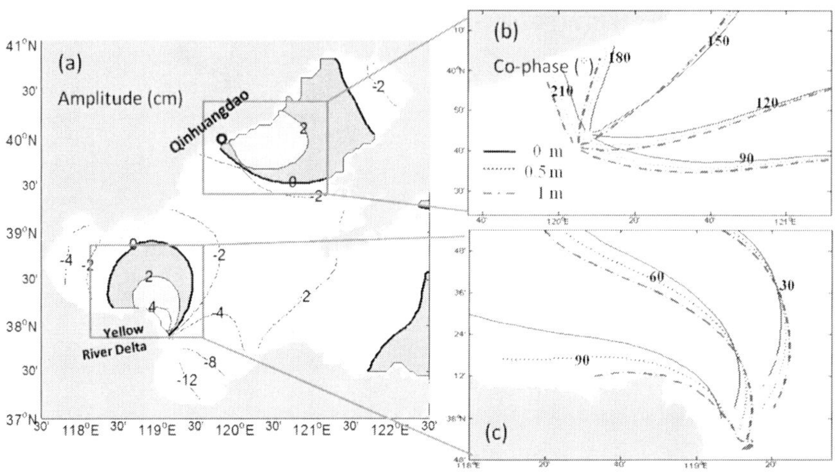

Figure 3.(a) Changes of amplitude of M_2 with SLR 1 meter (unit, cm), the shaded area are where amplitude increases. (b) Shift of co-phase lines near Qinhuangdao Island (black circle) due to different value of SLR and (c) changes near Yellow River Delta (unit: degree), black solid line is the present state, dashed line is SLR 0.5 m, and dot-line is SLR 1 m

Obviously SLR has changed the characteristics of tide, it can also influence the amphidromic point positions of tidal waves shift relatively to those of current tidal waves. In the shallow water, tidal wave propagated more quickly and wavelength increased with SLR increase, which would shift the amphidromic points. Table 1 showed the movements of amphidromic points under different SLR conditions. It was clear to see the migration of two amphidromic points, one amphidromic point near Yellow River Mouth moved to southeast due to SLR (Yao *et al.*, 2012; Yu *et al.*, 2007), and the other near Qinhuangdao moved to southwest (also shown in Figure 3b), which would change tidal phase lag.

Table 1. The location of amphidromic point of M_2 constituent under different SLR conditions

Location	Case	SLR (m)	Latitude (°)	Longitude (°)
Yellow River Mouth	0	0	37.975	119.15
	0.5	0.5 m	37.95	119.25
	1	1m	37.95	119.30
	2	2 m	37.80	119.40
	3	3m	37.60	119.80
	100y	<0.8 m	37.75	119.55
Qinhuangdao	0	0	39.775	120.15
	0.5	0.5 m	39.7	120.1
	1	1m	39.625	120.08
	2	2 m	39.55	120
	3	3m	39.5	119.95
	100y	<0.8 m	39.6	120.05

Note: '0' is the control run, which represents the present tidal state in the Bohai Sea. The number means the value of uniform sea level rise. '100y' represents scenario after one hundred years with the present sea level rising rate.

Change of tidal energy

According to formula (1) we calculated the tidal energy flux distribution in Bohai Sea in different SLR scenario. The simulated current state (no SLR) of M_2 tidal energy flux (Figure 4a) reached a good agreement with results from Song *et al.* (2013) and Yao *et al.* (2012). The tidal energy flux came into the Bohai Sea along the 50 m isobaths through the Bohai Strait, which broke into two branches: one branch turns northward into the Liaodong Bay and the remaining moves westward into the Bohai Sea interior (Zhu and Liu, 2012). And in the Bohai Sea, the maximum tidal energy appeared in the Laotieshan Channel, less than 5 GW, far less than that of Yellow Sea (Yao *et al.*, 2012). The mean tidal energy dissipation was also calculated using formula (2) and analyzed to study the changes of tidal energy dissipation in different SLR scenario. The spatial pattern of mean tidal energy dissipation of control run (0 m SLR) was also consistent with previous studies of tidal energy (Kang *et al.*, 2002; Song *et al.*, 2013; Yao *et al.*, 2012), with the largest energy dissipation occurred near coast, especially where coastline was irregular (not shown). Specifically, three sections were chosen to evaluate the effects of SLR on tidal energy (Figure 4b), it was clear to see that with SLR tidal energy entered into the Bohai Sea was decreased by increased energy dissipation, the tidal energy that entered into the Bohai Bay and Laizhou Bay also decreased. At the same condition of SLR increment, tidal energy through the Strait decreased more than those entering into the bays, which implied that with the propagation of tidal wave, the SLR effects on tidal energy decreased gradually.

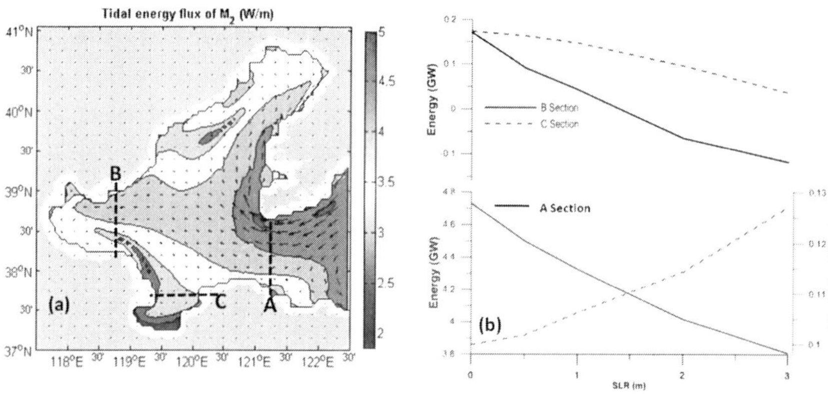

Figure 4.(a) Tidal energy flux of M_2 and energy flux vectors are plotted every three grid points (unit: W $m^{-1,}$ shown as logarithmic values base on 10). (b) Changes of tidal energy in three sections with different SLR, the positive values are energy flowing into Bohai Sea (blue solid line) or Bohai Bay (solid line) and Laizhou Bay (dashed line) with unit of GW, and tidal energy dissipation rate in A section (red dashed line, unit: W/$m^{2)}$

Change of tidal current

The characteristics of tidal currents can be depicted by the depth-averaged tidal current ellipses (Figure 5a). In the Bohai Sea, the strongest tidal currents are seen in the Bohai Bay with a maximum speed of about 0.6 m s^{-1} in west-east reversing flow. The weakest current occurs in the Laizhou Bay with a speed of only 0.1 ~ 0.2 m s^{-1}. The tidal currents in the center of Bohai Sea and Laizhou Bay behave as clockwise rotating currents rather than reversing ones appeared in Liaodong and Bohai Bays. And the tidal residual current has been recognized to play an important role influencing long-term sediment and other materials transport (Li *et al.*, 2005; Wei *et al.*, 2004). The mean residual currents in the interior of Bohai Sea are rather small, only of the order of a few cm $s^{-1,}$ and larger residual currents appeared near coast, no more than 5 cm s^{-1}. In the Bohai Bay there was a weak, cyclonic gyre, and there was an anticyclone gyre southwest to

Liaodong Peninsula, in the Bohai Strait, the mean tidal residual current flowed out the strait in the southern part and came into Bohai Sea in the northern (not shown). The simulation captured main characteristics of tidal currents.

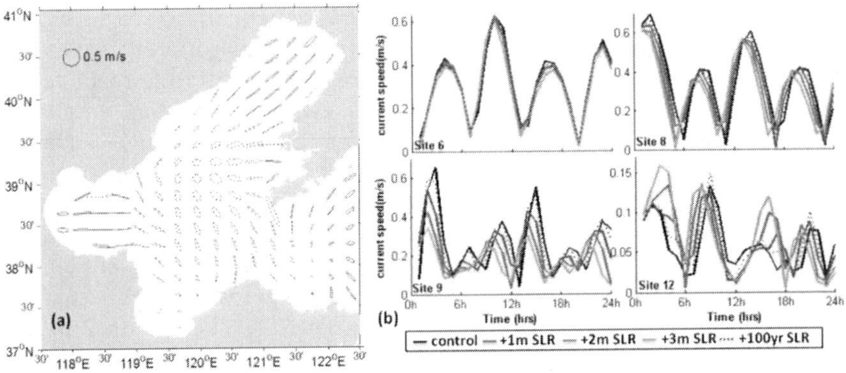

Figure 5.(a) Detail distribution of tidal current ellipsefor M_2, tidal ellipses are plotted every 6 grid points. The dashed (solid) ellipse shows the tidal ellipses rotate clockwise (counter-clockwise). (b) Tidal speed curves (m/s) from different SLR scenario at 4 ports. The five sea level scenarios are shown 0 m (black line), 1 m (red line), 2 m (mauve line), 3 m (green line), and 100 years (blue dashed line). Note that the ports are marked in Figure 1a

The changes in tidal energy act to influence tidal current as well. Four sites (where there was large human activity) were chosen to evaluate the changes caused by SLR (Figure 5b). It was clear to see that changes of tidal current was different in areas, in the Bohai Bay (site 8 and 9), influenced by sea level rise, tidal current speed was weakened, and there was a minor shift in time of maximum tidal current. In the Laizhou Bay (site 12) tidal current increased influenced by sea level-rise, and with SLR 3m scenario, tidal current could increased 5 cm/s, and in site 6 which located near Liaodong Bay, the effect of SLR on tidal current was small with little change in current speed.

DISCUSSION

Recently quick coastline changes caused by natural process and anthropogenic activity had influenced tidal system in the coastal zone

(Huang *et al.*, 2015; Pelling *et al.*, 2013c; Wang *et al.*, 2014). Compared to coastline changes SLR is a relative slow and imperceptible process, however, the long-time affection of sea level rise can not only increase the risk of extreme water level but also influence tidal dynamic system. As coastline change was mainly induced by anthropogenic reclamation, recent coastline was kept unchanged for future projection when evaluating the affection induced by SLR. In the shallow water equation, sea level rise increased water depth, which led to intensify the propagation of waves and modify the reflected and incident waves. The tidal amplitude responds to SLR in a spatially non-uniform manner, with substantial amplitude decreases and increases in different SLR scenarios. To get details we choose four sites (marked in Figure 1a with 6, 8, 9 and 12, respectively) to compare the effects of different SLR scenario (Figure 6). Detailed examinations show that the amplitude of M_2 responded appropriately linearly to uniform SLR (Figure 6a), but was not proportional to SLR. Influenced by 0.5 m SLR, the maximum of increased (decreased) M_2amplitude was 3 (8) cm, the maximum increment changed to 5 cm with 1m SLR, 11 cm with 2 m SLR scenario, and 15 cm under 3 m SLR condition. However, with the combination effects of water depth and bottom friction, the tidal energy dissipation (Figure 6b) showed a nonlinear response to different increment of sea level, which is not consistent with amplitude changes.

The way of SLR implemented in the models could lead to qualitatively different results (Pelling *et al.*, 2013a; Pelling and Green, 2013; Ward *et al.*, 2012). Pickering *et al.* (2012) using a drying and flooding cell to study the effects of SLR with 2m and 10m on European shelf, found that the tidal amplitude effects was complex and non-linear to SLR. Most previous results used uniform SLR to study and predict its effects on coast in the future, neglected the spatial pattern of SLR in the domain. It was clearly to see that the effects of uniform SLR were much more different compared to effects of spatial pattern SLR (Figure 6a and b, marked with filled symbols). As Figure 1b showed, maximum rising trend of sea level was in Bohai Bay, from 5 mm/yr to 8 mm/yr, much larger than other areas of Bohai Sea. We assumed that the sea level trend keep constant, in the future after one hundred years the mean sea level in the Bohai Sea increased only 0.2 ~ 0.8 m and distributed spatially, simulation result showed that the effect was stronger than that of uniform SLR with 1 meter, even equivalent to effects of SLR with 2 meter, only considering amplitude changes (Figure 6a); and also in the migration of amphidromic points (Table 1), in the future the movement caused by spatial-distribution of SLR variation was farther than that caused by uniform SLR. Because tides in the Bohai Sea was forced by ocean tides from the Pacific, the wave frequency remains constant, as depth increased with the SLR the tidal wave propagates more quickly in the domain. With the spatial distribution of sea level-rise, tidal wave propagated into the Bohai Sea from adjoining sea, influenced by local changed depth, finally the accumulated results led to a complex and non-linear response to SLR (Figure 6c).

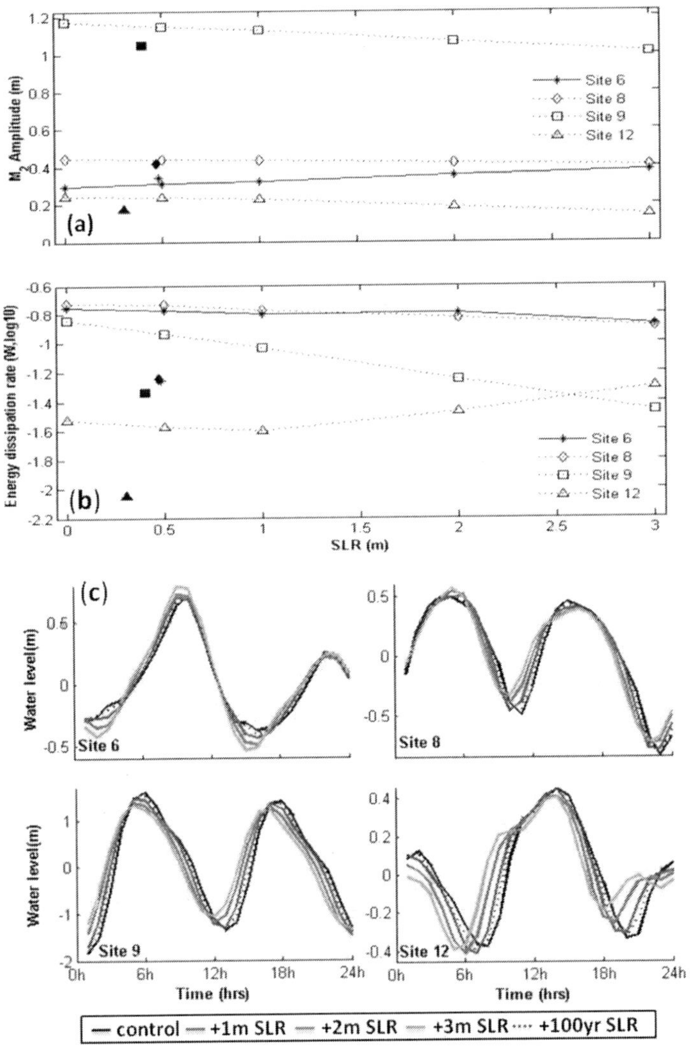

Figure 6.(a) M_2 amplitude response curves for station 6, 8, 9 and 12 (positions see Figure 1a); (b) The tidal energy dissipation at these four stations (shown as logarithmic values base on 10). The filled signals in the above two panels are values influenced by spatial distribution of SLR after one hundred years. (c) Water level curves (m) from different SLR scenario at 4 ports. The five sea level scenarios are shown 0 m (black line), 1 m (red line), 2 m (mauve line), 3 m (green line), and 100 years (blue dashed line)

The changes caused by sea level-rise act to change tidal currents as well. As Figure 7a showed, there are three domains that residual currents changed dramatically, though maximum variance values are of the order of a few cm s^{-1}. The distribution of residual current change was similar to that of tidal energy change (not shown), which illustrated that the increment of tidal energy increased residual current. Compared Figure 7a and 7b, SLR with 1 meter mainly strengthened the effects caused by SLR 0.5 m, however, compared to the results between uniform SLR with 0.5m and spatial distrubiton-SLR after one hundred years (Figure 7a and 7c), the effects were intensified by spatial distribution of sea level rise, especially in Bohai Bay, where quicker rising rate of sea level increased local water level much more compared to other areas (Figure 1b), the local increased water depth would reduce incident waves and change reflected waves, which led to the increment of the residual current. Considering the spatial distribution of sea level variation, the influence of SLR was intenser than that of uniform variation. The migration of amphidromic points also showed the non-linear response between experiment case 0.5 and case100y (Table 1). As Zhou *et al.* (2015) confirmed that tidal current flow is indeed the main agent for transporting the fine sediment to the mud patch in the central South Yellow Sea, the changed tidal and residual currents caused by SLR were also related to the long-term transport of sediment in the Bohai Sea (Ji and Jiang, 1994; Li *et al.*, 2005; Wei *et al.*, 2004), for example near the Yellow River Delta weakened residual current may decrease the sediment transport outside and accumulate more sediment in the Delta. And in the Bohai Bay SLR decreased the residual currents, would increase the risk of port siltation.

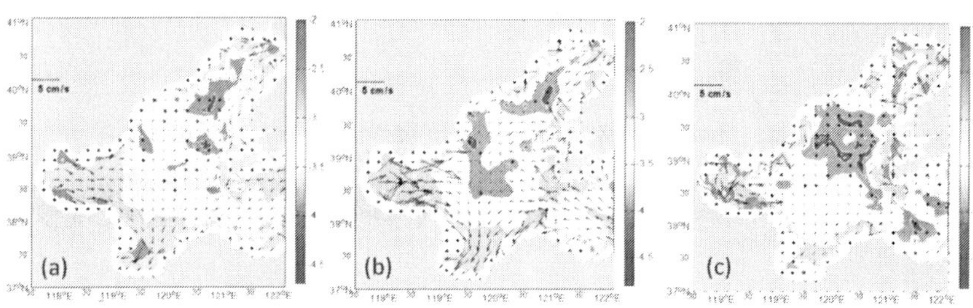

Figure 7.(a) Changes of tidal residual current between control run and uniform 0.5 m SLR run. (b) Changes of tidal residual current between control run and uniform 1 m SLR run. (c) Difference of effects from spatial distribution SLR run after one hundred years and uniform SLR 0.5 m. Values of both images are scaled by $\log_{10}$ with unit of m s^{-1}

CONCLUSION

The influence of sea level rise on tides in the Bohai Sea was assessed by a series of numerical experiments using ROMS model. The scenario of spatial distribution of SLR case was projected in the future keeping the present sea level rising trend derived from altimeter data after one hundred years, and the results were compared with other cases under different SLR conditions of uniform increment of water level from 0.5 m to 3 m.

The modeled results show that with synchronous increment of sea level, changes of M_2 amplitude responded to SLR non-proportionally, and showed a non-linear response. The M_2 amplitudes increased only around the two amphidromic points influenced by sea level rise. Due to the shape of coastline (headland etc.), changed bottom friction affected tidal energy flux and moved the amphidromic points to deep water under sea level rise conditions. Rising sea level increased water depth, combined local effect of sea level, it indeed reduced/increased tidal currents and residual currents, especially along coasts, which would change long-term sediment transport and influence the coastal erosion (Li *et al.*, 2005; Wei *et al.*, 2004). Simulated results showed that the response of tide regime was complex to spatial distribution of sea level rise and stronger than that from uniform rise of sea level, which illustrated the importance of considering spatial distribution of sea level rising when evaluating the risk of SLR in the future.

ACKNOWLEDGMENTS

This study was supported by the Strategic Priority Research Program of the Chinese Academy of Sciences (grant no. XDA11020305) and the Key Deployment Project of Chinese Academy of Sciences (grant no. KZZD-EW-14). The altimeter products were produced by Ssalto/Duacs and distributed by Aviso, with support from Cnes (http://www.aviso.altimetry.fr/duacs/). We thank Dr. Jan Harff for encouragement and support.

LITERATURE CITED

1. Bao, X.; Gao, G., and Ju, Y., 2001. Three dimensional simulation of tide and tidal current characteristics in the *East China Sea. Oceanologica Acta*, 24 (2), 135–149.
2. Chen, H.; Jiang, S., and Wang, F., 2011. Application of ADCIRC in M2 tide modeling in the Bohai Sea. *Chinese Marine Forcasts*, 28(4), 70–75. (in Chinese with English abstract)
3. Chu-Agor, M. L.; Munoz-Carpena, R.; Kiker, G.; Emanuelsson, A., and Linkov, I., 2011. Exploring vulnerability of coastal habitats to sea level rise through global

sensitivity and uncertainty analyses. *Environmental Modeling & Software*, 26(5), 593–604.
4. Church, J. A. and White, N. J., 2006. A 20th century acceleration in global sea-level rise. *Geophysical Research Letters*, 33, L01602.
5. Church, J. A. and White, N. J., 2011. Sea-level rise from the late 19th to the early 21st century. *Surveys in Geophysics*, 32 (4), 585–602.
6. Cooper, J. A. G. and Pilkey, O. H., 2004. Sea-level rise and shoreline retreat: time to abandon the Bruun Rule. *Global and Planetary Change*, 43, 157–171.
7. Fang, G.; Wang, Y.; Wei, Z.; Choi, B. H.; Wang, X. and Wang, J., 2004. Empirical cotidal charts of the Bohai, Yellow, and East China Seas from 10 years of TOPEX/Poseidon altimetry. *Journal of Geophysical Research*, 109, C11006,
8. Feng, G.; Jin, S., and Zhang, T., 2013. Coastal sea level changes in Europe from GPS, tide gauge, satellite altimetry and GRACE, 1993–2011. *Advances in Space Research*, 51(6), 1019–1029.
9. FitzGerald, D. M.; Fenster, M. S.; Argow, B. A., and Buynevich, I. V., 2008. Coastal impacts due to sea-level rise. *Annual Review of Earth and Planetary Sciences*, 36, 601–647.
10. Fu, L. L. and Cazenave, A. (Eds.), 2001. Satellite Altimetry and Earth Sciences: A Handbook of Techniques and Applications. *Academic Press, International Geophysics Series*.
11. Gong, Z.; Zhang, C.; Wan, L., and Zuo J., 2012. Tidal level response to sea-level rise in the yangtze estuary. *China Ocean Engineering*, 26(1), 109–122.
12. Greenberg, D. A., 1979. A numerical model investigation of tidal phenomena in the Bay of Fundy and Gulf of Maine. *Marine Geodesy*, 2(2), 161–187.
13. Hao, Y.; Le, K. T., and Liu, X. Q., 2010. A numerical prediction of the tidal characteristics in 2010 of Yellow River Delta. *Chinese Marine Science*, 24(6), 43–46.
14. Haidvogel, D. B.; Arango, H. G.; Hedstrom, K.; Beckmann, A.; Malanotte-Rizzoli, P., and Shchepetkin, A. F., 2000. Model evaluation experiments in the North Atlantic Basin: Simulations in nonlinear terrain-following coordinates. *Dynamics of Atmospheres and Oceans*, 32, 239–281.
15. Hu, Z.; Guo, J.; Tan, Z., and Chang, X., 2014. Sea level variation in Hong Kong determined with Topex/Poseidon and tide gauge. *Journal of Geodesy and Geodynamics*, 34(4), 56, 60.
16. Huang, J.; Xu, J.; Gao, S.; Lian, X., and Li, J., 2015. Analysis of influence on the Bohai Sea tidal system induced by coastline modification. *Journal of Coastal Research*, 73 (sp1), 359–363.
17. Ji, Z. and Jiang, Z., 1994. Impacts of sea level rise on coastal erosion in the Changjiang river delta and north Jiangsu coastal plain. *Chinese Geographical Science*, 4(4), 310–321.
18. Kang, S. K.; Foreman, M. G. G.; Lie, H. J.; Lee, J. H.; Cherniawsky, J., and Yum, K. D., 2002. Two-layer tidal modeling of the Yellow and East China Seas with application to seasonal variability of the M2 tide. *Journal of Geophysical Research*, 107(*C*3).
19. Kuang, C. P.; Chen, W.; Gu, J.; Zhu, D. Z.; He, L. L., and Huang, H. C., 2014. Numerical Assessment of the Impacts of Potential Future Sea-Level Rise on Hydrodynamics of the Yangtze River Estuary, China. *Journal of Coastal Research*, 30(3), 586–597.
20. Leatherman, S. P.; Zhang, K., and Douglas, B. C., 2000. Sea-level rise shown to drive coastal erosion. *EOS*, 81 (6), 55–57.
21. Lefèvre, F.; Le Provost, C., and Lyard, F. H., 2000. How can we improve a global ocean tide model at a regional scale? A test on the Yellow Sea and the East China Sea. *Journal of Geophysical Research*, 105(C4), 8707–8725.
22. Li, G. S.; Wang, H. L., and Li, B. L., 2005. A model study on seasonal spatial-temporal variability of the Lagrangian residual circulations in the Bohai Sea. *Journal of Geographical Sciences*, 15(3), 273–285.
23. Li, X.; Sun, X.; Wang, S.; Ye, F.; Li, Y., and Li, X., 2011. Characteristic analysis of Tianjin offshore tide. *Marine Science Bulletin*, 13, 40–49.

24. Lu, X. and Zhang, J., 2006. Numerical study on spatially varying bottom friction coefficient of a 2D tidal model with adjoint method. *Continental Shelf Research*, 26(16), 1905–1923.

25. Matsumoto, K.; Takashi T., and Masatsugu, O., 2000. Ocean tide models developed by assimilating Topex/Poseidon altimeter data into hydrodynamical model: A global model and a regional model around Japan. *Journal of Oceanography*, 56, 567–581.

26. Müller, M.; Arbic, B. K., and Mitrovica, J. X., 2011. Secular trends in ocean tides: Observations and model results. *Journal of Geophysical research*, 116, C05013,

27. Munk, W., 1997. Once again: once again-tidal friction. *Progress in Oceanography*, 40 (1–4), 7–35.

28. Pelling, H. E. and Green, J. A. M., 2013. Sea level rise and tidal power plants in the Gulf of Maine. *Journal of Geophysical Research Oceans*, 118, 2863–2873,

29. Pelling, H. E.; Green, J. A. M., and Ward, S. L., 2013a. Modeling tides and sea level rise: to flood or not to flood. *Ocean Modeling*, 63, 21–29.

30. Pelling, H. E.; Uehara, K., and Green, J. A. M., 2013b. The impact of rapid coastline changes and sea level rise on the tides in the Bohai Sea, China. *Journal of Geophysical Research: Oceans*, 118(7), 3462–3472.

31. Pickering, M. D.; Wells, N. C.; Horsburgh, K. J., and Green, J. A. M., 2012. The impact of future sea-level rise on the European Shelf tides. *Continental Shelf Research*, 35, 1–15.

32. Poulos, S. E.; Ghionis, G., and Maroukian, H., 2009. The consequences of a futureecstatic sea-level rise on the deltaic coasts of Inner Thermaikos Gulf (Aegean Sea) and Kyparissiakos Gulf (Ionian Sea), Greece. *Geomorphology*, 107(1–2), 18–24.

33. Snoussi, M.; Ouchani, T.; Khouakhi, A., and Niang–Diop, I., 2009. Impacts of sea-level rise on the Moroccan coastal zone: Quantifying coastal erosion and flooding in the Tangier Bay. *Geomorphology*, 107(1–2), 32–40.

34. Song, D. H.; Wang, H. X.; Zhu, X. M., and Bao, X. W., 2013. Modeling studies of the far-field effects of tidal flat reclamation on tidal dynamics in the East China Seas. Estuarine, *Coastal and Shelf Science*, 133, 147–160.

35. SOAPRC (State Oceanic Administration People's Republic of China), 2009. Communiqué on Marine Environmental Quality of Bohai Sea 2008 (in Chinese). SOAPRC (State Oceanic Administration People's Republic of China).

36. Taylor, G. I., 1919. Tidal friction in the Irish Sea. *Proc R Soc London Ser A* 96(678), 330.

37. Valentim, J. M.; Vaz, L.; Vaz, N.; Silva, H.; Duarte, B.; Cacador, I., and Dias, J. M., 2013. Sea level rise impact in residual circulation in Tagus estuary and Ria de Aveiro lagoon. *Journal of Coastal Research*, 65, 1981–1986.

38. Wang, Y.; Wei, Z.; Fang, G.; Chen, H., and Gao, X., 2014. A numerical study on the effect of changes in water depth and coastline on M_2 tidal component near the Yellow River estuary. *Chinese Advances in Marine Science*, 32 (2), 141–147. (in Chinese with English abstract)

39. Ward, S. L.; Green, J. A. M., and Pelling, H. E., 2012. Tides, sea-level rise and tidal power extraction on the European Shelf. *Ocean Dynamics*, 62, 1153–1167, Warner, J. C.; Sherwood, C. R.; Signell, R. P.; Harris, C. K., and Arango, H. G., 2008. Development of a three-dimensional, regional, coupled wave, current, and sediment-transport model. *Computers & Geosciences*, 34, 284–1306.

40. Wei, H.; Hainbucher, D.; Pohlmann, T.; Feng, S. Z., and Suendermann, J., 2004. Tidal-induced Lagrangian and Eulerian mean circulation in the Bohai Sea. *Journal of Marine Systems*, 44(3–4), 141–151.

41. Wei, H.; Su, J.; Wan, R.; Wang, L. and Lin, Y., 2003. Tidal front and the convergence of anchovy (Engraulis japonicus) eggs in the Yellow Sea. *Fish Oceanography*, 12 (45), 434–442.

42. Yao, Z.; He, R.; Bao, X.; Wu, D., and Song, J., 2012. M_2 tidal dynamics in Bohai and Yellow Seas: a hybrid data assimilative modeling study. *Ocean Dynamics*, 62 (5), 753–769.

43. Yu, Y.; Yu, Y., and Zuo, J., 2003. Effect of sea level variation on tidal characteristic values for the East China Sea. *China Ocean Engineering*, 17, 369–382.
44. Yu, Y.; Liu, L., and Guo, M. K., 2007. Numerical research on tidal waves changes due to mean sea level rise in the Bohai Sea, the Huanghai Sea and the East China Sea II: Numerical modeling of tidal waves after mean sea level rise in the areas. *Periodical of Ocean University of China*, 37(1), 7–14. (in Chinese with English abstract)
45. Zhang, W.; Zhang, J.; Lin, R., and Zong, H., 2013. Tidal response of sea level rise in marginal seas near China. *Advance in water science*, 24, 243–250. (in Chinese with English abstract)
46. Zhou, C. Y.; Dong, P., and Li, G. X., 2015. Hydrodynamic processes and their impacts on the mud deposit in the Southern Yellow Sea. *Marine Geology*, 360, 1–16.
47. Zhu, X. and Liu, G., 2012. Numerical study on the tidal currents, tidal energy fluxes and dissipation in the Chinese Sea. *Oceanologia et Limnologia Sinica*, 43(3), 669–677. (in Chinese with English abstract)

CHAPTER 11

Coastal Ecosystem Effects of Increased Summer Temperature and Contamination by the Flame Retardant HBCDD

Clare Bradshaw [1,2,*], Anna-Lea Golz [1] and Kerstin Gustafsson [2]

[1]*Department of Ecology, Environment and Plant Sciences, Stockholm University, 10691 Stockholm, Sweden*

[2]*School of Natural Sciences, Technology and Environmental Studies, Södertörn University, Alfred Nobels allé 7, SE-141 89 Huddinge, Sweden*

ABSTRACT

The combined effects of ocean warming and contaminants on marine ecosystems are poorly understood. In this study, we exposed model ecosystems comprising typical shallow coastal Baltic Sea communities to elevated temperature (+5 °C) and the flame retardant hexabromocyclododecane (HBCDD), both singly and in combination, for 13 days. Higher temperatures caused the release of PO_4 from the sediment, which in turn stimulated the growth of the cyanobacteria *Dolichospermum* sp. This in turn led to an increase in the copepod *Acartia bifilosa* and other indirect effects in the plankton, interpreted as being caused by changes in predation, grazing, and competition. Elevated temperatures also stimulated benthic primary production and increased production of benthic mollusk larvae. Although increased temperature was the dominant driver of effects in these systems, HBCDD also appeared to have some effects, mainly in the zooplankton (both direct and indirect effects) and benthic meiofauna (an interactive effect with temperature). Although the study used model ecosystems, which are an approximation of field conditions, it highlights that interactive ecosystem effects between two stressors are possible and demonstrates the ecological and temporal complexity of such responses. Such unpredictable responses to warming and contaminants are a major challenge for ecosystem management to deal with multistressor situations in the Baltic Sea.

Keywords: global warming; flame retardant; Baltic Sea; brackish; multiple stressors; benthic-pelagic coupling; community ecology; indirect effects; *Macoma balthica*; cosm

1. INTRODUCTION

Marine ecosystems are exposed to a wide range of anthropogenic stressors, such as eutrophication, global warming, overfishing, and chemical pollution. Substantial research has been carried out on the threats of each of these stressors on marine ecosystems, but the potential effects of these threats in combination are understudied and highly uncertain. There are strong possibilities for unforeseen interactive effects between multiple stressors [1], non-linear responses, feedback loops, and overall erosion of ecosystem resilience [2]. During the last decade there has been an increased awareness of the potential interactions between multiple stressors caused by climate change (e.g., [3]) and specifically between climate change and contaminants [4,5,6], but knowledge is lacking, particularly at an ecosystem level.

Increased temperature may alter ecosystem effects of contaminants in a number of ways. Physicochemical properties of the contaminant, and thus contaminants' environmental behavior, bioavailability, and toxicity, may be changed. The rates and efficiency of metabolic processes (including homeostasis, growth, contaminant uptake, and elimination) in organisms may also be altered [4,5,6]. Increased temperature may alter biogeochemical cycles, further impacting marine species through changes in, e.g., nutrient availability. Both factors may cause physiological stress but relative sensitivities of species differ, so exposure to these may change the relative fitness of species and indirectly alter the ecosystem or food web composition [7], in turn influencing exposure pathways. Species living at the edge of their range of physiological tolerance, such as many Baltic Sea species, may be particularly susceptible to such interactive effects [4,6].

Global warming and chemical pollution are recognized as two of the major threats to Baltic Sea ecosystems in general [8] and also to Baltic Sea red-listed biotopes [9]. In the Baltic Sea, climate models have predicted increased water temperatures as a result of global warming. The exact increase is of course difficult to predict, and will vary spatially and with season, but estimates are generally in the range of 2–5 °C by the end of the century [10,11,12]. In all sub-basins, the surface layer is projected to warm more than the deep-water [12]. It should, however, be noted that although these models have a good resolution (down to 10–50 m; [11]), these models are at best regional rather than local and shallow coastal bays could be expected to have more extreme temperature fluctuations. The Helsinki Commission (HELCOM) [13] identified temperature as an anthropogenic pressure likely to affect several Baltic Sea "core indicators" such as zooplankton size and abundance and a number of benthic indicators. Temperature is acknowledged by HELCOM to have potentially large impacts on the abundance and distributions of species in Baltic ecosystems, but potential interactions with contaminant pollution are not mentioned.

The Baltic Sea also experiences a number of other environmental stresses. Its limited water exchange with the North Sea and short geological history have resulted in a permanent north-south salinity gradient (3–15 psu) and species-poor flora and fauna of both marine and freshwater origins. Parts of the Baltic coastline are spatially complex, with multiple islands (i.e., archipelagos) or a large number of shallow inlets, lagoons, and shallow soft bottom ecosystems (e.g., [14]). Coastal areas such as these are often areas with high levels of anthropogenic pressures, and several of these types of biotopes have been identified as "vulnerable" or "near-threatened" in a recent evaluation by HELCOM [9]. In addition, the Baltic Sea has a large and heavily populated drainage basin which, in combination with the slow water exchange rate, leads to the accumulation of nutrients and contaminants. These include legacy contaminants such as polychlorinated biphenyls (PCBs), dioxins, dichlorodiphenyltrichloroethane (DDT), and metals, but also emerging contaminants such as pharmaceuticals, perfluoroalkyl substances, and flame retardants. Increasing our currently poor understanding of the potential effects of these emerging contaminants is essential for making robust assessments of environmental quality and regulatory decisions.

In the Baltic Sea, HELCOM's list of eleven substances of "specific concern" includes HBCDD (hexabromocyclododecane) and it is one of the substances in the set of HELCOM core indicators for hazardous substances [9], which are used to assess the state of the Baltic marine environment and the EU Marine Strategy Framework Directive (MSFD) qualitative descriptors ("contaminants in environment"). HBCDD is a brominated flame retardant (BFR) used widely in building materials, polystyrene, textiles, and electronics. It is used additively, i.e., it is not chemically bound in the matrix, and is thus gradually released from the products to which it has been added or from waste. It may therefore enter the marine environment directly, via atmospheric fallout or runoff from land. It is classified as a Persistent, Bioaccumulative, and Toxic (PBT) substance and a "Substance of very high concern" [15]. In 2013, HBCDD was added to Annex A of the Stockholm Convention on Persistent Organic Pollutants (POPs), a global treaty to protect human health and the environment from environmentally persistent chemicals by for example reducing production, import and export, use, and release to the environment. HBCDD is also listed in Annex XIV list of the European Union Regulation REACH (Registration, Evaluation, Authorisation and Restriction of Chemicals).

The main source of HBCDD to the Baltic is "leakage from products (e.g., insulation, electrical products, and textiles) via rivers, atmosphere, or wastewater treatment plants" [9]. HBCDD concentrations are measured in a range of species as part of the Swedish national monitoring program and temporal trends vary spatially and temporally and with the organism measured [16]. For example: concentrations in herring muscle and cod liver are generally higher in the Baltic Proper than on the Swedish West coast; concentrations in guillemot eggs have been decreasing over the past decade from their peak in the mid-2000s, but steadily increasing since 1980 in cod liver from SE Gotland. Concentrations may also vary widely between samples from the same species, site, and time point [16].

The effects of HBCDD used in its classification as a PBT were determined using standard ecotoxicological tests (for a summary see [15]). For aquatic species, reduced size and reproductive output was seen in *Daphnia magna* (21-day exposure, Lowest Observed Effect Concentration (LOEC) 5.6 μg L^{-1}), and reduced growth rates were seen in three microalgae (72-h 50% Effect Concentration (EC_{50}) c. 40–50 μg L^{-1}). No effects were seen in growth of the amphipod *Hyalella azteca* at concentrations up to 1000 mg·$kgdw^{-1}$ but numbers of the worm *Lumbriculus variegatus* were reduced at 50 mg·$kgdw^{-1}$. No measureable effect was seen on eggs, larvae, or adult rainbow trout at concentrations up to 6.8 μg L^{-1}. Assessments are often complicated by the low solubility of HBCDD and the different solubilities and toxicities of the different diastereomers present in the technical mixture of HBCDD. A few experiments have also been done using Baltic species, applying HBCDD in a more realistic manner, associated to suspended organic matter; Smolarz and Berger reported cytotoxic and genotoxic effects (significant increases in nuclear and nucleolar abnormalities and frequency of dead cells) of HBCDD in the bivalve *Limecola balthica* [17]. Pirzadeh et al. suggested that copepods may be a particularly sensitive part of the Baltic plankton [18]. Bradshaw et al. used an 8-month mesocosm study to demonstrate that HBCDD caused effects on benthic population structure and ecosystem function and possible changes in benthic-pelagic coupling of nutrients with knock-on effects in the plankton [19].

The main aim of this study was to determine if increased temperature could modify the ecological response of a shallow coastal Baltic ecosystem to a contaminant, HBCDD. This was done with an experiment using model ecosystems ("cosms") to approximate the natural environment whilst maintaining control over experimental conditions. The cosms were exposed to the two stressors either singly or in combination. The study found that increased temperature had the largest effect, mainly through stimulating the release of P from the sediment into the water column. HBCDD had a less dominant, though detectable effect in some parts of the ecosystem.

2. MATERIALS AND METHODS

2.1. General Experimental Design

This paper presents the results of a two-week experiment using 5 L model ecosystems (cosms) constructed from water, sediment, and naturally occurring species at relevant densities, from a coastal bay in the Baltic Proper. A crossed two-factorial design was used, with the factors of temperature (ambient and warm (+5 °C)) and contaminant (with or without HBCDD). The experiment was designed to be as environmentally relevant as possible, within the constraints of working with cosms. Thus, a concentration of HBCDD was used that may be found in the environment and at which effects have previously been seen in similar systems. The exposure pathway was spiked dead phytoplankton, such as might occur after a spring bloom settles out to the seabed. In addition, outdoor systems were used that reflected natural summer temperatures in a shallow coastal bay, as well as a realistic increased temperature (+5 °C) that might be expected with global warming in the Baltic region (see Section 1). An array of relevant structural and functional endpoints was measured in both the pelagic and benthic parts of the ecosystem.

2.2. Collection of Components of the Experimental System

Sediment, water, and organisms (the benthic bivalves *L. balthica* (previously *Macoma balthica*) and *Cerastoderma glaucum*, Hydrobiidae snails) were collected between March and May 2007 from c. 1 m water depth in a bay near Stockholm University's Askö Laboratory (58° 49.4′ N, 17° 38.2′ E) (see Figure 1 for the experimental time line). The sediment was sieved on a 1 mm mesh to remove all macrofauna, which was saved in separate sediment and kept aerated, cool, and dark until the start of the experiment. The sieved sediment and water was transferred to a large plastic tank (100 × 100 × 50 cm) which was placed outdoors to allow warming of the sediment and hatching of plankton from the sediment. After one week, the water was visibly cloudy with plankton. Extra zooplankton was also collected just before the start of the experiment by passing 450 L of water from the same bay through a 25 μm mesh plankton net. This sample was dominated by rotifers, with some cladocerans and copepods.

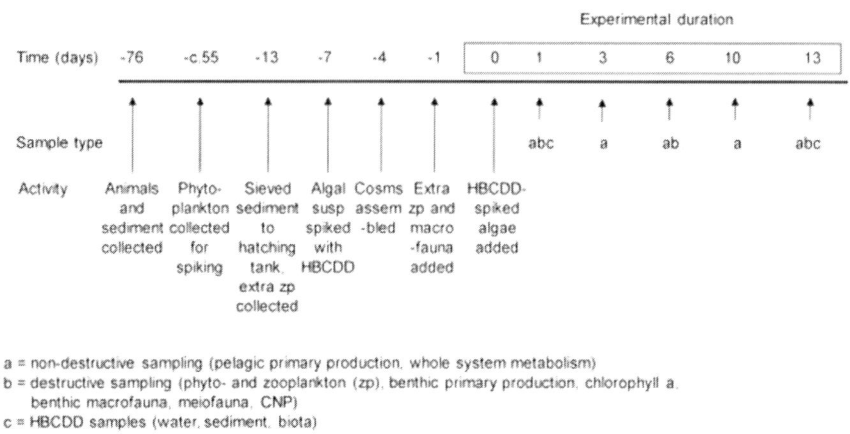

Figure 1. Timeline of cosm construction and sampling. HBCDD—Hexabromocyclododecane; CNP—carbon, nitrogen, phosphorus.

2.3. Spiking of Organic Material with HBCDD

Phytoplankton was collected in April 2007, partly during a monitoring program in the area, and partly by trawling from a small boat in the immediate vicinity of the Askö Laboratory, in both cases using a 90 μm plankton net. These samples were concentrated and stored at −20 °C until later spiking with HBCDD. At that time, they were thawed out, thoroughly mixed, and divided into two stocks, one that was spiked with a solution of HBCDD in an acetone carrier in an amount calculated to achieve a final nominal amount per cosm of 78.75 μg, the other unspiked but with an equal amount of acetone. These two batches were kept cold (4–5 °C) on magnetic stirrers for 1 week before being added in equal amounts to each of the experimental containers.

Later chemical analyses (see Section 2.5.4) showed that mean HBCDD concentration in the spiked suspension was 12.27 μg L^{-1} (±4.2 standard deviation (SE), n = 2), and that most of this was in the solid fraction (mean concentration: 3.21×10^3μg gdw^{-1} ($\pm 3.60 \times 10^2$ SE, n = 2) [20].

2.4. Experimental Set-Up and Cosm Construction

The experiment was conducted outdoors under an open-sided transparent plastic shelter that allowed sunlight through while preventing rain or debris entering the cosms. The cosms were randomly distributed between 4 troughs. Two of these troughs had cool water flow-through pumped directly from the sea. The other two were maintained at c. 5 °C above ambient using aquaria water heaters (2 × 300 W heaters per trough). Two water pumps ensured thorough mixing of the warmed water in the troughs. Photographs of the set-up can be found in Figure S1. Distilled water was added to the cosms five times during the experiment to compensate for evaporation and maintain constant salinity.

During the experiment, the temperature of the ambient treatments varied between c. 12–20 °C, depending on the time of day and the weather. The warm treatments varied between c. 17–26 °C. The maximum daily variation was 7.6 °C and 9.1 °C in the ambient and warm treatments, respectively. The salinity in the cosms was c. 7.5 psu and oxygen water concentrations were consistently >8 mg·L^{-1}.

The cosms were prepared as follows. Water from the hatching tank was drained off into holding containers. The sediment was then thoroughly mixed and divided equally (610 mL wet sediment each) into 58 separate 5 L buckets (cosms). These were factory-new, made of polypropylene, and approved for food use. Two of these cosms were taken immediately for analysis of CNP and HBCDD in the water and CNP content of the sediment (see Section 2.5.4). Hatching tank water was then equally distributed (3.75 L each) between the remaining 56 buckets, which were then placed in the troughs and allowed to stand for 3 days. After this time the sediment had settled out to form a layer 2 cm thick. The number of buckets allowed for 4–6 replicates of each of the four treatments (ambient control (A–), warm control (W–), ambient with HBCDD (A+), and warm with HBCDD (W+)), and for destructive sampling at three time points (1, 6, 13 days after HBCDD addition; t1, t6 and t13).

The previously collected organisms were sieved out of their holding sediment and four *L. balthica* (11.5 ± 1.2 mm), two *C. glaucum* (shell length c. 16 mm and c. 8.5 mm), and 10 Hydrobiidae snails were added to each of the treatments. This represented natural field densities of 187, 93, and 468 individuals per m^2, respectively. The *L. balthica* burrowed quickly into the sediment.

Finally, 4.5 mL HBCDD-spiked plankton suspension was added to half of the cosms (A+ and W+ treatments) to give a nominal amount in each cosm of 78.75 µg HBCDD. The same amount of unspiked suspension was added to the A– and W– cosms to ensure equal amounts of extra organic material were added to all cosms (63 mg dw or 21.4 mg TOC (total organic carbon) per cosm).

The experiment was run for 13 days. During this time the weather was in general warm and sunny.

2.5. Sampling

Non-destructive sampling was used for measurements of whole system metabolism, and pelagic primary production. Destructive sampling (i.e., disassembling the contents of the cosms) was done for the rest of the endpoints (pelagic chlorophyll a, phyto- and zooplankton community composition), and sediment fauna community composition (macrofauna, meiofauna), CNP, and HBCDD analysis. This was done three times during the experiment, at 1, 6, and 13 days after the HBCDD was added (t1, t6, t13) (see Figure 1).

2.5.1. Whole System Metabolism

The whole system metabolism was determined five times during the experiment from O_2 measurements using a WTW Multi 350i (WTW GmbH, Weilheim, Germany), though only data from t1, t6, and t13 is used here. Each time, O_2concentrations in the water were measured twice in the dark (starting c. 22:00) and twice in the light (starting c. 10:30). The difference between the dark measurements was used to calculate the whole system respiration (R, in mg O_2 h^{-1}) and the difference between the light measurements was used to calculate the net primary production (NPP, in mg O_2 h^{-1}). Gross primary production (GPP, in mg O_2 h^{-1}) was calculated as NPP-R and the production/respiration ratio (P/R) as GPP/R.

2.5.2. Pelagic Endpoints

Pelagic primary production was measured five times during the experiment (t1, t3, t6, t10, t13), though only data from t1, t6, and t13 is used in this paper. Two 10 mL water samples were taken (in the morning) from each microcosm and placed in two 20 mL scintillation vials. One hundred μL ^{14}C was added as buffered $NaH^{14}CO_3$ (specific activity 185 kBq mL^{-1}) and the samples were incubated at ambient temperature for a few hours, one of the samples in the light and the other in the dark (covered in black plastic). Three drops of 10% HCl was then added to stop primary production, excess ^{14}C bubbled off, and 10 mL Lumagel Safe scintillation cocktail (Perkin Elmer, Waltham, MA, USA) was added. ^{14}C activity in the samples was measured in a 1214 Rackbeta scintillation counter (LKB Wallac, Turku, Finland) and decays per minute (dpm) converted to $mg^{14}C\ L^{-1}\ h^{-1}$. Net primary production was calculated as the difference between ^{14}C uptake in the light minus that in the dark.

All other pelagic endpoints were measured at t1, t6, and t13. In order to measure pelagic chlorophyll a, phyto- and zooplankton community composition, different fractions of the overlying water were obtained. Firstly, the overlying water was siphoned off into a bucket from which a well-mixed water sample of known volume was filtered through a 90 μm sieve. The fraction retained on the sieve (zooplankton) and the fraction that had passed through the sieve (largely phytoplankton) were fixed separately in Lugol solution and saved in the dark until quantification and identification to the lowest possible taxonomic level. A further 1–2 L was filtered through a 90 μm sieve to remove the zooplankton and the filtrate was further filtered on 0.7 μm cellulose-nitrate filters and stored at −20 °C until the analysis of chlorophyll a, a measure of phytoplankton production/biomass. At that time, frozen filters were thawed and extracted in 5 mL acetone; the extract was analyzed with a U-2000 spectrophotometer (Hitachi, Tokyo, Japan) according to [21].

2.5.3. Benthic Endpoints

Directly after removing the water for the plankton and pelagic chlorophyll a measurements, a 1 mL (4 mm diameter) cut-off plastic pipette was then used to take 4 small sediment cores from the sediment. The top 2 mm of each mini-core (c. 25 mm^3) was saved and all four pooled to a single sample in a Winkler flask. 20 mL filtered seawater and 100 μL buffered $NaH^{14}CO_3$ solution (specific activity 185 kBq mL^{-1}) was then added to each. Double samples of this type were taken from each cosm for light and dark incubations and the flasks were incubated for several hours in the troughs at the temperature from which they had come. Twenty-five mL filtered seawater were then added and 10 mL samples of well-shaken suspension were taken. Three drops 10% HCl were added to stop all production, excess ^{14}C bubbled off, and 10 mL Lumagel Safe (Perkin Elmer, Waltham, MA, USA) was added. ^{14}C activity was measured in a 1214 Rackbeta scintillation counter (LKB Wallac, Turku, Finland) and dpm counts per sample were converted to mg ^{14}C cm^{-2}.

Macrofauna were sieved out of the remaining sediment on 1 mm and 5 mm sieves. *L. balthica* and *C. glaucum* shell lengths were measured and all species were counted, allowed to clear their guts for several hours, and were then frozen at −20 °C for later HBCDD analysis. The number of dead individuals was noted. At t13, 100 mL of the top 1 cm sediment was scraped off using a spatula and preserved with 40 mL 95% ethanol for identification and quantification of meiofauna. This was done by sieving the sediment on a 63 μm sieve and quantifying the organisms to the lowest possible taxon level.

2.5.4. CNP and HBCDD Sampling and Analysis

Water samples for the analysis of C, N, and P were taken at t1, t6, and t13; C was analyzed in unfiltered water and N and P in water filtered through a 0.45 μm filter. Nitrogen (NO_2^-, NO_3^-, and NH_4^+) was analyzed with a Leco-CHNS-932 analyzer (Isomass Scientific Inc., Calgary, AB, Canada) (EDTA and acetanilide as standards) and phosphorus (PO_4^{3-}) with a segmented flow system (ALPKEM Flow Solution IV) following combustion at 500 °C and acidic persulfate oxidation, at the accredited laboratory at the Department of Ecology, Environment, and Plant Sciences, Stockholm University, Sweden.

To determine the fate of the HBCDD in the experimental systems, samples of the spiked organic material were taken, as well as water and sediment samples at t1 and t13, from both A+ and W+ treatments. The soft tissues of *L. balthica* and *C. glaucum* collected at t13 (see above) were also analysed. All samples were frozen at −20 °C until analysis using gas chromatography–mass spectrometry (GC–MS/MS, Agilent Technologies, Santa Clara, CA, USA) and liquid chromatography–mass spectrometry (LC–MS/MS, Shimadzu Corp., Kyoto, Japan) at the Department of Environmental Chemistry, Stockholm University. The relative amounts of the three diastereomers (α, β, γ) were also determined. Full details of these chemical analyses are available in Bradshaw et al. [20] (Supplementary Information).

2.6. Data Handling and Statistics

Single and interactive effects of warming (ambient and warm), HBCDD addition (with HBCDD and without HBCDD), abiotic variables, and phyto-

and zooplankton species abundances, were assessed using constrained analysis of principle coordinates (CAP), based on Bray-Curtis dissimilarities. Prior to testing the effects of temperature and HBCDD addition and abiotic variables on species abundances of both phyto-, and zooplankton, species matrices were fourth root transformed to account for homoscedasticity and to downweigh the importance of taxa with very high abundances. Additionally, the abiotic variables were tested for autocorrelation between predictor variables and factors and no autocorrelation was found. Furthermore, the effect of the blocks (i.e., troughs) was tested, but since it did not have an effect, it was removed from the analyses. All multivariate analyses were performed in R version 3.2. [22].

The model for phytoplankton included both factors of warming and HBCDD addition as single and interactive effects, as well as all abiotic variables chlorophyll a, phosphate (PO_4), total nitrogen, P/R ratios, and pelagic primary production, with the holding times (t6 and t13) constant. Significance of the model, terms, and axis were assessed using a permutation procedure with 999 permutations. The minimal adequate model (containing only significant parameters $\alpha = 0.05$) was identified through step-wise deletion of non-significant parameters [23]. The analyses were performed using the "capscale" function from the {vegan} R package [24]. A similarity percentage test (SIMPER) was used to identify which phytoplankton species drove the community-level effects, using significant factor(s). For the zooplankton species matrix, both the abiotic variables (same as above) as well as all phytoplankton species were used as predictors. The analysis was then performed the same way as described for the phytoplankton species abundance.

Three-way ANOVA (factors: time, temperature, HBCDD) was used to test for: differences in pelagic and benthic primary production (standardized to average A– at each time point to remove the effect of differences in light intensities during incubations at the different time points), total N, and PO_4^- concentrations in the water. Homogeneity of variance was tested using Cochran's test and non-homogeneous data was Box-Cox transformed where necessary. The analyses were performed in Statistica version 12.5 (Dell Inc., Tulsa, OK, USA).

Since too few benthic community data were available for a meaningful benthic CAP analysis, and several benthic endpoints were only measured at one time point, benthic parameters were analyzed using correlation analysis and ANOVA. Two-way ANOVA (factors: temperature, HBCDD) was used to test for differences in: abundance of each meiofauna taxa at t13; abundance of Hydrobiidae at t6; *L. balthica* biomass per cosm at t6 and t13. Homogeneity of variance was tested using Cochran's test and non-homogeneous data were square-root transformed where necessary.

To investigate potential connections between the benthic and pelagic components of the ecosystem, two multiple correlations were performed (t6 and t13 separately) on the following data: abundances of the 5 phytoplankton and 6 zooplankton groups identified by SIMPER (see above) to cumulatively contribute to 70% of the dissimilarity between treatments in the pelagic analysis; pelagic and benthic primary production; chlorophyll a, total N, and PO_4 concentrations. For t6, the number of Hydrobiidae was also available. For the t13 data correlation, the following variables were also available: total number of meiofauna, numbers of Turbellaria and *Microarthridion littorale*, and total biomass of *L. balthica*. Although the focus of this analysis was benthic-pelagic linkages, it was also used to help explain the relationships between plankton species. The significance of the correlation coefficients r was tested using a two-tailed *t* test. The analyses were performed in Statistica v.12.5.

A significance level of 0.05 was used in all statistical tests.

3. RESULTS

3.1. HBCDD Partitioning and Exposure of the Model Ecosystem

After only one day, >90% of the HBCDD was detected in the sediment [20] and this increased to >99% by t13. Concentrations in the water were very low (Table 1), with >80% of the HBCDD in the water being present in the particulate fraction at t1, decreasing to 66.5%–78.7% by the end of the experiment, depending on the treatment (Table 1). Concentrations in both water fractions decreased throughout the experiment, with the warmer treatment (W+) having lower HBCDD concentrations than the ambient treatment (W−). No significant effect of either temperature or time was seen on sediment concentrations, although HBCDD concentrations tended to be lower at higher temperatures. There was no obvious difference between treatments for HBCDD body concentrations in *L. balthica* at the end of the experiment. For full details of partitioning, including that of HBCDD diastereomers, see [20].

Table 1. Measured concentrations of hexabromocyclododecane (HBCDD) in the various compartments and treatments one day after the start and at the end of the experiment. For more details, see [20].

Sample	Mean	SE	*n*	% Present in the Dissolved Phase
Water (unfiltered) (μg L^{-1})				
t1, A+	2.12×10^{-3}	1.14×10^{-5}	2	83.7
t1, W+	1.08×10^{-3}	1.09×10^{-5}	2	82.8
t13, A+	2.92×10^{-5}	1.66×10^{-6}	2	66.5
t13, W+	3.98×10^{-5}		1	78.7
Surface sediment (μg gdw^{-1})				
t1, A+	2.46×10^{-1}	7.16×10^{-2}	4	
t1, W+	1.29×10^{-1}	3.87×10^{-2}	4	
t13, A+	3.82×10^{-1}	1.46×10^{-1}	3	
t13, W+	1.71×10^{-1}	3.02×10^{-2}	3	
Limecola (μg g lipid^{-1})				
t13, A+	1.18×10^{2}			
t13, W+	9.63×10^{1}			

3.2. Pelagic Part of the Ecosystem

Pelagic primary production was consistently higher in the warm treatments (W− and W+) (3-way ANOVA, significant main effect of temperature, $F_{1,60}$, $p < 0.0001$), but there was no effect of HBCDD (Figure 2, Table S12).

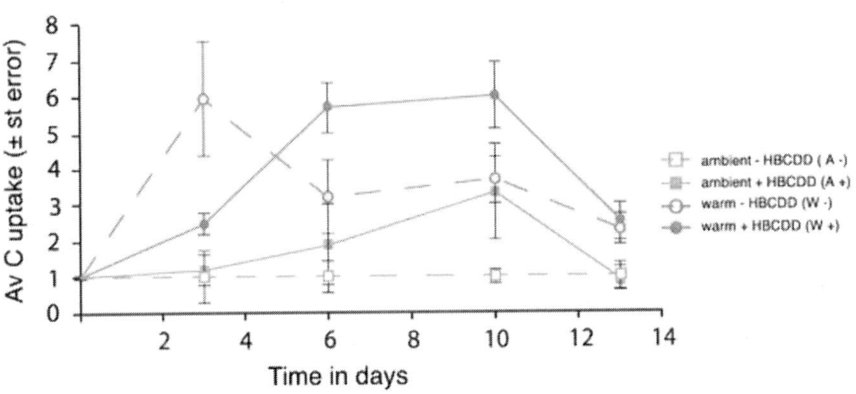

Figure 2. Pelagic primary production over time standardized to A− treatment. Values are means and SE of each treatment.

The total nitrogen concentration (sum of NO_2^-, NO_3^- and NH_4^+) decreased in all four treatments at t6 compared to t1 (3-way ANOVA, significant main effect of time, $F_{2,47}$, $p = 0.0001$). The decrease was more pronounced in the warm treatments compared to ambient treatments, though this was not statistically significant (3-way ANOVA, interaction of time and temperature, $F_{2,47}$, $p = 0.072$) (Figure 3a, Table S13). At t13, the total N concentrations further decreased in the ambient temperature treatments, while a slight increase could be observed in the warm treatments. Phosphate concentration slightly increased at t6 in the ambient treatments, while the increase in the warm treatments was significantly more pronounced (Figure 3b), as was the decrease in PO_4 by t13 (3-way ANOVA, interaction of time and temperature, $F_{2,47}$, $p < 0.0001$; Table S14).

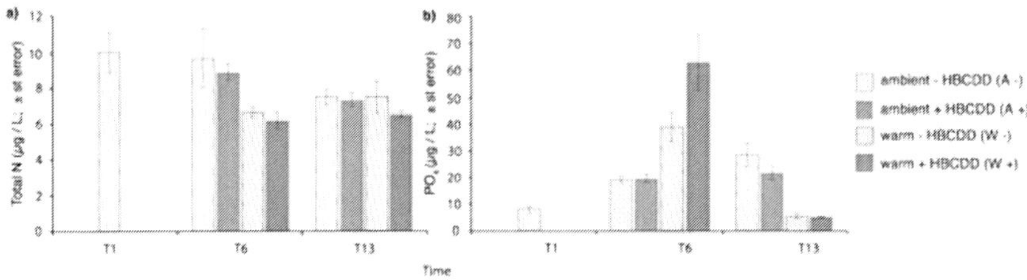

Figure 3. (**a**) Total N (sum of NO_2^-, NO_3^-, and NH_4^+); (**b**) PO_4 concentration in the water at t1, t6, and t13. Values are means and SE of each treatment; $n = 5$, except for t6 W− where $n = 4$.

3.2.1. Phytoplankton

The multivariate CAP analysis showed that the phytoplankton community structure was influenced by: warming ($F_{1,31} = 21.08$, $p = 0.001$), chlorophyll a ($F_{1,31} = 5.60$, $p = 0.002$), PO_4 ($F_{1,31} = 7.88$, $p = 0.002$), total N ($F_{1,31} = 3.58$, $p = 0.024$), and pelagic primary production (PPP) ($F_{1,31} = 3.88$, $p = 0.016$; Figure 4). Warming divided the phytoplankton community into two clusters, where the warm temperature treatments were mainly related to an increase in pelagic primary production as well as the chlorophyll a increase, which was driven by *Dolichospermum* filaments and heterocysts (Figure 4 and Figure 5a,b,d; Table S1). Increasing PO_4 concentrations influenced centric diatom species and *Pyramimonas* sp., which characterized ambient temperature treatments (Figure 4 and Figure 5c,e; Table S1). *Melosira* sp. abundances were influenced by increasing total N concentrations (Figure 4).

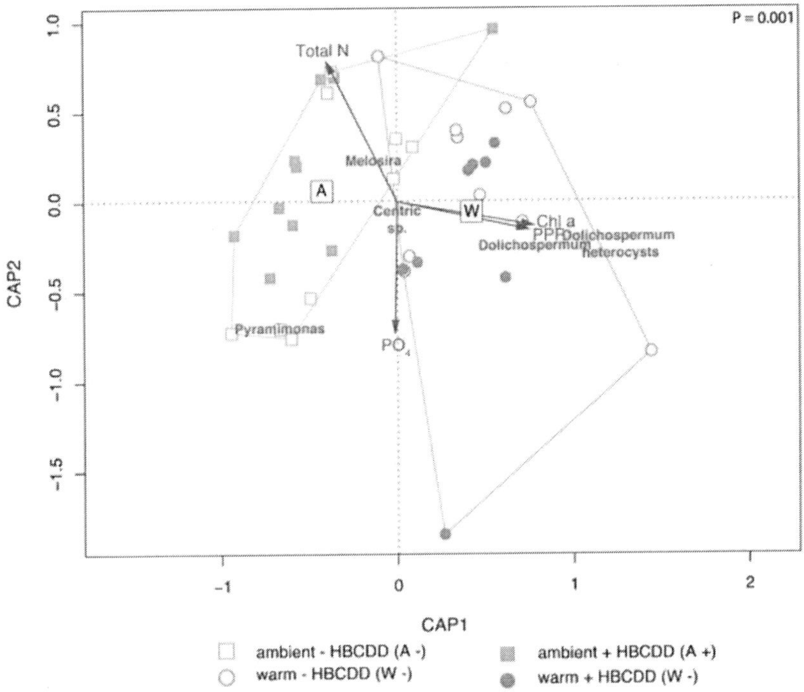

Figure 4. CAP (constrained analysis of principle coordinates) biplot, based on the phytoplankton community matrix, and constrained by time, where each symbol represents each cosm with the individual treatment. The centroids are ambient (A) and warm (W) temperature treatments, also indicated by different colors. The arrows represent the different predictor variables and point into the direction of increasing value. The species shown represent the taxa that contribute 70% to the dissimilarity identified in the similarity percentages (SIMPER) analysis (Tables S1–S3).

3.2.2. Zooplankton

The multivariate CAP analysis showed that the zooplankton community structure was influenced by: warming ($F_{1,31}$ = 4.91, p = 0.002), HBCDD addition ($F_{1,31}$ = 2.65, p = 0.035), PO_4 concentrations ($F_{1,31}$ = 7.88, p = 0.001), chlorophyll a concentrations ($F_{1,31}$ = 4.09, p = 0.009), and the abundance of *Melosira* sp. ($F_{1,31}$ = 3.72, p = 0.008; Figure 6). Overall, the warm treatments were characterized by bivalve larvae, copepod nauplii, and *Acartia bifilosa* (Figure 6 and Figure 7b,d,e), and differences between the warm and ambient treatments were driven mainly by higher

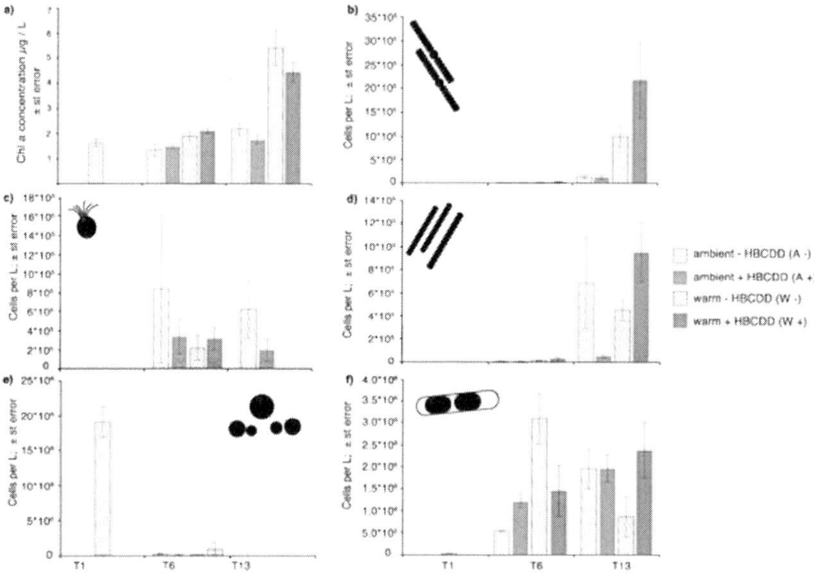

Figure 5. Time series of (**a**) average chlorophyll a concentrations in µg L^{-1} (± standard error (SE)) (n = 5, except t6 A− where n = 4); (**b**) average density of *Dolichospermum*heterocysts (±SE); (**c**) average cell density of *Pyramimonas* species (±SE); (**d**) average density of *Dolichospermum* filaments (±SE); (**e**) average cell density of centric diatom species (±SE); (**f**) average cell density of *Melosira* species (±SE). For phytoplankton graphs: n = 5, except for t6 W− where n = 6 and t6 W+ where n = 3. The taxa shown are those identified in the SIMPER analysis as together contributing to 70% of the dissimilarity between treatments/times (Tables S1–S3).

chlorophyll a values, while the main driving factors in ambient treatments were total N and *Melosira* sp. abundance (Figure 6). However, the relative abundance of these taxa varied with time point (Figure 7, Tables S6–S9). Overall, across all time and temperature treatments, HBCDD treatments were characterized by higher numbers of rotifers, bivalve larvae, gastropod larvae, and *A. bifilosa*, and lower numbers of copepod nauplii (Figure 7, Table S5).

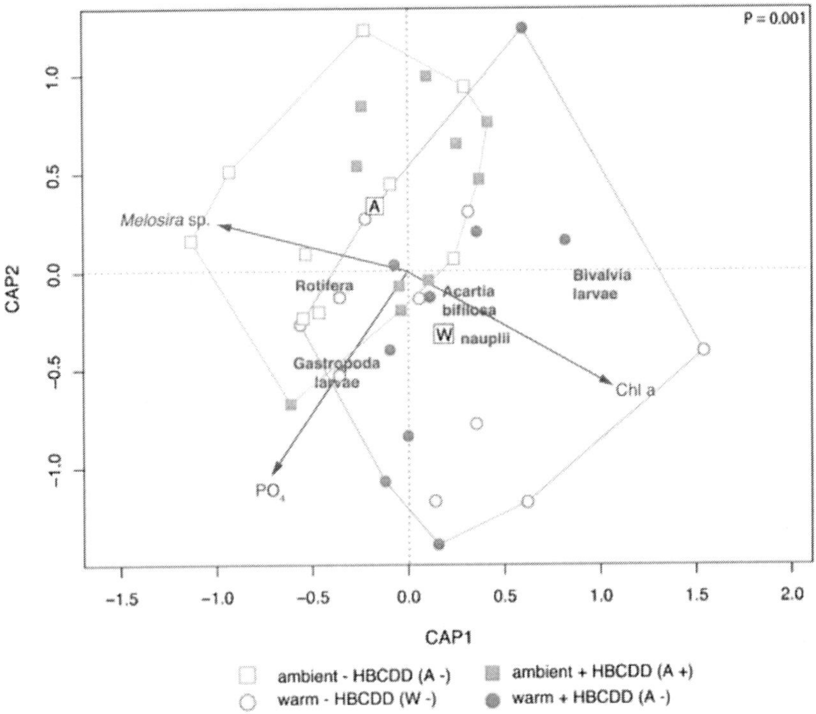

Figure 6. CAP biplot based on the zooplankton community matrix, and constrained by time, where each symbol represents each cosm with the individual treatment. The arrows represent the different predictor variables and point in the direction of increasing value of each variable and their lengths indicate their relative importance. The centroids displayed are ambient (A) and warm (W) temperature treatments, which were based on a higher significance of the temperature effect compared to the HBCDD effect. The species shown represent the taxa that contribute 70% to the dissimilarity identified in the SIMPER analysis (Tables S4–S9).

3.3. Benthic Part of the Ecosystem

Increased temperature significantly increased benthic primary production (3-way ANOVA, $F_{1,32}$, $p < 0.0001$), particularly at t6 (3-way ANOVA, significant interaction between time and temperature, $F_{1,32}$, $p = 0.0002$) (Table S15). There was no effect of HBCDD (Figure 8).

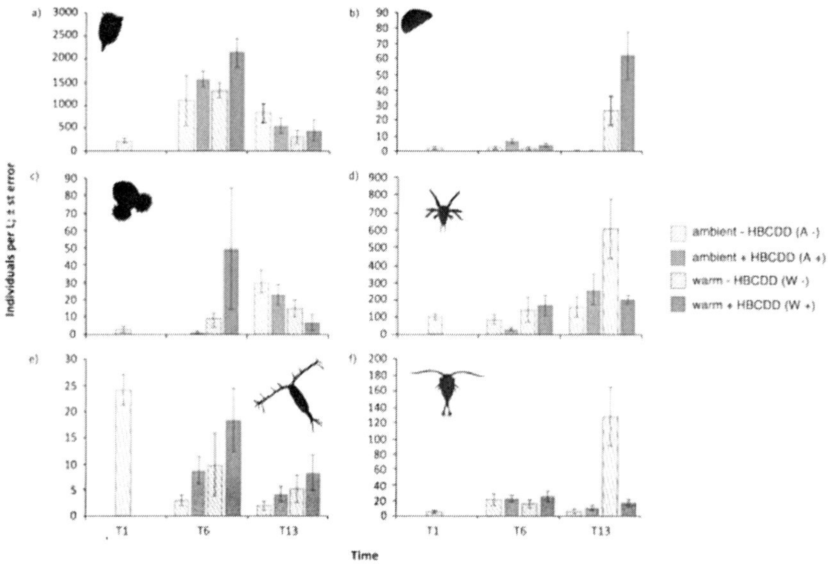

Figure 7. Time series of individuals per L (**a**) Rotifera (±SE); (**b**) Bivalvia larvae (±SE); (**c**) Gastropoda larvae (±SE); (**d**) Copepoda nauplii (±SE); (**e**) *Acartia bifilosa* (±SE); (**f**) *Eurytemora affinis* (±SE). n = 5, except for t13 W+ where n = 3. The taxa shown are those identified in the SIMPER analysis as together contributing to 70% of the dissimilarity between treatments/times (Tables S4–S9).

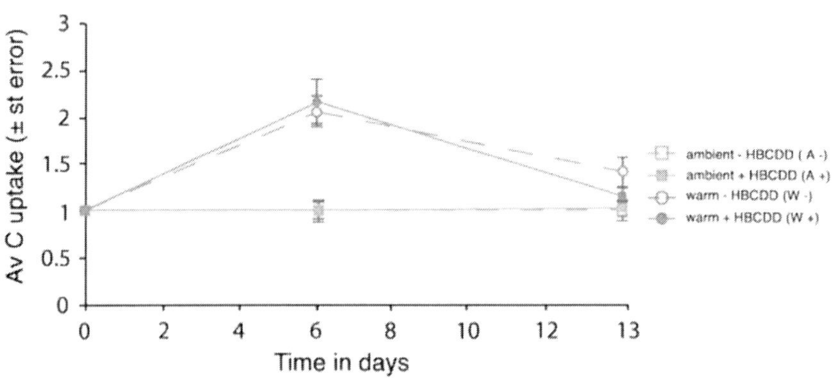

Figure 8. Benthic primary production, standardized to the A− treatment to enable valid comparisons between time points (mean ± SE).

3.3.1. Meiofauna

Total meiofauna (Figure 9a) abundances at the end of the experiment were dominated by the abundance of nematodes. Although total nematode numbers were lower in the warm treatments than in the ambient ones (Figure 9b), this was not significant (2-way ANOVA, $F_{1,15}$, $p = 0.12$; Table S16) and there was no significant effect of HBCDD (2-way ANOVA, $F_{1,15}$, $p = 0.88$), nor any significant interaction between the two factors (2-way ANOVA, $F_{1,15}$, $p = 0.44$). There was no overall effect of HBCDD or temperature on Turbellaria or *M. littorale*, but there was a significant interaction between these stressors (2-way ANOVA, $F_{1,15}$, $p < 0.05$) for both species (Figure 9c,d, Tables S17 and S18).

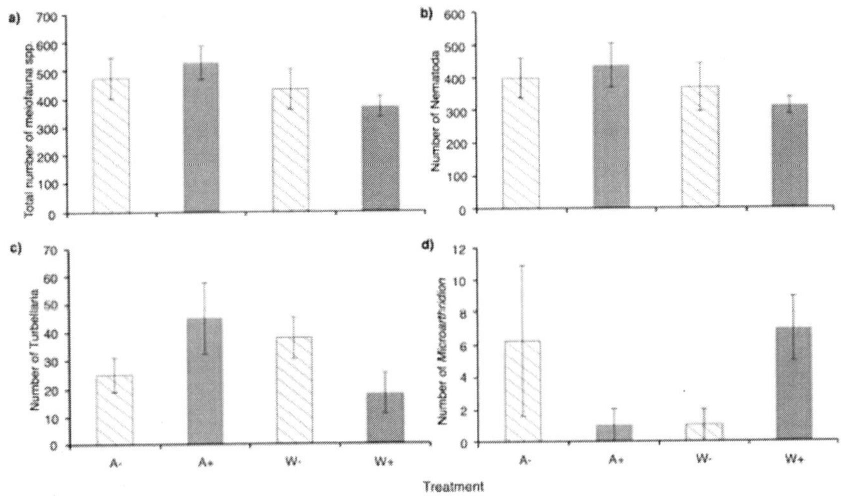

Figure 9. Abundances of (**a**) total meiofauna, (**b**) nematodes, (**c**) Turbellaria and (**d**) *Microarthridion littorale* at the end of the experiment (t13). Graphs show mean abundance per 100 mL sediment ± SE for each treatment; $n = 5$ except for A– where $n = 4$.

3.3.2. Benthic-Pelagic Coupling

Benthic and pelagic variables seemed in general to be only weakly correlated (Tables S10 and S11). We looked especially carefully at the potential influence of *L. balthica* biomass, hydrobid numbers, and benthic primary production in the systems and at the two zooplankton taxa that are the larvae of benthic species (gastropod and bivalve larvae).

Benthic primary production was significantly correlated to PO_4, pelagic primary production, and *Dolichospermum* and *Melosira* sp. abundance at t6 (Table 2). At t13 it was not correlated to any of the variables. *L. balthica* biomass was only significantly correlated ($r = 0.48, p = 0.05$) with one of the other 19 variables, *Melosira* sp., (Table 2 and Table S11) a result that is most likely not causative. In addition, *L. balthica* biomass per microcosm showed no significant difference between treatments, either at t6 or t13 (2-way ANOVA, $F_{1,15}, p > 0.05$; Tables S19 and S20), though there was a tendency for lower biomass in HBCDD treatments at t13. The number of Hydrobiidae increased from 10 at the start of the experiment to 15–20 by t6. There was a tendency for more Hydrobiidae in the HBCDD treatments at t6 (Figure S2), but this was not significant (Table S21). The number of Hydrobiidae at t6 only correlated significantly ($r = 0.45, p = 0.05$) with total N concentrations in the water (Table 2).

Table 2. Summary of significant results from the multiple correlations of the benthic-pelagic parameters. *r* is the correlation coefficient, *p* is the level of significance.

t6			t13		
Correlated Variable	*r*	*p*	Correlated Variable	*r*	*p*
Benthic primary production					
Dolichospermum heterocysts	0.670	0.001			
PO_4	0.657	0.002	no significant correlations		
Melosira sp.	0.627	0.003			
Dolichospermum filaments	0.596	0.006			
Chlorophyll a	0.478	0.033			
Bivalve larvae					
Pelagic PP	0.694	0.002	no significant correlations		
PO_4	−0.631	0.007			
Dolichospermum heterocysts	0.583	0.014			
Gastropod larvae			**Gastropod larvae**		
PO_4	0.718	0.000	Total meiofauna	0.620	0.008
Chlorophyll a	0.511	0.021	Chlorophyll a	−0.566	0.018
Dolichospermum heterocysts	0.501	0.024	*Dolichospermum* heterocysts	−0.505	0.039
			PO_4	0.475	0.054
Hydrobiidae			no data available		
Total N	0.449	0.047			
no data available			**Total *Limecola balthica* biomass (gdw)**		
			Melosira sp.	−0.481	0.050

Abundances of gastropod larvae were significantly positively correlated with PO_4 at both time points, positively correlated with pelagic chlorophyll a and *Dolichospermum* heterocysts at t6, but negatively at t13. There was also a significant correlation ($r = 0.62, p = 0.008$) between the number of gastropod larvae and the total number of meiofauna, but this result is probably coincidental.

4. DISCUSSION

4.1. Benthic-Pelagic Coupling: Nutrient Release from Sediment and Effects on the Plankton

In previous work, with larger cosms run over eight months [19], we showed that HBCDD negatively affected the biomass of "large" (>8 mm) *L. balthica* in the same size class as in this experiment. This in turn affected benthic-pelagic coupling, reducing the recirculation of nutrients from the sediment to the water. In the current study, no changes in *L. balthica*biomass were observed, probably due to the short duration of this study. Trends in benthic and pelagic parameters were not linked to differences in *L. balthica* biomass between treatments, but to overall effects of temperature, and to a lesser degree HBCDD (see Section 4.3).

It is well-known from freshwater systems that increased temperature causes the release of PO_4 from sediments [25,26,27]. The few studies that have been performed in marine ecosystems suggest the same (e.g., [28,29,30]). This PO_4 release is thought to occur due to increased microbial activity increasing the decomposition of organic matter at the sediment surface [31,32,33] and by consuming oxygen at the sediment-water interface [25], leading to a redox-mediated release of P [27]. Studies of anoxic marine benthic ecosystems also confirm that low oxygen conditions facilitate fluxes of nutrients from the sediment to the water [28,30,34]. Although overall oxygen levels in the water column remained high in this experiment, it is possible that local anoxia/hypoxia at the sediment surface occurred in the warm treatments. We did not measure benthic bacterial activity specifically in this study, but measurements of benthic primary productivity support the idea that benthic processes were higher at higher temperatures, particularly at t6. By t13, P levels in the sediment were probably quite low, leading to the reduction in benthic primary productivity at this time point. It should be noted that by adding organic matter to the cosms at t0, we also added a source of P, which potentially could have increased water PO_4 concentrations. However, we estimate that this addition was negligible compared to the P pool in the sediment. Calculations using measured TOC and C:P ratios in the added suspended organic matter and the sediment suggest that the organic matter provided c. 0.35 mgP per cosm compared c. 1200 mgP in the sediment.

This nutrient regime in the warm treatments therefore has similarities with the Baltic Sea nutrient regime after the spring bloom, where N and P are depleted and the sedimentation of organic material leads to increased microbial activity, leading to anoxia, which in turn results in an increased P release by the anoxic sediment [35]. While our microcosms did not turn anoxic, filamentous nitrogen-fixing cyanobacteria *Dolichospermum* sp. increased from relatively low numbers at t6 to the most abundant species in the warm treatments at t13. As P is the primary limiting nutrient for N-fixation, the increase in phosphorus concentrations, through PO_4 release from the sediment, triggered a rapid increase in *Dolichospermum* sp. abundances [36] and in the number of nitrogen-fixing heterocysts present. While cyanobacteria blooms are generally considered to be harmful, recent studies indicate that the N fixed and released as NH_4 by cyanobacteria might actually be beneficial for other primary producers (i.e., [35]). Stable isotope analysis and nanoscale secondary ion mass spectrometry (nanoSIMS) have shown that N derived from cyanobacterial N-fixation can be taken up by diatoms in the form of NH_4. Furthermore, recent studies indicate that cyanobacteria can be a direct food source for meso- and microzooplankton. Stable isotope analyses show that cyanobacteria are a major food source of copepods species such as *Acartia* spp. [37]. This could explain the positive relationship of *Acartia* with increasing chlorophyll a concentrations, as this increase was most likely driven by *Dolichospermum* filaments and therefore represents an increase in food items. The increase in rotifers was most likely due to an increase in abundance of a potential food species, *Melosira* sp., in ambient temperature treatments, or could be explained by predator-prey release. Copepods are known to not only graze on phytoplankton but also prey on other zooplankton species including rotifers [38]. Due to an increase in PO_4 and a consecutive cyanobacteria increase in the warm treatments, *Acartia* increased in the warm treatments, while abundances in the ambient treatments stayed relatively low, which in turn resulted in lower predation pressure of rotifers in the ambient treatments.

The overwhelming effect of temperature-induced PO_4 release, and the resulting changes in the phytoplankton and pelagic primary production, was probably responsible for the significant correlations observed at t6 between the benthic primary production and PO_4, chlorophyll a, and some phytoplankton species. It is unlikely that these correlations are a cause-effect relationship; rather, they are probably reflecting the common increase in rates of ecological processes in the warm treatments at that time point (Figure 10).

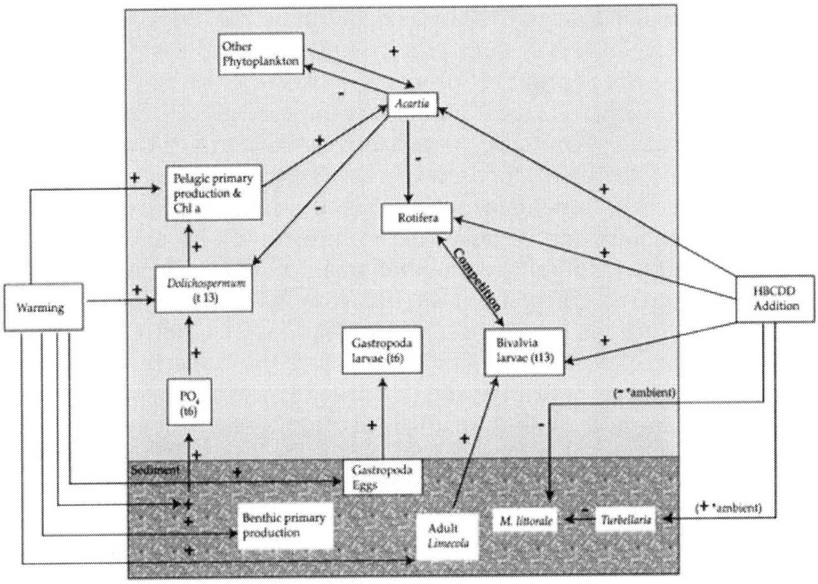

Figure 10. Conceptual diagram of direct and indirect effects of increased temperature and HBCDD addition in model ecosystems.

4.2. Benthic-Pelagic Coupling: Benthic Species' Larvae in the Zooplankton

Benthic species were also present in the water column as larvae; bivalve and gastropod larvae were identified as two of the planktonic taxa that were responsible for differences in the zooplankton community structure. Although bivalve larvae may survive in the water column for many weeks, their general increase in numbers in the warm treatments throughout the experiment suggests that these were released from adults in the cosms, rather than being present in the water column from the start. In this case, they were most probably *L. balthica* (10 adults present in each cosm), or possibly *C. glaucum* (2 per cosm), both of which are known to spawn at the time of year that the experiment was performed (e.g., [39,40]).

There were no clear effects on adult *L. balthica* or *C. glaucum* in terms of mortality or size, so the clear effects on their larvae must be either due to direct effects on adult spawning or larval survival, or indirect effects on the larvae related to foodweb dynamics. Drent showed that *L. balthica* larvae grew faster and developed more quickly at higher temperatures (experimental range 10–20 °C) [41], and several other studies have shown that spawning is triggered by the temperature rising to a critical level [42,43], albeit at lower temperatures than in this experiment. Benthic primary production at t6 was also positively correlated with bivalve larval abundance at t13; adult *L. balthica* may have been able to spawn more successfully in the warmer treatments by t13 due to increased food levels on the sediment surface. *L. balthica* larvae are thought to feed on small unicellular phytoplankton, bacteria, flagellates, and organic matter [44,45,46,47], and will likely compete for these food items with other small zooplankton, such as other larvae and rotifers. They are also likely to be preyed on by larger zooplankton, such as copepods [38]. Relationships between bivalve larvae and these other taxa could therefore be expected. However, multiple correlations of bivalve and gastropod larval abundances with all possible prey species, competitors, and predators revealed no significant correlations with the abundances of any of these species. The possibility of indirect effects should, however, not be ruled out. It is difficult in studies such as these to capture the fine scale temporal changes with a limited number of "snapshot" sampling times. Processes or events that have happened in between sampling points may well be the most important in determining the effects that are subsequently measured.

Most of the common gastropod species present in shallow Baltic bays (e.g., *Hydrobia ventrosa*, *H. neglecta*, *Potamopyrgus antipodarum*, *Theodoxus fluviatilis*, *Bithynia tentaculata*, *Radix* spp.) either produce live young, or lay eggs that hatch into juveniles. The only common species that has a larval stage, hatching from eggs laid on the seabed, is *Hydrobia ulvae* [48], which was most probably the species introduced to the cosms ("Hydrobiidae") at the start of the experiment. Adult Hydrobiidae increased in number from 10 at the start to c. 25 at t6 and >35 (not quantified) at t13, with no significant differences between treatments at t6. However, the larval abundance showed a clear treatment difference, with warm treatments having higher numbers at t6 and ambient treatments having higher numbers at t13. Given that the time to hatching takes at least 5 days, it seems probable that the larvae seen in this experiment hatched from eggs already present in the sediment at the start, a process that was stimulated (or occurred earlier) in the warm treatments. Fish and Fish [49] showed that *H. ulvae* from a Welsh estuary had a clear optimum temperature with respect to embryonic development and survival as well as the time taken for eggs to hatch, as is the case for most invertebrates [3].

H. ulvae are generalist feeders that graze on organic matter and biofilm on the sediment surface [50]. They also consume more biomass at higher temperatures [51]. They might therefore be expected to take up a considerable amount of HBCDD from the spiked algal material added to the cosms, particularly at increased temperature, but if this was the case, it did not seem to have adversely affected adult survival in any way. One might also expect to see a correlation between adult or larval hydrobid numbers and benthic primary production (i.e., food supply), but this was not the case. Again, this may due to comparing these parameters at the same time point, whereas in reality there would be a lag between the increased food supply and the increase in population size.

4.3. HBCDD Effects

Increased temperature was the factor that had the largest effect on these systems. However, HBCDD was also partly responsible for changes in the zooplankton community: higher numbers of gastropod larvae, rotifers, bivalve larvae, and *A. bifilosa* at t6, and higher numbers of the latter two at t13; fewer nauplii, rotifers, and *Eurytemora affinis* at t13 (and rotifers at t6). The decrease in numbers of some taxa may be direct negative effects of the contaminant, but the increase in numbers of some taxa in the presence of HBCDD are likely indirect effects caused by the release of these taxa from predation pressure or competition, or the presence of more food species in these treatments. However, exploration of the potential correlations between competitors, food species, and predators, both within and between time points, revealed little. This is in contrast to Pirzadeh et al. [18], who found copepods (all life stages) to be the most sensitive to HBCDD, resulting in increases in their prey taxa, rotifers, in HBCDD treatments. However, this apparent discrepancy may be explained by the time and frequency of sampling, HBCDD dosage, and other experimental factors. These types of experimental systems are highly dynamic and the temporal resolution of sampling is not sufficient to capture the fine-scale and complex temporal changes in ecosystem structure and function that occur.

It is surprising that more effects of HBCDD were not seen in the benthos, since this is where the majority of the contaminant was present (see Section 3.1). Benthic primary production, *L. balthica* biomass, and the abundance of most meiofauna species was unaffected by HBCDD. The harpacticoid copepod *M. littorale* feeds preferentially on diatom biofilm [52] and fresh organic matter settling on the sediment surface, e.g., after phytoplankton blooms [53,54]. In this experiment, it is therefore likely that it ingested a considerable amount of HBCDD as the spiked algae settled out quickly on to the sediment. However, HBCDD negatively affected *M. littorale*'s abundance only in the ambient treatments; in warm treatments, abundances were higher in the presence of HBCDD. It is possible that this is due to release from predation by Turbellaria, whose abundances showed approximately the opposite trend to *M. littorale*.

Turbellaria are hard to identify and little is known about their ecology [55], but some species are known to predators on other meiofauna. Turbellaria were least abundant in the treatments A– and W+ where *L. balthica* were the largest (although these size differences were non-significant). *L. balthica*'s deposit-feeding activity at the sediment surface is known to disturb the feeding activity of some meiofauna groups [56] and this may have been the case for Turbellaria in this study, although there was no correlation between these variables.

5. CONCLUSIONS

This model ecosystem study experimentally demonstrated the complex nature of shallow Baltic ecosystem responses to a combination of two stressors; the increase in temperature and the presence of a contaminant (HBCDD). Increased temperature was the dominant driver of ecosystem changes, mainly through stimulating the release of PO_4 from the sediment to the water column, resulting in a cyanobacteria bloom and knock-on effects in the plankton community. Higher temperature also enhanced benthic primary production and the production of benthic mollusk larvae. HBCDD effects were detectable in the zooplankton and meiofauna but were more complex, in some cases being different depending on the temperature (i.e., an interactive effect; Figure 10).

Both stressors were applied at levels that are not unreasonable to expect in nature. Climate models for the Baltic Sea predict warming in the range of 2–5 °C by the end of the century, and this is also within the natural fluctuation of a shallow bay in the summer. HBCDD was added at a realistic concentration and applied to the system in an ecologically relevant way (associated with suspended organic matter). Although the study was based on model ecosystems, which are an approximation of field conditions, it has shown that interactive ecosystem effects between two stressors are indeed possible in shallow Baltic ecosystems, as well as demonstrating the ecological and temporal complexity of such responses.

This is only a first step to increasing our understanding of such complex interactions, using only two stressors. In reality, ecosystems may be exposed to multiple chemicals, temperature stress, salinity stress, overfishing, and eutrophication. A major challenge for future management of Baltic Sea ecosystems is how to ensure "good ecological status" [13] in such multistressor situations.

ACKNOWLEDGMENTS

We thank: Bo Sundström and Marc Andersen for identifying the phytoplankton and zooplankton taxa; Annika Cala, Mattias Murphy, Susann Ericsson, Michael Gilek, Leif Lundgren, and Kasia Smolarz for experimental assistance; Anna Strid and Hans von Stedingk (former Department of Materials and Environmental Chemistry, Stockholm University) for HBCDD analyses; Emil Olafsson for meiofauna analyses; and the DEEP Chemistry Laboratory at Stockholm University for nutrient analyses. This study was funded by grants from The Foundation for Baltic and East European Studies to C.B. and K.G., His Majesty Carl XVI Gustaf's Foundation for Science and Education to K.G. and the BEAM program at Stockholm University. Symbols and illustrations in Figure 7 are courtesy of the Integration and Applications Network, University of Maryland Centre for Environmental Science (ian.umces.edu/symbols/).

AUTHOR CONTRIBUTIONS

C.B. and K.G. conceived, designed, and carried out the experiments; C.B. and A.-L.G. analyzed the data and wrote the paper; K.G. also contributed to the interpretation of the results.

CONFLICTS OF INTEREST

The authors declare no conflict of interest. The founding sponsors had no role in the design of the study; in the collection, analyses, or interpretation of data; in the writing of the manuscript, and in the decision to publish the results.

REFERENCES

1. Darling, E.S.; Côté, I.M. Quantifying the evidence for ecological synergies. Ecol. Lett. 2008, 11, 1278–1286.
2. Rocha, J.; Yletyinen, J.; Biggs, R.; Blenckner, T.; Peterson, G. Marine regime shifts: Drivers and impacts on ecosystems services. Philos. Trans. R. Soc. B Biol. Sci. 2014, 370, 20130273.
3. Byrne, M. Global change ecotoxicology: Identification of early life history bottlenecks in marine invertebrates, variable species responses and variable experimental approaches. Mar. Environ. Res. 2012, 76, 3–15.

4. Noyes, P.D.; McElwee, M.K.; Miller, H.D.; Clark, B.W.; van Tiem, L.A.; Walcott, K.C.; Erwin, K.N.; Levin, E.D. The toxicology of climate change: Environmental contaminants in a warming world. Environ. Int. 2009, 35, 971–986.
5. UNEP-AMAP Climate Change and POPs: Predicting the Impacts; Report of the UnePAMAP Expert Group; Secretariat of the Stockholm Convention: Geneva, Switzerland, 2011; pp. 1–66.
6. Schiedek, D.; Sundelin, B.; Readman, J.W.; Macdonald, R.W. Interactions between climate change and contaminants. Mar. Pollut. Bull. 2007, 54, 1845–1856.
7. Fleeger, J.W.; Carman, K.R.; Nisbet, R.M. Indirect effects of contaminants in aquatic ecosystems. Sci. Total Environ.2003, 317, 207–233.
8. Jutterström, S.; Andersson, H.C.; Omstedt, A.; Malmaeus, J.M. Multiple stressors threatening the future of the Baltic Sea-Kattegat marine ecosystem: Implications for policy and management actions. Mar. Pollut. Bull. 2014, 86, 468–480.
9. Helsinki Commission (HELCOM). Red List of Baltic Sea underwater biotopes, habitats and biotope complexes. In Baltic Sea Environment Proceedings No. 138; HELCOM: Helsinki, Finland, 2013; pp. 1–74.
10. Helsinki Commission (HELCOM). Climate Change in the Baltic Sea Area—HELCOM Thematic Assessment. In Baltic Sea Environment Proceedings No. 111; HELCOM: Helsinki, Finland, 2007; pp. 1–54.
11. Helsinki Commission (HELCOM). Climate Change in the Baltic Sea Area Draft HELCOM Thematic Assessment. In Baltic Sea Environment Proceedings No. 137; HELCOM: Helsinki, Finland, 2013.
12. BACC II Author Team (Ed.) Second Assessment of Climate Change for the Baltic Sea Basin. Regional Climate Studies; Springer: Cham, Germany, 2015.
13. Helsinki Commission (HELCOM). HELCOM core indicators: Final report of the HELCOM CORESET Project. In Baltic Sea Environment Proceedings No. 136; HELCOM: Helsinki, Finland, 2013; pp. 1–74.
14. Hansen, J. Benthic Vegetation in Shallow Inlets of the Baltic Sea: Analyses of Human Influences and Proposal of a Method for Assessment of Ecological Status; Report from the EU Central Baltic Interreg IVA Project NANNUT; Department of Botany, Stockholm University: Stockholm, Sweden, 2012.
15. European Chemicals Agency ECHA (European Chemicals Agency). Risk Assessment: Hexabromocyclododecane; Final Report; ECB: Luxembourg, 2008.
16. Bignert, A.; Danielsson, S.; Nyberg, E.; Asplund, L. Comments Concerning the National Swedish Contaminant Monitoring Programme in Marine Biota; Swedish Museum of Natural History: Stockholm, Sweden, 2016; Volume 5.
17. Smolarz, K.; Berger, A. Long-term toxicity of hexabromocyclododecane (HBCDD) to the benthic clam Macoma balthica(L.) from the Baltic Sea. Aquat. Toxicol. 2009, 95, 239–247.
18. Pirzadeh, P.; Gustafsson, K.; Woin, P. Effects of the Brominated Flame Retardant HBCDD on the Structure of Baltic Sea plankton Communities—A Microcosm Study. In Gustafsson K (2004) Uptake and Toxicity of Brominated Flame Retardants and Pesticides. Ph.D. Thesis, Stockholm University, Stockholm, Sweden, 2004.
19. Bradshaw, C.; Näslund, J.; Hansen, J.; Kozlowsky-Suzuki, B.; Sundström, B.; Gustafsson, K. Hexabromocyclododecane affects benthic-pelagic coupling in an experimental ecosystem. Environ. Pollut. 2015, 206, 306–314.
20. Bradshaw, C.; Strid, A.; von Stedingk, H.; Gustafsson, K. Effects of benthos, temperature, and dose on the fate of hexabromocyclododecane in experimental coastal ecosystems. Environ. Toxicol. Chem. 2015, 34, 1246–1257.
21. Jespersen, A.M.; Christoffersen, K. Measurements of chlorophyll—A from phytoplankton using ethanol as extraction solvent. Arch. Hydrobiol. 1987, 109, 445–454.
22. R Core Team. R: A Language and Environment for Statistical Computing; R Foundation for Statistical Computing: Vienna, Austria, 2016; Available online: https://www.R-project.org (accessed on 3 October 2016).

23. Eklöf, J.S.; Havenhand, J.N.; Alsterberg, C.; Gamfeldt, L. Community-level effects of rapid experimental warming and consumer loss outweigh effects of rapid ocean acidification. Oikos 2015, 124, 1040–1049.
24. Oksanen, J.; Blanchet, F.G.; Friendly, M.; Kindt, R.; Legendre, P.; McGlinn, D.; Minchin, P.R.; O'Hara, R.B.; Simpson, G.L.; Solymos, P.; et al. Vegan: Community Ecology Package. R package version 2.4-1. Available online: https://CRAN.R-project.org/package=vegan (accessed on 3 October 2016).
25. Boström, B.; Andersen, J.M.; Fleischer, S.; Jansson, M. Exchange of Phosphorus Across the Sediment—Water Interface. Hydrobiologia 1988, 170, 229–244.
26. Jensen, H.S.; Andersen, F. Importance of temp etc for PO4 release from lake seds. Limnol. Oceanogr. 1992, 37, 577–589.
27. Mortimer, C.H. Chemical Exchanges Between Sediments and Water in the Great Lakes-Speculations on Probable Regulatory Mechanisms. Limnol. Oceangr. 1971, 16, 387–404.
28. Lukkari, K.; Leivuori, M.; Kotilainen, A. The chemical character and behaviour of phosphorus in poorly oxygenated sediments from open sea to organic-rich inner bay in the Baltic Sea. Biogeochemistry 2009, 96, 25–48.
29. Sanz-Lázaro, C.; Valdemarsen, T.; Holmer, M. Effects of temperature and organic pollution on nutrient cycling in marine sediments. Biogeosciences 2015, 12, 4565–4575.
30. Viktorsson, L.; Almroth-Rosell, E.; Tengberg, A.; Vankevich, R.; Neelov, I.; Isaev, A.; Kravtsov, V.; Hall, P.O.J. Benthic Phosphorus Dynamics in the Gulf of Finland, Baltic Sea. Aquat. Geochem. 2012, 18, 543–564.
31. Gächter, R.; Meyer, J.S. The Role of Microorganisms in Mobilization and Fixation of Phosphorus in Sediments. Hydrobiologia 1993, 253, 103–121.
32. Holdren, G.C.; Armstrong, D.E. Factors Affecting Phosphorus Release From Intact Lake Sediment Cores. Environ. Sci. Technol. 1980, 14, 79–87.
33. Jensen, H.S.; Mortensen, P.B.; Andersen, F.O.; Rasmussen, E.; Jensen, A. Phosphorus Cycling in a Coastal Marine Sediment, Aarhus Bay, Denmark. Limnol. Oceangr. 1995, 40, 908–917.
34. Koop, K.; Boynton, W.R.; Wulff, F.; Carman, R. Sediment-Water Oxygen and Nutrient Exchanges Along a Depth Gradient in the Baltic Sea. Mar. Ecol. Prog. Ser. 1990, 63, 65–77.
35. Karlson, A.M.L.; Duberg, J.; Motwani, N.H.; Hogfors, H.; Klawonn, I.; Ploug, H.; Svedén, J.B.; Garbaras, A.; Sundelin, B.; Hajdu, S.; et al. Nitrogen fixation by cyanobacteria stimulates production in Baltic food webs. Ambio 2015, 44, 413–426.
36. Walve, J.; Larsson, U. Seasonal changes in Baltic Sea seston stoichiometry: The influence of diazotrophic cyanobacteria. Mar. Ecol. Prog. Ser. 2010, 407, 13–25.
37. Nielsen, J.M.; Winder, M. Seasonal dynamics of zooplankton resource use revealed by carbon amino acid stable isotope values. Mar. Ecol. Prog. Ser. 2015, 531, 143–154.
38. Brandl, Z. Freshwater Copepods and Rotifers: Predators and their Prey. Hydrobiologia 2005, 546, 475–489.
39. Bonsdorff, E.; Norkko, A.; Boström, C. Recruitment and population maintenance of the bivalve Macoma balthica (L.)—factors affecting settling success and early survival on shallow sandy bottoms. In Biology and Ecology of Shallow Coastal Waters: Proceedings of the 28 EMBS Symposium; Olsen & Olsen: Fredensborg, Denmark, 1995; pp. 253–260.
40. Caddy, J.F. Development of Mantle Organs, Feeding, and Locomotion in Postlarval Macoma balthica (L) (Lamellibranchiata). Can. J. Zool. 1969, 47, 609–617.
41. Drent, J. Temperature responses in larvae of Macoma balthica from a northerly and southerly population of the European distribution range. J. Exp. Mar. Biol. Ecol. 2002, 275, 117–129.
42. De Goeij, P.; Honkoop, P.J. Experimental effects of immersion time and water temperature on body condition, burying depth and timing of spawning of the tellinid bivalve Macoma balthica. Helgol. Mar. Res. 2003, 57, 20–26.

43. Kennedy, V.S.; Mihursky, J.A. Upper temperature tolerances of some estuarine bivalves. Chesap. Sci. 1971, 12, 193–204.
44. Bos, O.G.; Philippart, C.; van der Meer, J. Effects of temporary food limitation on development and mortality of Macoma balthica larvae. Mar. Ecol. Prog. Ser. 2007, 330, 155–162.
45. Bos, O.G.; Hendriks, I.E.; Strasser, M.; Dolmer, P.; Kamermans, P. Estimation of food limitation of bivalve larvae in coastal waters of north-western Europe. J. Sea Res. 2006, 55, 191–206.
46. Bos, O.G.; Philippart, C.J.M.; Cadee, G.C.; van der Meer, J. Recruitment variation in Macoma balthica: A laboratory examination of the match/mismatch hypothesis. Mar. Ecol. Prog. Ser. 2006, 320, 207–214.
47. Raby, D.; Mingelbier, M.; Dodson, J.J.; Klein, B.; Lagadeuc, Y.; Legendre, L. Food-particle size and selection by bivalve larvae in a temperate embayment. Mar. Biol. 1997, 127, 665–672.
48. Lassen, H.H. Reproductive effort in Danish mudsnails (Hydrobiidae). Oecologia 1979, 40, 365–369.
49. Fish, J.D.; Fish, S. The effects of temperature and salinity on the embryonic development of Hydrobia ulvae (Pennant). J. Mar. Biol. Assoc. UK 1977, 57, 213–218.
50. Hagerthey, S.E.; Defew, E.C.; Paterson, D.M. Influence of Corophium volutator and Hydrobia ulvae on intertidal benthic diatom assemblages under different nutrient and temperature regimes. Mar. Ecol. Prog. Ser. 2002, 245, 47–59.
51. Blanchard, G.F.; Guarini, J.-M.; Provot, L.; Richard, P.; Sauriau, P.-G. Measurement of ingestion rate of Hydrobia ulvae(Pennant) on intertidal epipelic microalgae: The effect of mud snail density. J. Exp. Mar. Biol. Ecol. 2000, 255, 247–260.
52. De Troch, M.; Chepurnov, V.A.; Vincx, M.; Olafsson, E. The effect of Fucus vesiculosus on the grazing of harpacticoid copepods on diatom biofilms. J. Sea Res. 2008, 60, 139–143.
53. Nascimento, F.; Karlson, A. Settling blooms of filamentous cyanobacteria as food for meiofauna assemblages. Limnol. Oceanogr. 2008, 53, 2636–2643.
54. Olafsson, E.; Modig, H.; van de Bund, W.J. Species specific uptake of radio-labelled phyto-detritus by benthic meiofauna from the Baltic Sea. Mar. Ecol. Prog. Ser. 1999, 177, 63–72.
55. Martens, P.M.; Schockaert, E.R. The Importance of Turbellarians in the Marine Meiobenthos—A Review. Hydrobiologia1986, 132, 295–303.
56. Nascimento, F.J.A.; Karlson, A.M.L.; Näslund, J.; Elmgren, R. Diversity of larger consumers enhances interference competition effects on smaller competitors. Oecologia 2011, 166, 337–347.

CHAPTER 12

Multi-Layered Stratification in the Baltic Sea: Insight from a Modeling Study with Reference to Environmental Conditions

Bijan Dargahi [1,3,*], Venkat Kolluru [2,*] and Vladimir Cvetkovic [3]

[1]*Division of Hydraulic Engineering, School of Civil and Architectural Engineering, KTH Royal Institute of Technology, 10044 Stockholm, Sweden*

[2]*Environmental Resources Management, Inc., 75 Valley Stream Parkway, Suite 200, Malvern, PA 19355, USA*

[3]*Division of Water Resource Engineering, School of Civil and Built Environment, KTH Royal Institute of Technology, 10044 Stockholm, Sweden*

ABSTRACT

The hydrodynamic and transport characteristics of the Baltic Sea in the period 2000–2009 were studied using a fully calibrated and validated 3D hydrodynamic model with a horizontal resolution of 4.8 km. This study provided new insight into the type and dynamics of vertical structure in the Baltic Sea, not considered in previous studies. Thermal and salinity stratification are both addressed, with a focus on the structural properties of the layers. The detection of cooler regions (dicothermal) within the layer structure is an important finding. The detailed investigation of thermal stratification for a 10-year period (i.e., 2000–2009) revealed some new features. A multilayered structure that contains several thermocline and dicothermal layers was identified from this study. Statistical analysis of the simulation results made it possible to derive the mean thermal stratification properties, expressed as mean temperatures and the normalized layer thicknesses. The three-layered model proposed by previous investigators appears to be valid only during the winter periods; for other periods, a multi-layered structure with more than five layers has been identified during this investigation. This study provides detailed insight into thermal and salinity stratification in the Baltic Sea during a recent decade that can be used as a basis for diverse environmental assessments. It extends previous studies on stratification in the Baltic Sea regarding both the extent and the nature of stratification.

Keywords: Baltic Sea; hydrodynamics; modeling; vertical structure; stratification; dicothermal; GEMSS

1. INTRODUCTION

The Baltic Sea is a brackish sea located in northern Europe from 53° N to 66° N latitude and from 20° E to 26° E longitude. It is connected to the Atlantic Ocean via the Danish Straits. The Baltic Ice Lake was born 13,000 years ago and its present brackish state emerged 7000 years ago. For 2000 years, the salinity has been close to the present level (mean salinity: 7 parts per thousand). The Baltic Sea borders nine coastal countries with a total population of 85 million people (see Figure 1, [1]). The maximum length and width are 1600 km and 193 km, respectively. The surface area is 377,000 km^2, with an average depth of 55 m and a water volume of 20,000 km^3. Its maximum depth is 459 m, which is located between Stockholm and the Island of Gotland. The Baltic Sea is a shallow sea that consists of a series of basins interconnected through narrow sills (Figure 2).

In spite of the Baltic Sea HELCOM agreement signed in 1974, the state of the Baltic Sea has worsened (http://www.helcom.fi). Nutrient levels in the water and sediments are high, and poor oxygen conditions and “dead bottoms” exist in large archipelago areas of both Sweden and Finland [2]. However, during the past few decades there have been considerable efforts towards better and more sustainable management of the Baltic Sea. Here, the hydrodynamics of the Baltic Sea, which has been the subject of intensive research since the 1930s, plays a major role. The number of available journal articles and other publications exceeds several hundred. The foci of these studies are exchange processes, especially salt transport from the North Sea, and water age. Some of the main contributors are Meier [3], and Lehman and Hinrichsen [4]. Here, the work of Omstedt et al. [5] should be mentioned as it presents the state of knowledge on various hydrodynamic features of the Baltic Sea. There are also several excellent books covering many different aspects, including Feistel et al. [6], Leppäranta and Myrberg [7], and Harff et al. [8]. For a general literature review, the interested reader is referred to the comprehensive review given by Dargahi and Cvetkovic [9].

Figure 1. The Baltic model region and the model open boundary location [1].

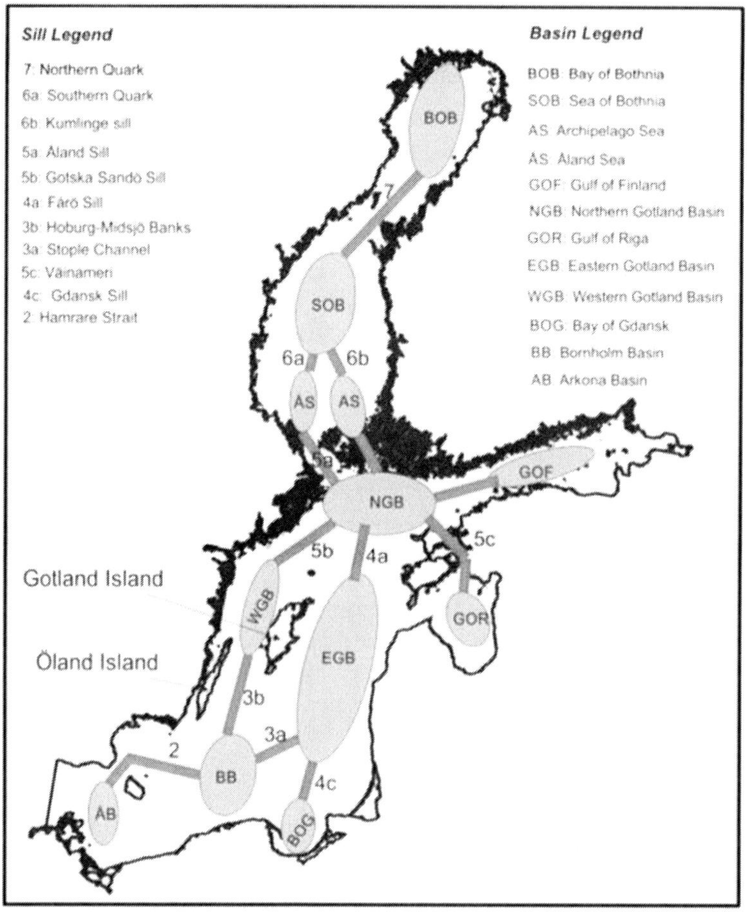

Figure 2. The Baltic Sea basin system, redrawn from an interpolated figure by Leppäranta & Myrberg [7].

The present study concerns the hydrodynamics of the Baltic, with a focus on the details of stratification, which is the primary feature of the sea. The novel features of the work are the relatively long simulation period of 10 years and the use of a complete set of external boundary conditions with maximum spatial and temporal resolution, in combination with highly accurate bathymetry. The main objectives were to create an accurate and validated 3D hydrodynamic model for investigating specific stratification characteristics that have not been addressed in the previous studies. In the following paragraph we present a short summary of previous research works on stratification.

The Baltic Sea is highly stratified by strong vertical salinity and temperature gradients. The stratification is commonly referred to as a two-layer structure that consists of an upper and a lower layer. A transitional middle layer exists between the upper and lower layers, which are known as halocline and thermocline, respectively. There is a significant variation in the depth of the halocline, from 40 m to 80 m in deeper regions to 10 m–30 m in shallower regions [7]. The lower values are found in the Gulf of Riga (mean value = 25 m), and Arkona Basin, with a mean value of 25 m for both regions [10,11]. The surface salinity varies in the north with a mean value of 3 ppt (parts per thousand) to 8 ppt to the south, i.e., the Arkona Basin. The corresponding mean value at the lower layer is in the range of 4–12.5 ppt. However, the salinities are considerably higher in the open sea. The mean values in the Kattegatt (the region in which the Baltic Sea drains through the Danish strait) are 22 ppt and 31 ppt, respectively (same references). Suominen et al. [12] studied surface salinity gradients and their temporal fluctuations in the Archipelago Sea of the northern Baltic Sea based on field salinity data for the time period July 2007–August 2008. They identified a broad scale gradient from low salinity in the shallow inner bays to the high salinity in the open sea areas towards the Baltic proper. The steepest gradients were observed in the semi-closed part of the archipelago. One important result was that the use of temporal mean values of salinity was insufficient for coastal management purposes in the region.

The halocline depth is controlled by wind-induced mixing and advection, which appears to change little over time. Väli et al. [11] looked into variations of the halocline during 1961–2007. Two periods were identified with shallow halocline during 1970–1975, and with deep halocline during 1990–1995. The main conclusion was that the freshwater content and absolute wind speed control the halocline depth in the Baltic Sea. However, they found the wind speed to have a moderate impact on the mean halocline depth in the Baltic proper due to the low impact of runoff.

An important issue is the effect of fresh water on stratification in the Baltic Sea. Hordoir and Meier [13] studied the dynamics of fresh water, which is released during spring into the Baltic proper. They showed that the fresh water only reaches the center of the Baltic proper after late summer. A small amount of fresh water may reach the entrance of the Baltic Sea during one season. The arrival of fresh water increases vertical stratification, which can in turn trigger the onset of the spring blooms. They also found that the seasonal changes in the fresh water outflow were closely connected with those of the zonal wind. An important result was the correlation of the annual variability of the seasonal freshwater outflow maximum with the North Atlantic Oscillation.

The temperature stratification in the Baltic Sea has a mean three-layer structure in analogy to the salinity stratification. The layers are commonly referred to as the epilimnion (upper layer), thermocline (middle layer), and hypolimnion (lower layer). In similarity with the other large water bodies, there is a seasonal stratification cycle, which is driven by the variations in the energy balance. However, in the case of the Baltic Sea there are two specific features. First, fall and spring overturns are not well defined, and second, the hypolimnion has a nearly constant temperature, with few seasonal variations. During the winter periods, the epilimnion layer has a lower temperature than the hypolimnion layer. The Baltic Sea is monitored using 22 stations, as shown in Figure 3. A typical temperature difference at the gauge station BY15 is about 5 °C. In the spring, following the ice melt, a thin warmer surface layer rapidly develops and sets up the thermocline. The thickness of the layer varies considerably from north to south, but a mean value of about 15 m can be used. The temperature gradient is high within the epilimnion layer. For instance, at BY15, the temperature can vary from 1.5 °C to 5 °C within a depth of 60 m. Below this depth, the temperatures increase rapidly to reach the constant temperature of the hypolimnion layer (about 5 °C). The layer with transitional temperature is known as the dicothermal layer, which is a cold layer sandwiched between two layers with higher temperatures (dicothermal). The dicothermal layer was first discovered by the Ekman expedition of 1877 to the Baltic Sea (see Fonselius [14]). The layer appears to originate from the vertical convection of the surface water in the winter. It is further explained that this cold surface water from the previous winter was preserved between the thermocline and halocline (see Fonselius [14]). The dicothermal layer develops at high latitudes with cold climates. Peter [15] reports the development of a layer with a thickness of 100 m in the Indian Ocean region of the Antarctic. The reported thickness in the Baltic Sea is in the range of 5–30 m, which persists during the summer and disappears during the autumn [7]. Here, we note that the measured temperature profiles in the southern basins confirm the formation of the dicothermal layer even during the spring.

The stratification is strongest during the summer due to high solar radiation input and warm air temperatures. The surface layer thickness increases to about 20 m during the summer period due to wind induced vertical mixing The temperature within the layer is nearly constant. Below the surface layer, a strong thermocline develops that has a sharp temperature drop of about 10 °C over a depth of about 10 m (e.g., temperature profile at BY15). There is also a dicothermal layer below the thermocline, which has a thickness of about 30 m at BY15. The hypolimnion layer has a relatively constant temperature of 4–5 °C, which is close to the temperature of maximum density of water. The layer thickness varies considerably, from about 30 m in shallow regions to 100 m in the deeper region. The negative effect of the strong stratification limits the exchange between the epilimnion and hypolimnion layers.

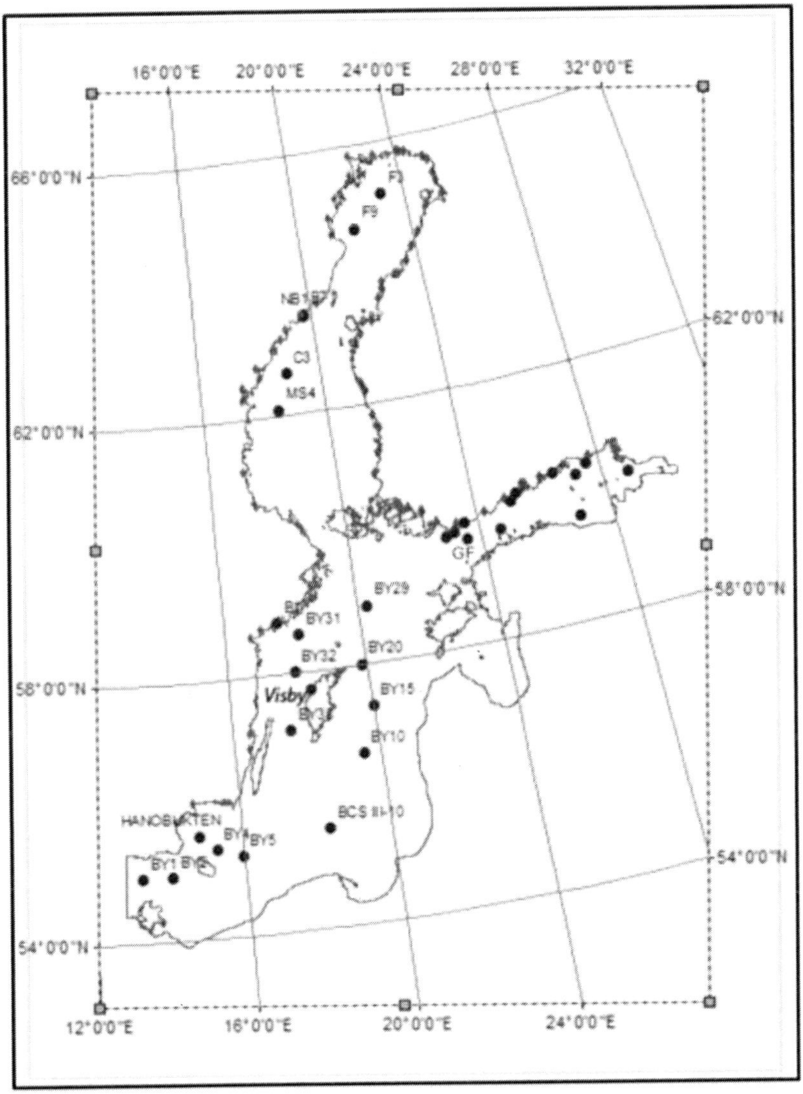

Figure 3. Map showing the monitoring stations in the Baltic Sea.

During autumn, the surface heat losses start to increase and the thermocline depth deepens. For instance, in the Eastern Gotland Basin, the thermocline reaches a depth of about 40 m. The lower temperatures cause the temperature gradient to decrease, which in turn weakens the thermocline. The temperature changes in the thermocline cause a weak positive temperature gradient to develop in the hypolimnion layer (e.g., 1 °C over 100 m).

2. MATERIALS AND METHODS

The materials used to model the Baltic Sea consisted of basic geometrical and various flow and meteorological data for the period 2000–2009 as listed below.

- The shoreline and the bathymetry in GIS format.
- Daily flow discharges for 24 Swedish rivers, 38 Finnish rivers, and five Eastern European rivers (i.e., the Daugava, Neman, Neva, Odra, and Vistula).
- Monthly mean flow discharge for four Eastern European rivers (i.e., Lielupe, Narva, Pärnu, and Narva). The daily records were not available.
- Water temperature for all the rivers.
- The forcing meteorological data (air temperature, dew point, cloud cover, pressure, wind speed, and wind direction) at 3-h intervals as grid data.
- Precipitation as rain intensity at 19 stations at daily intervals.
- Water quality data at 15-day intervals for 22 different stations spread across the sea. The data included water temperature, salinity, dissolved oxygen, and phosphorous.
- Wave heights and sea and water levels at several stations across the sea.

The main data sources were the Swedish Meteorological and Hydrological Institute, SMHI [16], and the Finnish Meteorological Institute, FMI [17]. The gridded meteorological data were obtained from http://www.smhi.se/en/research/research-departments/analysis-and-prediction, computed based on actual measurements. Among the many variables that affect the hydrodynamics of a large water body, the bathymetry and inflow of freshwater are the most important. So, special attention was paid to improving the quality and reliability of the bathymetry data.

The digitized bathymetry data for the Baltic Sea were obtained from the Leibniz Institute for Baltic Sea Research [18]. This dataset was frequently used in previous models of the Baltic Sea and also agrees well with published bathymetry maps. This dataset performed poorly in resolving the various channels along the coastlines of Finland, and in the Stockholm archipelago. The Åland Sea and its archipelago were also poorly reproduced. The latter problems caused a significant flow blockage in the forenamed areas. To resolve the foregoing issues, the bathymetry had to be refined using several different resolutions ranging from 50 m to 400 m that depended upon the model regions. The focus was on the areas along the coastlines and the interconnected channel systems. The modifications were done by a combined method using published maps and other databases in ARCGIS. The final bathymetry was analyzed using standard statistical methods to gain an understanding of the depth distribution and its relationship for resolving the vertical layers for the study domain. It was observed that depths less than 100 m cover 80% of the sea. This is an important result since it indicates the need for finer vertical grid resolution within the depth of 0–100 m.

2.1. Model Description

A three-dimensional, time-dependent hydrodynamic model, GEMSS® (Generalized Environmental Modelling System for Surface waters), was used. GEMSS was developed and maintained by Environmental Resources Mangement, Inc., Malvern, PA, USA. GEMSS is an integrated system of 3D hydrodynamic and transport models embedded in GIS. GEMSS is in the public domain [19] and has been used for similar studies throughout the United States and worldwide. Edinger and Buchak [20,21] first presented the theoretical basis of the model. The model was enhanced by implementing higher-order transport schemes, construction of various constituent modules, incorporation of various supporting software tools, GIS interoperability, visualization tools, graphical user interface (GUI), and post processors [22,23,24,25,26].

The hydrodynamic and transport relations are developed from the horizontal momentum balance, continuity, constituent transport and the equation of state. A detailed mathematical formulation of the model both in the hydrostatic and non-hydrostatic forms is described in [17,19] so will not be repeated here. The hydrodynamic equations are semi-implicit in time, have the advantage of computational stability, and are not limited by the Courant condition. The vertical momentum dispersion coefficient and vertical shear are evaluated from a Von Karman relationship modified by the Richardson number. Higher-order turbulence closure schemes (two-equation model, and second-moment closure model by Mellor and Yamada [27] are also included. The Two-Equation model used in GEMSS is based on the Generic Length Scale (GLS) model proposed by Umlauf and Burchard [28], and by Warner et al. [29]. The longitudinal and lateral coefficients are scaled to the dimensions of the grid cell using the dispersion relationships field developed by Okubo [30] and modified to include the velocity gradients using the Smagorinsky [31] relationship. The wind stress and bottom shear stress are computed using quadratic relationships with appropriate friction coefficients.

The transport module can run in fully explicit to fully implicit modes in a vertical direction while performing explicit computations in a horizontal direction [32]. GEMSS uses a curvilinear variably spacing horizontal staggered finite difference grid, which is based on control volume, with the elevation and constituent concentration computed at cell centers and velocities through a cell interface. Z-level with variable thickness is used for defining the grid in the vertical direction. Additional details of the model can be found in the technical documentation of GEMSS [32,33,34,35].

The wave dynamic module of GEMSS has two wave models, i.e., a steady state linear and a non-linear model. In the present investigation, the non-linear model was used. The model accounts for the wave influence on the bottom shear stresses by using the Madsen and Grant [36] equation. In the current study, a simplified linear ice model that relates the growth of ice thickness to the temperature differences between water and melting ice, and ice in equilibrium was used. Based on sensitivity studies, the linear ice model is a reasonably good model for the present study, which was run for a long period of time to understand the hydrodynamic characteristics. In addition, the particle tracking module of GEMSS was used for the current study to understand the travel time, water age [37] and vertical mixing processes at various basins in the Baltic Sea using the current persistency index defined in [38].

The GEMSS model was recently used by the first author to investigate the hydrodynamic and related water quality characteristics of Saltsjo [33] and Lake Tana in Ehiopia [34].

2.2. Model Setup

The model setup involved several main steps of creating the model grid, interpolating the bathymetry into the grid, defining the boundary conditions that include river inflow, water levels, precipitation, and forcing meteorological conditions.

The model grid was created using the grid generator tool of GEMSS®. A non-uniform boundary-fitted curvilinear grid in a horizontal plane (x-y) with a non-uniform z-layering in the vertical plane was created for the study domain and is shown in Figure 4. The grid dimensions are 195 × 200, with an approximate cell size of 4.8 km. The final bathymetry described in the previous section was used to interpolate the depths for each grid cell in the study domain. Since depths less than 100 m cover almost 80% of the Baltic Sea, the vertical layers were designed to have a finer resolution within this depth. The total number of vertical layers used in the current study was 47, with thicknesses varying from 1.5 m to 12 m.

2.2.1. Input Data

The model boundary conditions consisted of discharge, head (water levels), precipitation, and meteorological forcing conditions. The true dynamic character of the input data is an important issue that directly affects the results of any hydrodynamic simulations. Here, the input data are the river discharge hydrographs and the forcing meteorological conditions. The amplitude and frequency of the data control various hydrodynamic properties such as stratification and mixing processes. The control file generator tool of GEMSS was used to define all forcing data needed for the current study.

The discharge boundary conditions were defined for 69 rivers out of 72 that enter the modeled region. To define the exact locations of the rivers, a GIS file was used. The river data included the flow discharge, water temperature, and salinity. We assumed that the rivers enter through the water surface grid in the model.

The water level was set at the open boundary with the North Sea, as shown in Figure 1 with a red line (≈104 km wide). The GPS coordinates are 54°28′ N 12°50′ E and 55°22′ N 13°03′ E in the south and north directions, respectively. To set the water level, the data at the gauge station Skanör (55°26′ N 12°50′ E) were used. At the open boundary the water temperatures and salinity profiles from monitoring station BY1 (see Figure 3) were used.

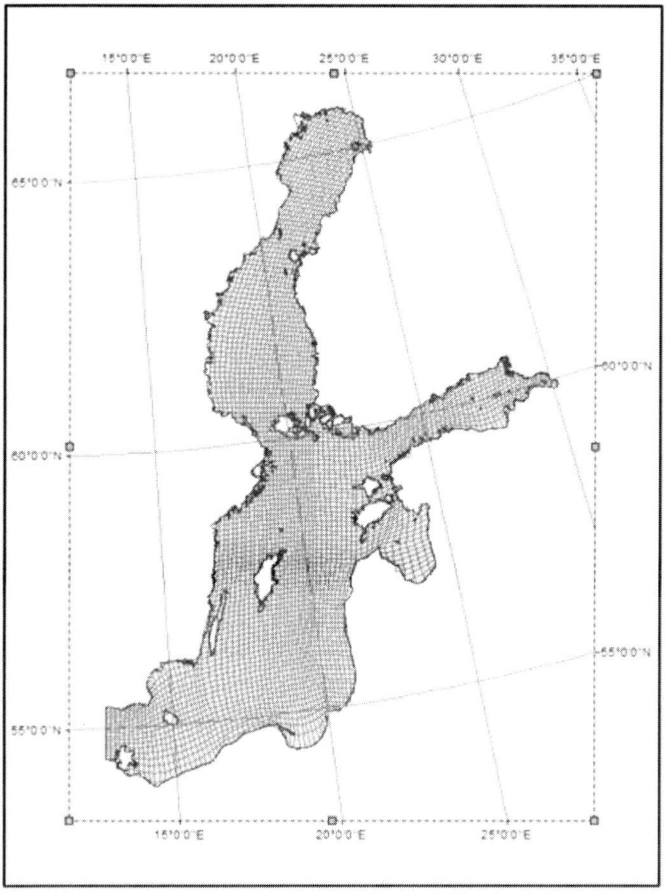

Figure 4. The Baltic Sea numerical model grid.

The precipitation data in mm/day were applied regionally by dividing the Baltic Sea surface into 19 regions, each with its corresponding rain intensity. Both point and gridded data were used for meteorological forcing conditions. The gridded data cover the whole Baltic drainage basin with a grid of (1° × 1°) squares. The grid extends over the area: Latitude 49.5°–71.5° N, Longitude 7.5°–39.5° E. The gridded data covers 32 years, starting in 1970. The gridded meteorological data represent geostrophic wind and were converted to the open surface wind speed needed for the model using the 2/3 power law. The second set was point data, which are available mainly along the coastlines.

2.2.2. Initialization

The model initialization is an important part of any hydrodynamic simulation, especially in the case of large water bodies such as the Baltic Sea. The long residence times make the model output sensitive to the choice of method. In the present study, we initiated the model using the data available at all the monitoring stations for water temperature and salinity. A total of 22 monitoring stations (see Figure 3) were used to interpolate temperature and salinity for the entire model grid at the start of the model simulation on 1 January 2000.

2.3. Simulations

In the present study we used the following setups for both partial and complete simulations, i.e., one year and 10 years.

- Vertical dispersion: Two-Equations with Mellor–Yamada formulation.
- Mixing dispersion: Okubo formulation.
- Transport diffusion: Prandtl method.
- Surface heat exchange: term by term, defining all the heat source and sink terms with atmospheric interaction.
- Transport model: Quick.
- Vertical momentum: Non-hydrostatic.
- Coriolis force: Using model grid.

The maximum time step used in the model simulations was 360 s. The auto time step feature available in GEMSS® was used so that the model time step goes only below the maximum time step due to extreme forcing conditions such as large wind speeds and river discharges. The calibration and verification simulations were carried out on a 4.8-km model grid shown in Figure 4. The model was run for the year 2000 for calibration. The model validation was done for the full 10 years using the restart files created by the calibration run.

3. RESULTS

3.1. Model Calibration & Validation

A detailed description of model calibration and verification is available in [9] and will not be repeated here. Instead, it will be briefly described here using some relevant plots and tables.

The model calibration for year 2000 was done using all the 22 monitoring stations shown in Figure 3. The temperature, salinity, and water levels field data were used. It involved a systematic two-step approach with a focus on temperature and salinity profiles. The first step was initiating the model using the data available at all the monitoring stations for water temperature and salinity. The RMS (Root Mean Square) errors of temperature and salinity were then evaluated. In Step 2 the model was re-run with new sets of initial data that were successively adjusted to lower the RMS values. The procedure significantly improved the agreement with the measurements, with a relative error range of 10%–20% for both temperature and salinity. During the calibration procedure, the bottom friction coefficient, wind drag coefficient, and coefficients related to surface heat exchange processes were adjusted to lower the RMS values with respect to field observations. The coefficient values used for the calibration matched reasonably well with similar modeling exercises completed for the Baltic Sea. Based on all the calibration adjustments we have done for various model parameters, we found that establishing the appropriate horizontally and vertically varying initial condition was the most important precursor to achieve low RMS values on model results. The model successfully captured the seasonal variations in water temperature and salinity profiles as well as the stratification across the entire Baltic Sea. Here, we have chosen only some typical results for a few stations. Figure 5 and Figure 6 show the combined temperature and salinity profiles at the monitoring stations F3, GF, BY15, and BY5. Table 1summarizes the absolute and mean relative errors at these monitoring stations. The range of the error is 4%–10%, which is acceptable considering the large volume of the Baltic Sea and its complex hydrodynamics.

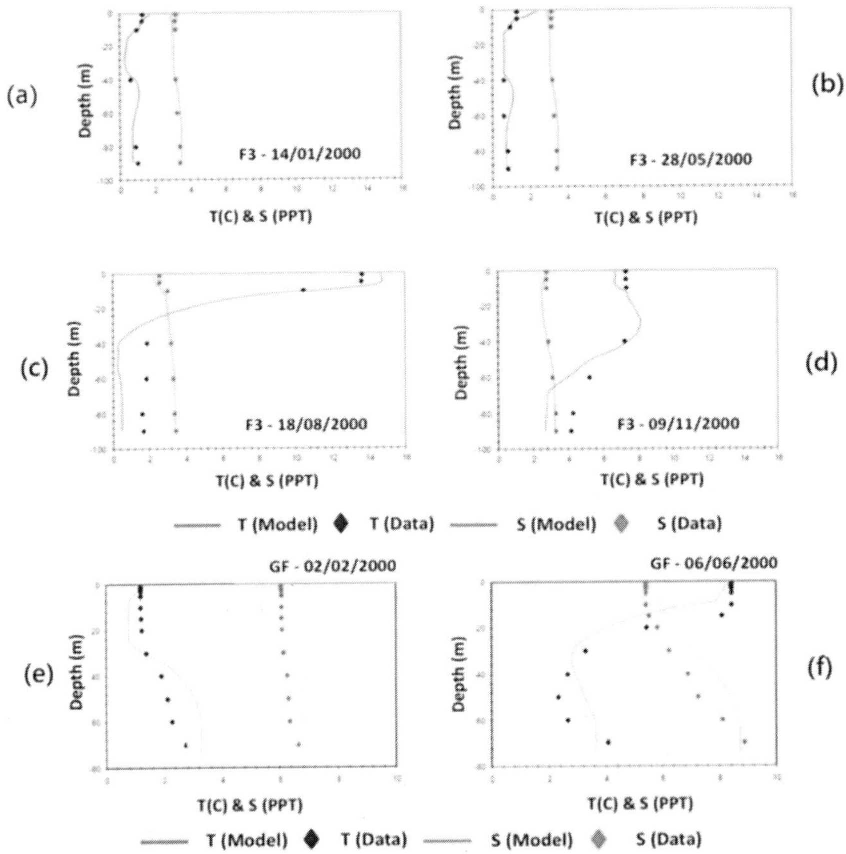

Figure 5. Comparison of model predicted vertical temperature and salinity profiles at the monitoring stations F3 (Bay of Bothnia) and GF (Gulf of Finland) with field measurements for the year 2000. (**a**) Station F3—14/01/2000; (**b**) Station F3—28/05/2000; (**c**) Station F3—18/08/2000; (**d**) Station F3—09/11/2000; (**e**) Station GF—02/02/2000; Station (**f**)—06/06/2000.

Table 1. Calibration error summary for temperature and salinity.

	Absolute Error								Mean Relative Error %	
Time	14 January 2000		28 May 2000		18 August 2000		9 November 2000			
Station	*T* (°C)	*S* (ppt)	*T* (°C)	*S* (ppt)	*T* (°C)	*S* (ppt)	*T* (°C)	*S* (ppt)	*T*	*S*
F3	0.15–0.6	0.1–0.2	0.1–1.5	0.1–0.2	0.1–0.2	0.1–0.2	0.1–0.2	0.1–0.2	10	3
Time	14 January 2000		28 May 2000		22 August 2000		25 October 2000			
BY15	0.1–0.57	0.1–0.8	0.1–1.3	0.1–0.8	0.1–0.9	0.1–0.7	0.1–1.6	0.1–1.2	4	4
Time	14 January 2000		28 May 2000		22 August 2000		25 October 2000			
BY5	0.1–0.2	0.1–0.2	0.1–0.5	0.1–0.3	0.1–0.5	0.1–0.3	0.1–1.6	0.1–2	5	6
Time	2 January 2000		6 June 2000							
GF	0.1–1	0.1–0.7	0.1–1	0.1–0.2					9	8

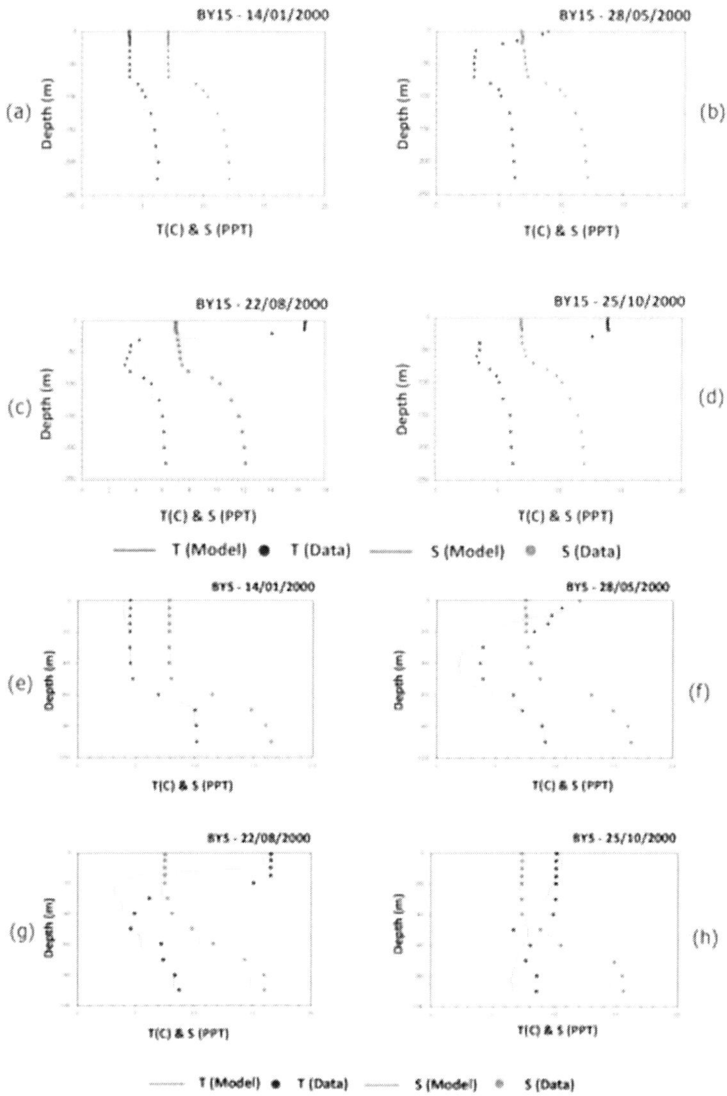

Figure 6. Comparison of model predicted vertical temperature and salinity profiles at the monitoring stations BY15 (Eastern Gotland Basin) and BY5 (Arkona Basin) with field measurements for the year 2000. (**a**) Station BY15—14/01/2000; (**b**) Station BY15—28/05/2000; (**c**) Station BY15—22/08/2000; (**d**) Station BY15—25/10/2000; (**e**) Station BY5—14/01/2000; (**f**) Station BY5—28/05/2000; (**g**) Station BY5—22/08/2000; (**h**) Station BY5—25/10/2000.

Model validation was done for the whole 10 years using the restart files created by the calibration simulation for the year 2000. For this purpose, the complete sets of field data at all the monitoring stations were used (Figure 3). These data were considered independently from the simulation results as only the first day records were used in the calibration simulations.

The model predictions of water temperatures and salinities were satisfactory at all the monitoring stations. Here, we have chosen some representative results for only a few stations. The time history plots in Figure 7 and Figure 8 compare the model predicted temperature and salinity with the field data at stations F3 and BY29 at water surface and 40 m depth.

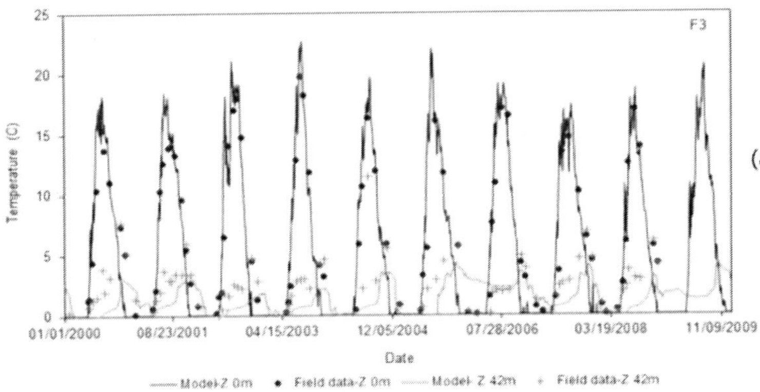

(a)

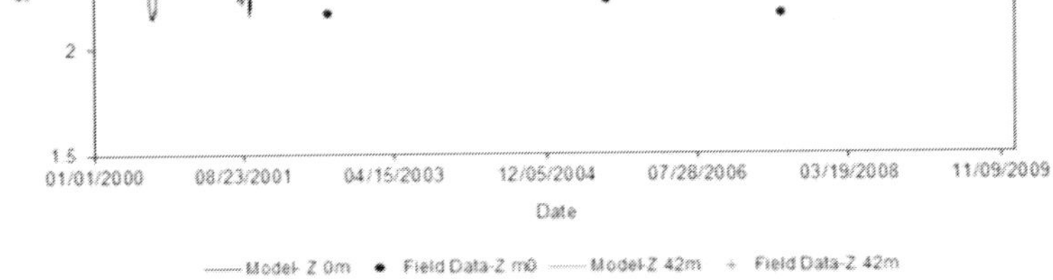

(b)

Figure 7. (**a**) Comparison of model predicted and measurements of temperature in the Baltic Sea at the Station F3 surface and 42 m depth for the time period 2000–2009; (**b**) Comparison of model predicted and measurements of salinity in the Baltic Sea at the Station F3 surface and 42 m depth for the time period 2000–2009.

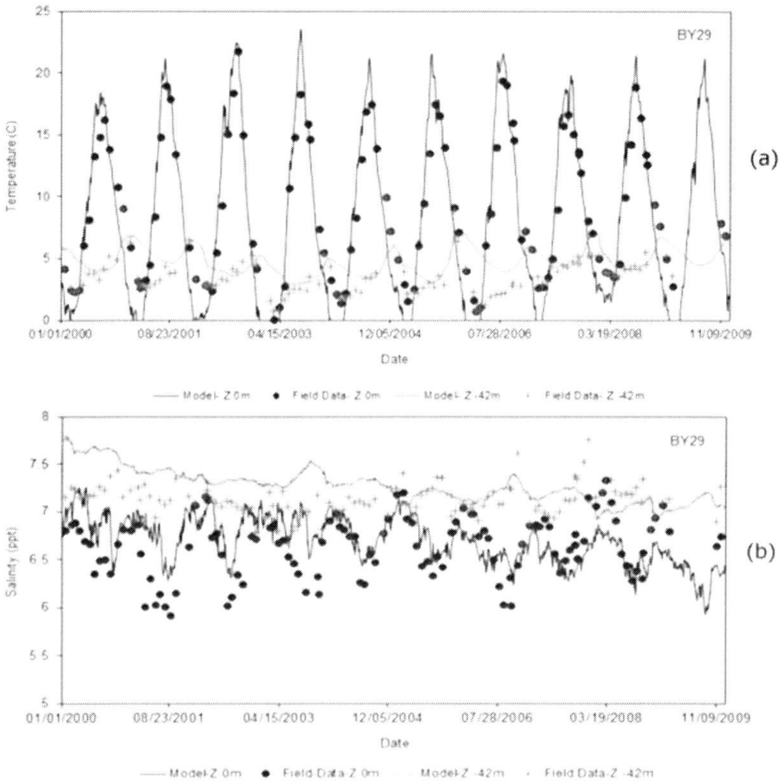

Figure 8. (**a**) Comparison of model predicted and measurements of temperature in the Baltic Sea at the Station BY29 surface and 42 m depth for the time period 2000-2009; (**b**) Comparison of model predicted and measurements of salinity in the Baltic Sea at the Station BY29 surface and 42 m depth for the time period 2000–2009.

The time history plots in Figure 9 compare the model predicted salinity with the field data at stations BY15 and BY5. The comparisons are shown at the water surface and at the deepest points in the basins. The agreements for these stations are better than the other stations as they are closer to station BY1, which defines the model boundary condition.

The range of Nash–Sutcliffe coefficients [39] calculated for all the monitoring stations was 0.72–0.83, respectively, also suggesting close agreement. One other important feature is the ability of the model to capture the 10-year seasonal variations of water temperatures and salinities. The conclusion is that the model is reasonably validated over a considerable period, with a maximum relative error of 10%.

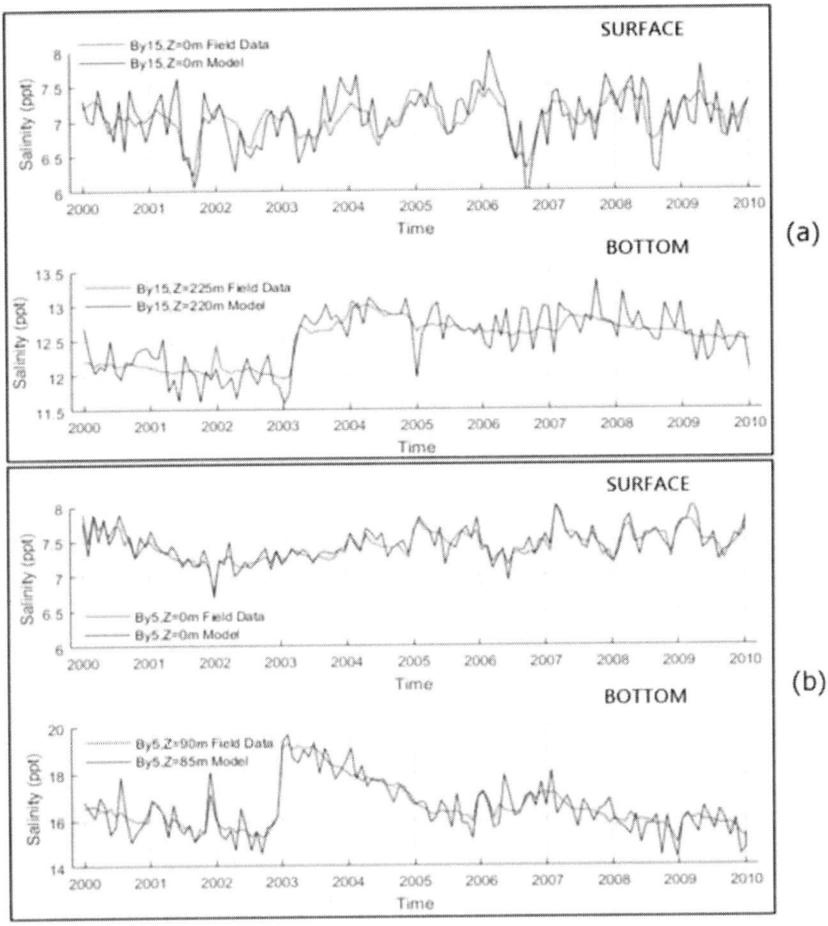

Figure 9. (**a**) Comparison of model predicted and measurements of salinity in the Baltic Sea at the Station BY15 surface and bottom for the time period 2000–2009; (**b**) Comparison of model predicted and measurements of salinity in the Baltic Sea at the Station BY5 surface and bottom for the time period 2000–2009.

3.2. Stratified Vertical Structure

We used the validated model to investigate the stratified structure within all the basins of the Baltic Sea for the time period 2000–2009. The complete sets of numerical results are too extensive to be reported here. We will present and discuss some general results that are illustrated using a few plots (i.e., temperature and salinity) in both horizontal and vertical sections. The latter were investigated along several cross sections shown in Figure 10 that were selected to cut through all the basins (see Figure 2 for basin names) included in this study. The structure concerns thermocline and halocline stratifications, which are controlled by the hydrodynamic variables, forcing meteorological parameters, the topography, the shorelines, and the exchange processes with the North Sea.

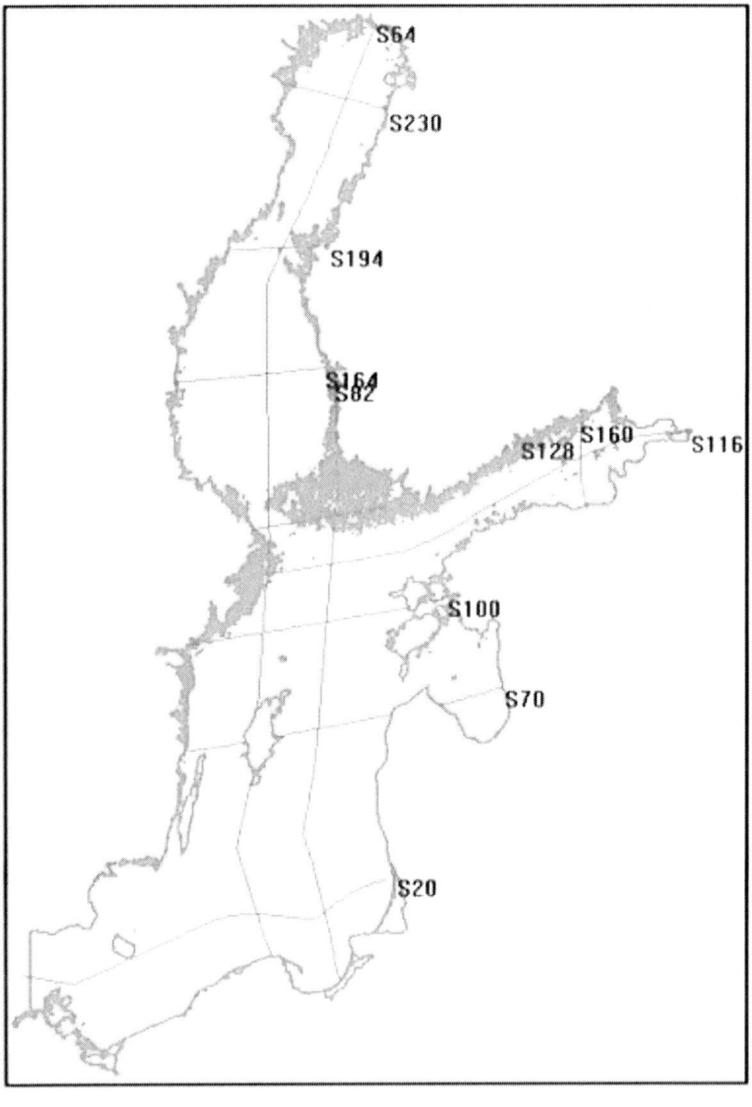

Figure 10. Selected cross sections in the Baltic Sea for vertical structure analysis.

Examples of simulation results for temperature and salinity are presented seasonally for winter (15 January), spring (15 May), summer (15 August), and autumn (15 October). Figure 11 and Figure 12 show the surface and bottom contour plots of temperature for the year 2000, respectively. The typical cross-sectional plots of temperature are given for sections S64 and S116 for the year 2000 are given in Figure 13 and Figure 14. The basins' abbreviated names are also given in the foregoing figures. It should be noted that the plotting scales were adjusted to show the full range of the hydrodynamic variables. This is particularly important for insuring that the stratification features are not obscured by the choice of the color scales. The summary results for all nine basins are presented in Table 2 and Table 3. In these tables the layer thicknesses are normalized (L_n) with mean depth.

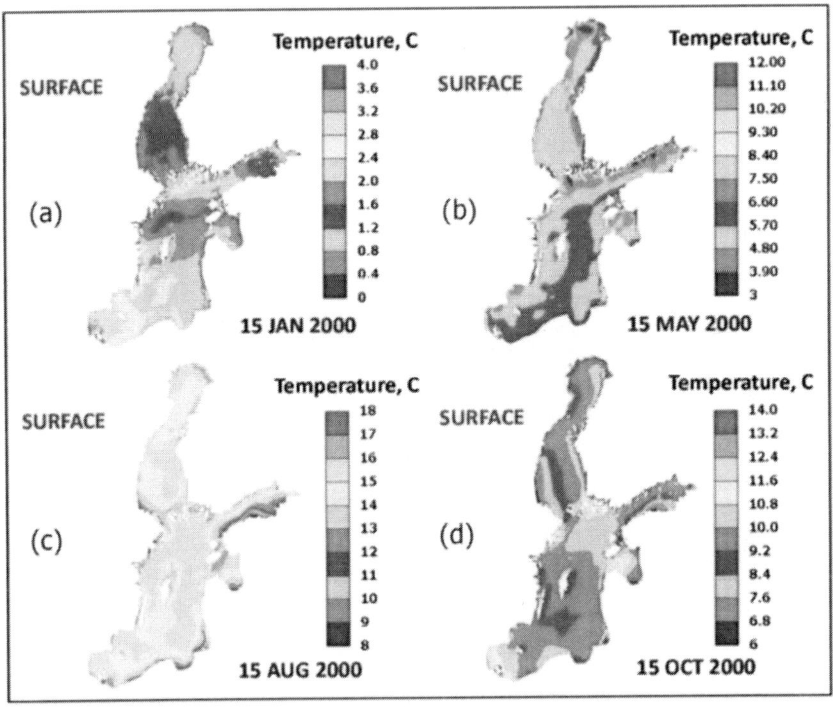

Figure 11. Seasonal surface temperature contour plots for the year 2000. (**a**) Winter; (**b**) Spring; (**c**) Summer; (**d**) Autumn.

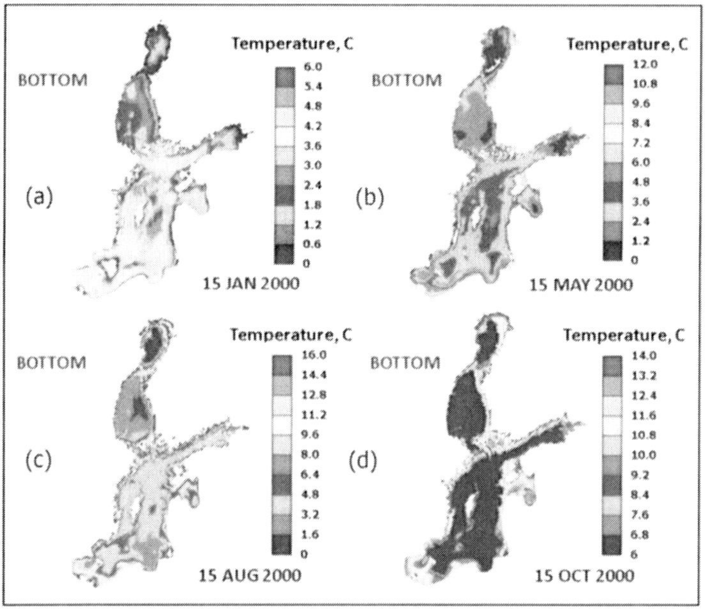

Figure 12. Seasonal bottom temperature contour plots for the year 2000. (**a**) Winter; (**b**) Spring; (**c**) Summer; (**d**) Autumn.

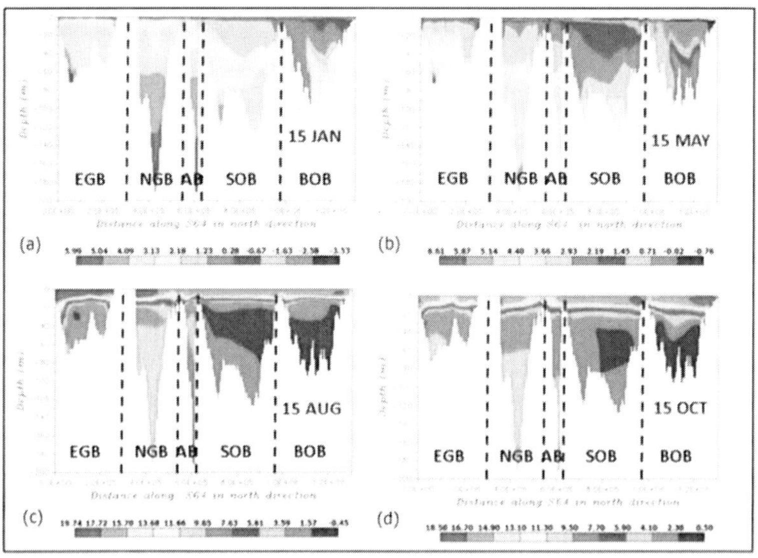

Figure 13. Seasonal thermal stratification plots in the Baltic Sea at cross section S64 for the year 2000. (**a**) Winter; (**b**) Spring; (**c**) Summer; (**d**) Autumn.

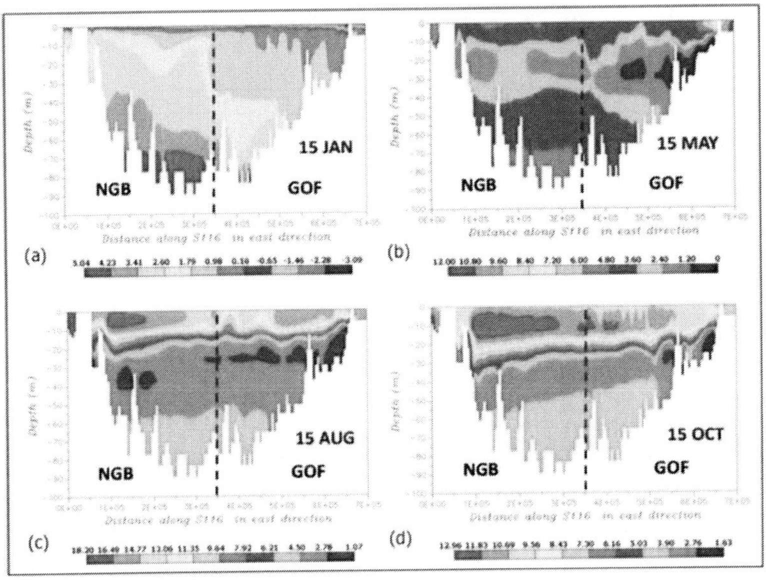

Figure 14. Seasonal thermal stratification plots in the Baltic Sea at cross section S116 for the year 2000. (**a**) Winter; (**b**) Spring; (**c**) Summer; (**d**) Autumn.

Table 2. Characteristics of thermal stratification in the basin system of the Baltic Sea, 2000–2009.

Basins	Layers	Winter		Spring		Summer		Autumn	
		T °C	L_K %	T °C	L_K %	T °C	L_K %	T °C	L_K %
Bay of Bothnia	1	0.5	65	1.2	5	11	10	13.5	5
	2	0.5–1.5	23	1.2–0.8	5	11–3	30	13.5–3	35
	3	1.5	12	0.8	80	3–2	50	3–2	50
Sea of Bothnia	1	1.5	12	4.5	4.5	14–11.2	3	12	10
	2	1.5–3	33	4.5–1	13.5	11.2–3.2	25	12–2.5	20
	3	3	55	1.3	82	3.2–1.5	9	2.5–2	10
	4					1.5–3	40	2–3	47
	5					3	23	3	13
Gulf of Finland	1	0.2–0.25	42	7.5–1.5	42	16.2	21	13–12.8	21
	2	0.25–2.6	42	1.5	16	16.2–2.6	50	12.8–3.5	50
	3	2.6	16	1.5–4.2	42	2.6–4.2	29	3.5	29
Sea of Åland	1	0.5–4	25	4	10	14.2	10	10	10
	2	4–4.3	35	4–2.4	27	14.2–6.5	27	10–9	18
	3	4.3–4.8	40	2.4	27	6.5–3.9	63	9–5	72
	4			2.4–2.9	36				
Northern Gotland	1	4.2	43	6	8	16.2	5	13.5	10
	2	4.2–5.2	10	6–3.2	12	16.2–16.8	5	13.5–14.2	10
	3	5.2	47	3.2–2.8	11	16.8–16	10	14.2–3.5	20
	4			2.8–4.8	29	16–3	20	3.5–5	15
	5			5.5	40	3–4.8	16	5	35
	6					4.8	42		
Western Gotland	1	2	3	7–2.6	12	16–3.2	12	13.5	8
	2	2–5	20	2.6–5	18	3.2–5	11	3.5	4
	3	5–5.2	77	5–5.2	70	5–5.2	77	3.5–5	11
	4							5–5.2	77
Eastern Gotland	1	3.8	28	9–3	19	16.5	10	14	9
	2	3.8–6	33	3	13	16.5–3.7	10	14–3.7	9
	3	6–6.3	39	3–6	35	3.7–3	12	3.7–3	9
	4			6–6.3	33	3–6	26	3–6	28
	5					6–6.2	42	6–6.2	45
Bornholm Basin	1	4.5	55	12–4	33	17–16.5	28	15–14.8	18
	2	4.5–10	22	4	22	16.5–4.7	28	14.8–4.7	27
	3	10	23	9	33	7–6.2	16	4.7–7.8	22
	4			9–9.2	12	6.2–8	28	7.8–8.5	33
Arkona Basin	1	4	25	13.2–5.2	76	16.5	25	17–16.2	50
	2	4–4.5	25	5.2–4.8	23	16.5–8.5	25	16.2–8.8	25
	3	4.5–6	50			8.5–10.7	50	8.8–10.5	25

Table 3. Characteristics of salinity stratification in the basin system of the Baltic Sea, 2000–2009.

Basins	Layers	Winter L_H %	Spring L_H %	Summer L_H %	Autumn L_H %	Salinity ‰
Bay of Bothnia	Upper	18	34	31	22	3
	Halocline	33	15	22	66	3.2
	Bottom	49	51	41	12	4
Sea of Bothnia	Upper	50	15	5	15	5
	Halocline	35	65	70	70	5.5
	Bottom	15	20	25	15	6.5
Sea of Åland	Upper	15	10	10	10	5.9
	Halocline	35	50	46	40	7.6
	Bottom	50	40	44	50	10
Gulf of Finland	Upper	25	32	25	30	4.7
	Halocline	30	38	50	40	6.5
	Bottom	45	40	25	30	8.3
Northern Gotland	Upper	35	38	20	25	5
	Halocline	30	32	55	45	42
	Bottom	35	30	25	30	28
Western Gotland	Upper	35	50	55	50	6.9
	Halocline	50	40	30	30	8.1
	Bottom	15	10	15	20	10
Eastern Gotland	Upper	30	20	25	20	6.8
	Halocline	35	50	50	55	9.3
	Bottom	35	30	25	25	11.5
Bornholm Basin	Upper	45	35	50	50	7.8
	Halocline	10	30	35	30	12.5
	Bottom	45	35	15	20	15.5
Arkona Basin	Upper	35	60	38	60	8
	Halocline	50	25	50	25	9
	Bottom	15	15	12	15	12

The temperature development at the Baltic Sea is affected by the intrusion of the warmer water masses from the North Sea and cold low salinity river water masses that flow into the Bay of Bothnia. The northern part of the Baltic Sea is covered by ice during winter and spring, which reduces the water temperature well into the summer. Figure 11 shows the seasonal variations of water temperature at the surface and bottom layers. The surface temperatures are in good agreement with the SMHI measurements available in [40]. The surface temperatures vary in the range of 0 °C to 18 °C. In the winter, the bottom temperatures are higher than the surface temperatures by several degrees. The spring is marked by higher surface temperatures, with the exception of the Bay of Bothnia and the eastern Gulf of Finland. The surface temperatures increase significantly during the summer and are about 2 °C higher than the bottom layer temperatures. The autumn shows a transitional behavior, with a decreasing temperature differences between the two layers. This marks the start of the cooling of surface water due to increased mechanical and thermal convection.

In summary, the thermal stratification in the Baltic Sea shows a considerable variation among the different basins with a clear seasonal feature. During the winter period, the lower water layers are warmer than the upper layers. The ice starts to melt in late spring but still maintains a higher water temperature in the bottom layers. The temperature of the upper water layers starts to increase as the end of the summer approaches. By then, the upper water layers have a higher temperature and the stratification is reversed.

The common thermal stratification features in the Baltic Sea are obtained by analyzing the transient plots along all the cross sections:

1. The number of the stratified layers varies from two to five.
2. There are two distinct layers, one with lower surface temperature, i.e., winter stratification and a reversed summer-type stratification with higher surface temperatures. These two types prevail during winter–spring and summer–autumn, respectively. Typical transitional behavior may begin in late autumn, when the wind speed increases, but does not last more than a couple of weeks. These stratification features dominantly occur in shallower regions of the Baltic Sea.
3. The surface layer has a transient structure composed of two to three minor thermoclines with a mean thickness of about 10 m. It is often difficult to distinguish between these layers. Here, the thermocline layer is considered as a layer with temperature variations more than 0.4 °C, which gives negligible density difference (i.e., 0.01 kg/m^3). The thermocline is located at a depth of 10–30 m.
4. The summer stratification has a dicothermal character in the northern basins of the Baltic Sea. This implies the existence of a colder layer sandwiched between two layers of higher temperatures. The layer is stable since the upper layer has a lower salinity than the deep underlying layer.

The foregoing features are shown in Figure 13 and Figure 14 at two typical cross sections of S64 and S116, respectively. The former crosses most of the major basins in the Baltic Sea. This plot does not do justice to the complexity of the stratification in the Baltic Sea. The intention is to provide support to the results and the discussion. For thermal stratification, the layers are numbered in descending order from the surface (Table 2). The division was needed due to the complex nature of the multilayered thermal stratification. Layer 1 is defined as the surface or the top layer and the last layer refers to the bottom layer. It is not very useful to list the absolute values due to the significant geometric variations within and among the basins (size, volume, and depth).

In the Bay of Bothnia, there is a significant seasonal variation regarding the number and thickness of the layers. The stratification in the Aland Sea is dominated by a three-layer structure, with the exception of a four-layer structure during the spring period. The stratified structure in the Gulf of Finland consists of three layers with significant variations in both thickness and temperature. The multilayered stratification structure for the Northern Gotland Basin is composed of top and bottom layers with constant temperatures. The Western Gotland Basin shows a sustained three-layered structure from the winter to summer period. The Eastern Gotland Basin stratification is similar to the northern basin except that the water temperatures are higher by 1 °C during the spring and autumn periods. The Bornholm Basin features are similar to the Eastern Gotland Basin. The Arkona Basin has a clear three-layer structure and during the winter and summer periods, the temperatures are constant in the 10-m thick upper layer, i.e., 4 °C and 16.5 °C, respectively. A short summary of each basin is reported in this paper. A detailed description of each basin's stratification is given in Dargahi and Cvektovic [9]. The foregoing results are summarized in Table 2.

The seasonal variation in surface and bottom layer salinities is shown in Figure 15 and Figure 16, respectively. Figure 17and Figure 18 show salinity contours for the years 2000 and 2006 as two typical years in the 10-year simulation period of 2000–2009. The general features are an outgoing surface flow with low salinity that is diluted by the fresh water from the rivers and a denser incoming bottom flow from the North Sea. The surface salinity varies from 1.6% to 9% and the bottom salinity in the range 3%–16%, moving southwards from the Bay of Bothnia to the Arkona Basin. The corresponding salinity gradients are 0.0042%/m and 0.0054%/m, respectively. The intruded high salinity water from the North Sea propagates like a plume as far as the Northern Gotland Basin, with a mean advection velocity of 6 cm/s. The feature is apparent from Figure 13, Figure 14, Figure 17 and Figure 18, which show the dense water settling mainly in the deep regions of the Northern and Eastern Gotland Basins. The salinity stratification has a multi-layer structure similar to the thermal stratification shown in Figure 13 and Figure 14. The general features are two thick (i.e., >30 m) surface and bottom layers with a transition layer in-between. The stratification is complicated by its temporal and spatial variations both within and among the basins. A detailed discussion on salinity stratification is provided by Dargahi and Cvektovic [9] for all basins and so will not be repeated here. As an example, salinity stratification is briefly discussed for the Gulf of Finland and the Northern Gotland Basin.

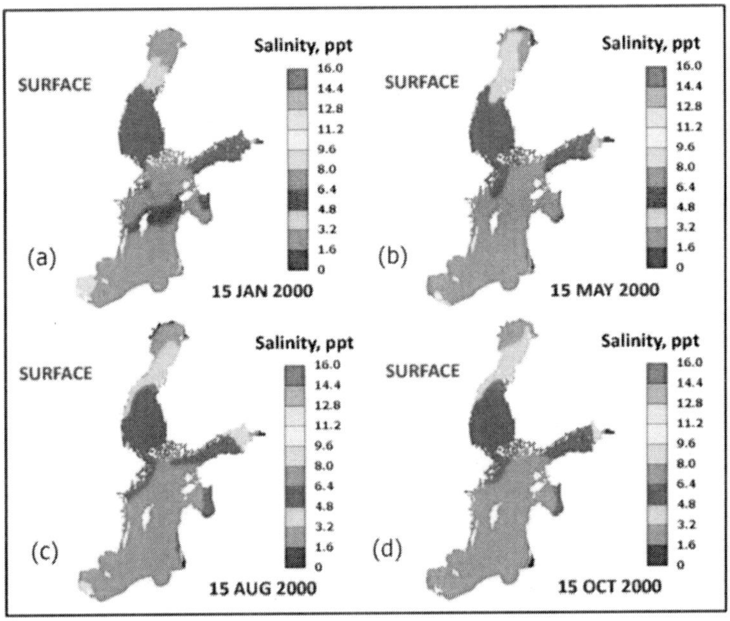

Figure 15. Seasonal plots of surface salinity in the Baltic Sea for the year 2000. (**a**) Winter; (**b**) Spring; (**c**) Summer; (**d**) Autumn.

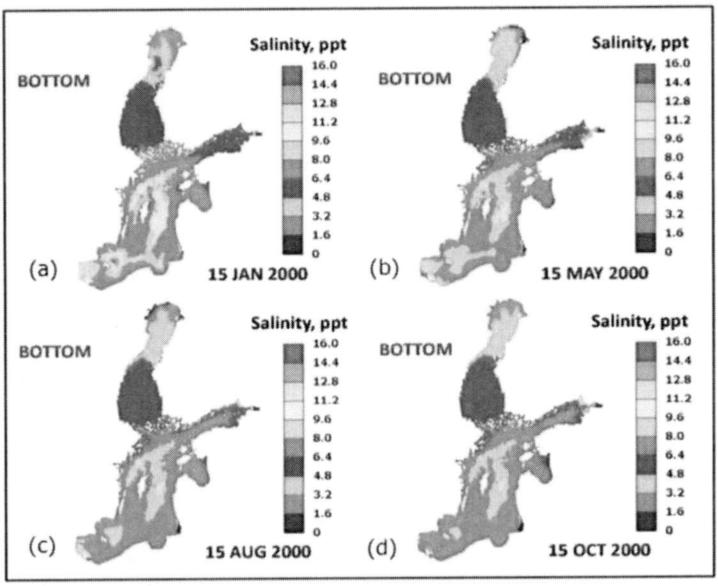

Figure 16. Seasonal plots of bottom salinity in the Baltic Sea for the year 2000. (**a**) Winter; (**b**) Spring; (**c**) Summer; (**d**) Autumn.

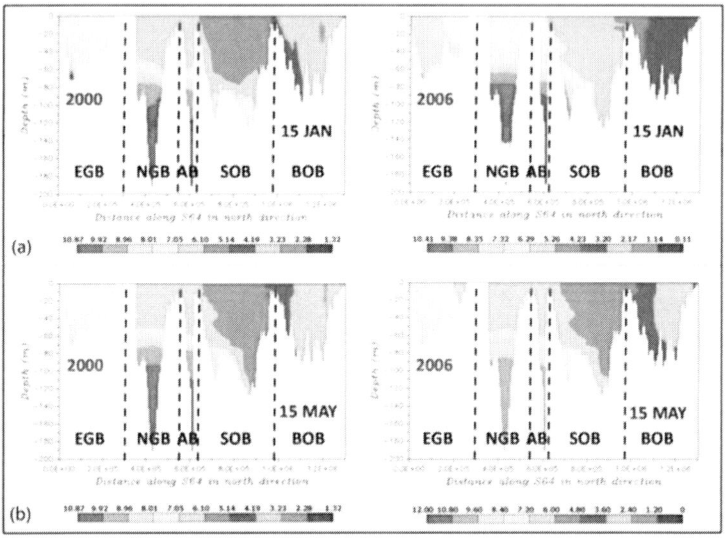

Figure 17. (**a**) Winter salinity stratification in the Baltic Sea at cross section S64 for the years 2000 and 2006; (**b**) Spring salinity stratification in the Baltic Sea at cross section S64 for the years 2000 and 2006.

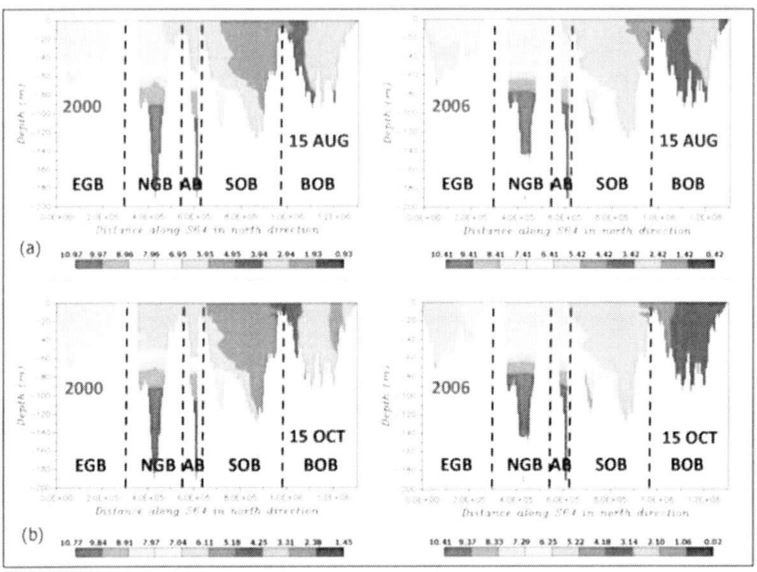

Figure 18. (**a**) Summer salinity stratification in the Baltic Sea at cross section S64 for the years 2000 and 2006; (**b**) Autumn salinity stratification in the Baltic Sea at cross section S64 for the years 2000 and 2006.

4. DISCUSSION

The understanding of stratification in large water bodies is of significant environmental importance due to its direct coupling with water quality dynamics. Severe thermal stratification can have a number of adverse effects, among which are reduced water quality and the spatial distribution of fish [41]. Any prolonged stratification can reduce oxygen solubility, which leads to oxygen depletion in deep water masses. The deeper water masses in the Baltic Sea are particularly vulnerable to eutrophication below the halocline or in regions affected by thermal stratification [42]. We know that eutrophication is the most serious and challenging environmental problem for the Baltic Sea [1]. Consequently, accurate knowledge of stratification dynamics is required for the management of eutrophication in the Baltic Sea.

The river Neva, the most voluminous river entering the Gulf of Finland (referred as Gulf in the foregoing discussion) from the eastern coastline (2500 m^3/s), significantly affects the stratification in the Gulf. The zero-salinity water from the river creates a permanent plume that propagates westwards to a distance of 50–150 km. The plume salinity varies from 0% to 5% and extends to a depth of about 15 m. The maximum extension occurs during early spring periods. The seasonal cross section plots are shown in Figure 14 at S116 for the year 2000. The stratification in the Neva region appears to have a permanent multi-layer structure with few seasonal variations. The characteristic three-layer structure is found in the inner Gulf region. The stratification characteristics in the Northern Gotland Basin are similar to those in the inner Gulf region. It is interesting to note the significant salinity increase to 9.5% that is confined to the Northern Gotland Basin and the bottom layer.

The model prediction of three-layered salinity stratification in all basins of the Baltic Sea is in general agreement with previous studies [8]. Salinity stratification has a strong seasonal variability but is much weaker than the corresponding variability in temperature. The summary of model-predicted salinity data for the entire 10-year period into seasonal normalized layer thicknesses is shown in Table 3. The normalized values are more applicable than the absolute values given in the literature as the layer thicknesses could be estimated as a function of depth. The latter vary significantly within and among the basins. To examine the validity of the generalization, we compared the model and measured salinities at all stations and found good agreement between the data. An example is given in Table 4, which compares model results with measurements at station C3 (located in the Sea of Bothnia; see Figure 3) for the year 2000. Evidently, the predicted mean thicknesses are within the range of the measured data that supports our generalization.

Table 4. Comparison of modeled and measured salinity stratification at station C3 (Sea of Bothnia), 2000.

Period	Upper Layer (m)		Halocline (m)		Bottom Layer (m)	
	Measured	Model	Measured	Model	Measured	Model
Winter	60–90	100	50–90	62	10–40	30
Spring	10–20	18	100–125	135	20–50	30
Summer	10–15	10	110–140	140	20–50	33
Autumn	15–30	22	100–130	127	20–45	22

The detailed investigation of thermal stratification for a 10-year period (i.e., 2000–2009) revealed some new features. The current study revealed a multilayered structure that contains several thermocline and dicothermal layers. The statistical analysis of all simulation results made it possible to derive the mean thermal stratification properties, expressed as mean temperatures and the normalized layer thicknesses (Table 2). The three-layered structure reported by Leppäranta and Myrberg [7] appears to be oversimplified.

The thermocline layer has a sharp temperature gradient that connects the upper and lower two layers. We have found the three-layer model to be valid only during the winter. In the northern basins, the vertical temperature gradients are significantly lower than for the southern basins (a factor 3). We attribute the difference to the formation of an ice cover during winter and spring in the northern basins (i.e., Northern Gotland Basin and above), which affects the surface heat transfer and its exchange with the overlying cold air. The ice layer acts as a thermal barrier preventing further heat losses due to the action of wind and convective transport. Consequently, the upper layer is rather thick and can occupy up to 65% of the depth in these basins. Following the winter, the three-layered structure is decomposed into several layers with increasing or decreasing temperature gradients. The process takes place in all the basins in the Baltic Sea. Here, we believe the primary driving force is the increased mixing processes between the basins during the ice-free periods. The high-momentum fresh water inflows from the rivers contribute to higher surface water temperatures, thus increasing the temperature gradients. On average, the thickest layer is the bottom layer (≈50%) with a small temperature gradient, except in the Bornholm and Arkona basins. We believe the low gradients are due to low-intensity exchange processes in the northern basin of the deep water regions.

The Bornholm and Arkona basins are shallower and more influenced by the exchange of warmer and more saline waters with the North Sea. This is reflected by the increased temperature gradients in these basins (from ≈0.02 °C/m to ≈0.5 °C/m). Our simulations indicate a wide spectrum in the layering properties among the basins. However, a few generalizations appear to be possible based on the results listed in Table 2. The thermocline occupies about 25% of the water depth in each basin. The deep bottom layer is about 40% of the water depth with temperatures of about 3 °C and 5 °C in the northern and southern basins, respectively. Our mean results on the properties of thermal stratification agree well with the results reported by Leppäranta and Myrberg [7]. Here, the reported annual averaged halocline thickness is 10–20 m. For example, in the Bay of Bothnia we get an averaged normalized depth of 35% from Table 2. The mean depth is 40 m, which yields a thickness of 14 m—well within the range given by Leppäranta and Myrberg [7].

There is also some evidence of upwelling and downwelling along the coastlines across the Baltic Sea [3,6,39,43]. The existence of the dicothermal layer is an indication of upwelling. The two features occur as wind causes surface water to diverge (Ekman transport) or converge (downwelling). During both processes, water is either replenished from the deep region or forced downwards [44]. Consequently, the salinity variations, which are mainly controlled by the water balance, are further modified.

We will also examine the validity of having normalized layer thicknesses (Table 2) in each basin that only vary seasonally. We start by examining the measured vertical temperature profiles for two extreme locations of F3 (Bay of Bothnia) and BY5 (Bornholm Sea). Figure 19 shows the measured vertical temperature profiles at F3, and BY5 for the years 2000, 2008, and 2009. Note that the data were rather limited and thus the lines are only illustrative and do not necessarily show the correct trends. The good agreement with the simulated summary results listed in Table 2 is apparent. For example, Table 2 predicts a thick top layer in the winter period (red line) at both stations that agrees well with the plots for F3, on 27 January 2000, and BY5, on 19 January 2000. We can also note that the profile shapes are preserved during the simulation period. However, the magnitudes of the temperatures show some significant variation. In conclusion, we believe the normalized layer thicknesses give reasonable estimates, with a standard deviation of ±15%.

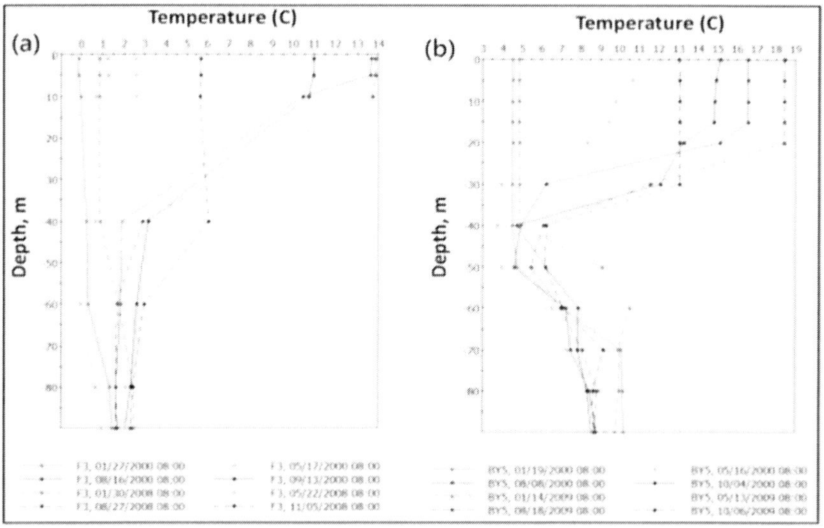

Figure 19. (**a**) Measured temperature profiles at F3 in 2000, 2008 and 2009; (**b**) Measured temperature profiles at BY5 in 2000, 2008 and 2009.

An interesting feature is the extreme inflow event of January 2003. According to Lehmann et al. [45], a massive salt intrusion of cold and oxygen-rich water from the North Sea took place at Darss Sill, which is a few kilometers west of the open sea boundary used in the model domain (see Figure 1). They consider the event as "the most important inflow from 1993". Their results indicate significant changes in both salinity and temperature distributions in the deeper basins of the Baltic Sea. In the present study, we have investigated the reported event by comparing the surface and bottom salinities at all the stations (see Figure 3). We present the results in Figure 20, illustrating the salinity time series at stations BY1, BY5, and BY15 at the surface and bottom layers. The curves at BY1 are measured field data but the curves at BY5 and BY15 are model results. The model results are plotted at seven-day intervals for ease of comparison with the field data. Several important features are evident from this figure:

1. The event took place on 18 January 2003 with a salinity of 29%.
2. It took nearly a month for the peak salinity to reach station BY5, which dropped to 19%.
3. The peak salinity was reduced to 12.5% at station BY15 after almost five months.

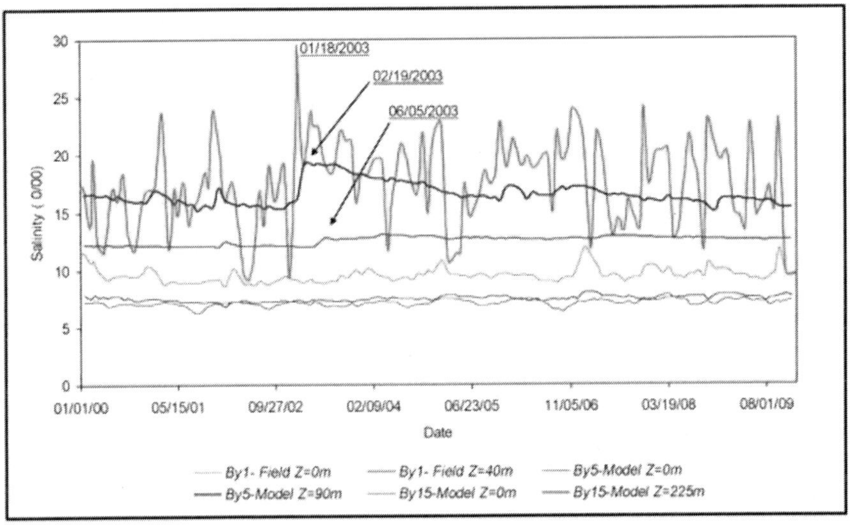

Figure 20. Time series of surface and bottom salinities at stations BY1, BY5, and BY15.

We computed the horizontal diffusive *[Math Processing Error]* and advective transports *[Math Processing Error]* using the foregoing information and the distances between the three stations (Table 5). A representative diffusion coefficient (*[Math Processing Error]*) for the Baltic Sea is 10^3 m^2/s [7]; the advective velocity is typically 0.1 m/s in the Arkona and Bornholm basins (present study). We can conclude that the diffusive transport is much faster than the advective transport. There is a significant reduction in the Eastern Gotland Basin in both modes of transport. It is interesting to note that the temporal variations in the bottom salinity appear to be random as opposed to the surface salinities. The surface salinities have lower amplitudes with no corresponding peaks as the bottom salinity. We believe the outgoing freshwater plays a significant role in lowering the salinity amplitude.

Table 5. Estimates of salinity transport in the southern basins.

Transport of Salt	BY1 (Arkona)-BY5 (Bornholm)	BY5-BY15(Eastern Gotland)
Diffusive (‰/day)	17×10^{-3}	5×10^{-3}
Advective (‰/day)	86×10^{-8}	1.12×10^{-8}

5. CONCLUSIONS

An integrated 3D modeling system was developed for the Baltic Sea using a public domain model called GEMSS®. The model was calibrated and verified using 10 years of data covering 2000–2009. We have incorporated boundary conditions and bathymetry with as high an accuracy as possible, to serve as an improvement over previous studies. The model was then used to investigate the vertical structure of the Baltic Sea to understand the stratification and exchange processes across various basins. This paper addressed in detail both the thermal and salinity stratifications, with a focus on the structural properties of the layers.

The hypothesis was that the layer properties could be expressed as dimensionless numbers valid for all seasons. In particular, the detection of cooler regions (dicothermal) within the layer structure has been an important finding. The detailed investigation of thermal stratification for a 10-year period (i.e., 2000–2009) revealed some new features. A multilayered structure that contains several thermocline and dicothermal layers prevails. Statistical analysis of the simulation results made it possible to derive the mean thermal str atification properties, expressed as mean temperatures and the normalized layer thicknesses.

The three-layered structure reported in the literature appears to be rather simplified. The current study found that the three-layer model is valid only during the winter.

The layering properties vary significantly among the basins whereby the layered structure could not be generalized. Nevertheless, a few generalizations appear to be possible. The thermocline occupies about 25% of the water depth in each basin. The deep bottom layer is about 40% of the water depth, with temperatures of about 3 °C and 5 °C in northern and southern basins, respectively.

Three-layered salinity stratification prevails in all basins of the Baltic Sea, in general agreement with previous studies. Salinity stratification has a strong seasonal variability but this is much weaker than the corresponding variability in temperature. We have succeeded in generalizing the seasonal normalized layer thicknesses for the entire 10-year period. The use of normalized values is advantageous compared to the absolute values given in the literature, enabling estimation of layer thickness as a function of depth.

This study provides detailed insight into thermal and salinity stratifications in the Baltic Sea during a recent decade and can be used as a basis for diverse environmental assessments (e.g., anoxia and reduced nutrient mixing between layers [1]). It extends previous studies on stratification in the Baltic Sea regarding both the extent and the nature of stratification.

ACKNOWLEDGMENTS

This study was initiated and the first stage financed by the Central Baltic Interreg IV Programme as part of the project "Phosphorous from the seabed and water quality in archipelagos—modeling attempt (SEABED)". The second stage of this study was funded by KTH Royal Institute of Technology, Division of Water Resources Engineering and Division of Hydraulic Engineering. The third and final stage of this study was funded by the project "The Baltic Sea Region System (BALSYS)".

AUTHOR CONTRIBUTIONS

The Baltic Sea hydrodynamic modeling study was done entirely by Bijan Dargahi. The original manuscript was prepared by Bijan Dargahi. The GEMSS model was developed by Venkat Kolluru. He also provided GEMSS model training tor this study. The subsequent manuscript revisions and refinement of graphics were prepared by Venkat Kolluru. The project management and funding from three different sources were managed by Vladimir Cvetkovic.

CONFLICTS OF INTEREST

The authors declare no conflict of interest.

REFERENCES

1. World Atlas. Available online: http://www.worldatlas.com/aatlas/infopage/balticsea.htm (accessed on 12 December 2016).
2. Elmgren, R.; Larsson, U. Himmerfjajrden: Changes in a Nutrient-Enriched Coastal Ecosystem in the Baltic Sea; Report 4565; Swedish Environment Protection Agency: Naturvårdsverket, Stockholm, Sweden, 1997.
3. Meier, M. On the parameterization of mixing in three-dimensional Baltic Sea models. J. Geophys. Res. 2001, 106, 30997–31061.
4. Lehmann, A.; Hinrichsen, H.H. Water, heat and salt exchange between the deep basins of the Baltic Sea. Boreal Environ. Res. 2002, 7, 405–415.

5. Omstedt, A.; Elken, J.; Lehmann, A.; Piechura, J. Knowledge of the Baltic Sea physics gained during the BALTEX and related programmes. Progress Oceanogr. 2004, 63, 1–28.
6. Feistel, R.; Nausch, G.; Wasmund, N. State and Evolution of the Baltic Sea, 1952–2005: A Detailed 50-Year Survey of Meteorology and Climate, Physics, Chemistry, Biology, and Marine Environment, 1st ed.; John Wiley & Sons, Inc.: Hoboken, NJ, USA, 2008.
7. Leppäranta, M.; Myrberg, K. Physical Oceanography of the Baltic Sea, 1st ed.; Springer: Berlin, Germany, 2009.
8. Harff, J.; Björck, S.; Hoth, P. The Baltic Sea Basin, 1st ed.; Central and Eastern European Development Studies; Springer: Berlin, Germany, 2011.
9. Dargahi, B.; Cvektovic, V. Hydrodynamics and Transport Characterization of the Baltic Sea 2000–2009, 1st ed.; TRITA-LWR REPORT 2014:03; KTH Royal Institute of Technology, School of Architecture and the Marine Environment: Stockholm, Sweden, 2014.
10. Bock, K.H. Monthly salinity maps of the Baltic Sea for different depths. Deustsche Hydrogr. Z. 1971, 12, 1–147.
11. Väli, G.; Meier, H.E.; Elken, J. Simulated Variations of the Baltic Sea Halocline during 1961–2007; Report Oceanography; Swedish Meteorological and Hydrological Institute (SMHI): Stockholm, Sweden, 2012; No. 44; pp. 1–37.
12. Suominen, T.; Tolvanen, H.; Kalliola, R. Geographical persistence of surface-layer water properties in the Archipelago Sea, SW Finland. Fennia 2010, 188, 179–196.
13. Hordoir, R.; Meier, H.E.M. Effect of climate change on the thermal stratification of the Baltic Sea: A sensitivity experiment. Clim. Dyn. 2012, 38, 1703–1713.
14. Fonselius, S. History of hydrographic research in Sweden. Proc. Estonian Acad. Sci. Biol. Ecol. 2001, 50, 110–129.
15. Peter, B. Dicothermal Layer Characteristics—A Case Study in the Antarctic Zone of Indian Ocean. Proc. Indian Natl. Acad. Sci. USA 1993, 59, 439–447.
16. SMHI. Swedish Meteorological and Hydrological Institute. Available online: http://www.smhi.se/en (accessed on 25 February 2012).
17. FMI. Finnish Meteorological Institute. Available online: http://en.ilmatieteenlaitos.fi/ (accessed on 10 April 2012).
18. Leibniz Institute for Baltic Sea Research Warnemunde. Available online: http://www.io-warnemuende.de/topography-of-the-baltic-sea.html (accessed on 20 September 2012).
19. GEMSS®. Available online: http://www.gemss.com (accessed on 10 August 2011).
20. Edinger, J.E.; Buchak, E.M. Numerical hydrodynamics of estuaries in estuarine and wetland processes with emphasis on modeling. In Estuarine and Wetland Processes with Emphasis on Modeling; Hamilton, P., Macdonald, K.B., Eds.; Plenum Press: New York, NY, USA, 1980; pp. 115–146.
21. Edinger, J.E.; Buchak, E.M. Numerical waterbody dynamics and small computers. In Proceedings of the ASCE Hydraulic Division Specialty Conference on Hydraulics and Hydrology in the Small Computer Age, Lake Buena Vista, FL, USA, 13–16 August 1985; American Society of Civil Engineers: Reston, VA, USA, 1985; pp. 13–16.
22. Kolluru, V.S.; Buchak, E.M.; Edinger, J.E. Integrated Model to Simulate the Transport and Fate of Mine Tailings in Deep Waters. In Proceedings of the Tailings and Mine Waste '98 Conference, Fort Collins, CO, USA, 26–29 January 1998; pp. 599–610.
23. Kolluru, V.S.; Buchak, E.M.; Wu, J. Use of Membrane Boundaries to Simulate Fixed and Floating Structures in GLLVHT. In Proceedings of the 6th International Conference on Estuarine and Coastal Modeling, New Orleans, LA, USA, 3–5 November 1999; Spaulding, M.L., Butler, H.L., Eds.; pp. 485–500.
24. Kolluru, V.S.; Edinger, J.E.; Buchak, E.M.; Brinkmann, P. Hydrodynamic Modeling of Coastal LNG Cooling Water Discharge. J. Energy Eng. 2003, 129, 16–31.
25. Kolluru, V.S.; Fichera, M. Development and Application of Combined 1-D and 3-D Modeling System for TMDL Studies. In Proceedings of the Eighth International

Conference on Estuarine and Coastal Modeling, Monterey, CA, USA, 3–5 November 2003; American Society of Civil Engineers: Reston, VA, USA, 2003; pp. 108–127.

26. Kolluru, V.S.; Prakash, S. Evaluation of Urbanization and Impacts on Water Quality in Nottawasaga Bay Using an Integrated 3-D Modeling Framework. In Proceedings of the 7th International Congress on Environmental Modelling and Software, San Diego, CA, USA, 15–19 June 2014; Ames, D.P., Quinn, N.W.T., Rizzoli, A.E., Eds.; Available online:http://www.iemss.org/society/index.php/iemss-2014-proceedings (accessed on 25 January 2013).
27. Mellor, G.L.; Yamada, T. Development of a turbulence closure model for geophysical fluid problems. Rev. Geophys. Space Phys. 1982, 20, 851–875.
28. Umlauf, L.; Burchard, H. A generic length-scale equation for geophysical turbulence models. J. Mar. Res. 2003, 61, 235–265.
29. Warner, J.C.; Sherwood, C.R. Performance of four turbulence closure methods implemented using a generic length scale method. Ocean Model. 2005, 8, 81–113.
30. Okubo, A. Oceanic diffusion diagrams. Deep-Sea Res. 1971, 18, 789–802.
31. Smagorinsky, J. General circulation experiments with the primitive equations. Mon. Weather Rev. 1963, 91, 99–164.
32. Prakash, S.; Kolluru, V.S. Implementation of higher order transport schemes with explicit and implicit formulations in a 3-D hydrodynamic and transport model. In Proceedings of the 7th International Conference on Hydroscience and Engineering, Philadelphia, PA, USA, 10–13 September 2006; Available online: http://web.abo.fi /seabed (accessed on 10 June 2013).
33. Dargahi, B.; Cvetkovic, V. Hydrodynamic and transport properties of Saltsjö Bay in the inner Stockholm Archipelago. J. Coast. Res. 2010, 27, 572–584.
34. Dargahi, B.; Setegn, S. Combined 3D hydrodynamic and watershed modelling of Lake Tana, Ethiopia. J. Hydrol. 2011, 398, 44–64.
35. Environmental Resources Management. GEMSS Software Technical Documentation; ERM: Malvern, PA, USA, 2006.
36. Madsen, O.S.; Grant, W.D. Quantitative description of sediment transport by waves. In Proceedings of the 15th International Conference on Coastal Engineering, ASCE, Honolulu, HI, USA, 11–17 July 1976; pp. 1093–1112.
37. Myrberg, K.; Andrejev, O. Modelling of the circulation, water exchange and water age properties of the Gulf of Bothnia. Oceanologia 2006, 48, 55–74.
38. Palmen, E. Untersuchungen über die Stromungen in den Finnland umgebenden Meeren. Soc. Sci. Fennica Comment. Phys. Math. 1930, 12, 1–94.
39. Harmel, R.D.; Smith, P.K.; Migliaccio, W.; Chaubey, I.; Douglas-Mankin, K.R.; Benham, B.; Shukla, S.; Muñoz-Carpena, R.; Robson, B.J. Evaluating, interpreting and communicating performance of hydrologic/water quality models considering intended use: A review and recommendations. Environ. Model. Softw. 2014, 57, 40–51.
40. SMHI Oceanography. Available online: http://www.smhi.se /klimatdata/ oceanografi/havsis (accessed on 7 March 2012).
41. Lackey, R.T. Response of physical and chemical parameters to eliminating thermal stratification in a reservoir. Water Resour. Bull. 1972, 8, 589–599.
42. Baltic Marine Environment Protection Commission. The Baltic Marine Environment 1999–2002; Baltic Sea Environment Proceedings, No. 87; HELCOM: Helsinki, Finland, 2003.
43. Myrberg, K.; Andrejev, O. Main upwelling regions in the Baltic Sea a statistical analysis based on three-dimensional modelling. Boreal Environ. Res. 2003, 8, 97–112.
44. Lehmann, A.; Krauss, W.; Hinrichsen, H.H. Effects of remote and local atmospheric forcing on circulation an upwelling in the Baltic Sea. Tellus 2002, 54, 299–316.
45. Lehmann, A.; Lorenz, P.; Jacob, D. Modelling the exceptional Baltic Sea inflow events in 2002–2003. Geophys. Res. Lett.2004, 31, 1–4.

CHAPTER 13

Modelling of Storm Impacts on the Shandong Peninsula Coast

Guilherme L. Dalledonne and Roberto Mayerle

Research and Technology Centre Westcoast, University of Kiel, 24118 Kiel, Germany

ABSTRACT

Dalledonne, G. L., and Mayerle, R., 2016. Modelling of storm impacts on the Shandong peninsula coast.

This paper presents results of the investigation of the effect of storms on the morphodynamics of the Yantai coastal areas in the Shandong Peninsula, China. A two-dimensional depth integrated morphodynamic model for prediction of the morphological changes during extreme events was developed and applied. The model covers the entire Bohai Sea with increasing grid resolution towards the coastal areas near Yantai. Calibration and validation using tidal constituents and measured water levels and current velocities showed that the model is capable of predicting the hydrodynamics in good agreement with observations. Emphasis was given to the proper description of the spatial distribution of the seabed sediments and bed stratigraphy. The 20 most severe storms observed in the region from 2000 to 2013 were identified and reconstructed. Storm surges with increases in mean sea level of up to 1.7 m and significant wave heights up to 4.1 m were considered. The results showed that during the storms the levels of energy are mainly due to waves and tidal currents are only relevant at the capes and some channels and straits along the Shandong Peninsula. The resulting morphological changes indicated that although there is an overall loss of seabed material, particularly during the peak of the storms, the beaches near Yantai recover quite quickly just a few days after the extreme events.

Keywords: Bohai Sea, morphodynamics, sediment transport.

INTRODUCTION

China is annually harmed by a large number of storm surges. The majority of the storms are induced by tropical cyclones, which usually strike, at the southern coasts in summer, while during winter periods storm surges as a result of cold-air outbreak can hit the northern coastal regions. Like other Chinese Seas, the Bohai Sea is susceptible to storm surges induced by both tropical and extra-tropical cyclones. However, only a few tropical cyclones manage to move to such high latitudes. Despite of the fact that in the Bohai Sea the increases in water levels during such events are not significant, in general they take place every season causing serious damage. On the other hand, winter storm surges generated by cold-air outbreaks usually lead to major increases in sea level in the southern part of the Bohai Sea coast. As more than 90% of the strong winds over the Bohai Sea are induced by cold-air outbreaks, they are more likely to occur at the time of the high tide than tropical cyclones and to bring damage along the coast (Zhao and Jiang, 2011).

The Bohai Sea is a semi-enclosed shallow water body situated in the northern part of China. The sea and its surrounding provinces are rich in natural resources such as oil, coal and fish thus playing an important economic role. Hence it is subject to the strongest marine environment preservation requirements currently in place in China. The sea covers an area of approximately 79 000 km^2 with average depths of about 18 m. It is rather shallow except for the submerged valley near the north connecting the Bohai Sea with the Yellow Sea. Figure 1 shows the bathymetry of the Bohai/Yellow Sea. The seas are bounded in the west by Hebei Province; in the south by Shandong Peninsula, and in the east partly by the Liaodong Peninsula. The Bohai Strait in the east, also known as Huanghai Sea (Yellow Sea), is the only connection to the open ocean. According to Huatong and Wang (1985) the flow circulation in the area is mainly due to the combined effect of tidal forcing and winds. Jiang *et al.* (2004) states that wind generated waves play an important role on the temporal and spatial distribution of the suspended particulate matter concentration. Northerly winds prevail in winter. In the flood seasons of summer and fall river discharges affect tidal currents to a certain extent.

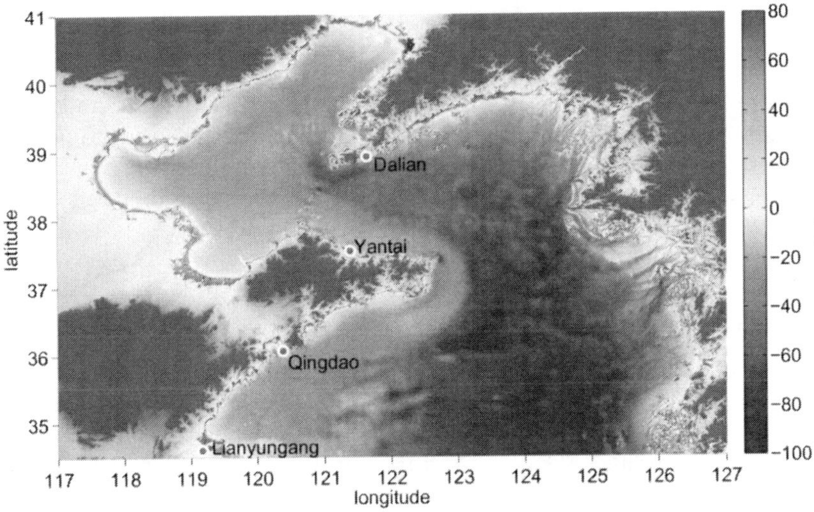

Figure 1. Topography of the Bohai/Yellow Sea region. (Source: GEBCO(OC, IHO, and BODC, 2003))

Storm surges in the Bohai Sea were investigated using dimensional analysis by Chin and Feng (1975). With the development of computational capabilities numerical models have been developed and applied to study storm surges in the region. Shi *et al.* (1997) introduced an effective numerical model for real-time forecasting of storm surge flooding using a wet-dry grid point method with generalized curvilinear computational grids. Li *et al.* (2003) coupled a wave model (WAVEWATCH III) with a three-dimensional ocean model (POM) to simulate storm surges. A strong impact on the elevation and current was found by considering the wave effect.

Despite of the large number of investigations, little is known about the impacts of storms on the sediment dynamics and morphodynamics in the near coastal areas of the Bohai Sea. Except for the mouth of the Yellow River, there is no much information about the dynamics of storms and its relevance on the resulting sediment transport in other coastal regions. In the framework of a sino-german cooperation, the relevance of storms on the coast of the Shandong Peninsula in the southeast of the Bohai Sea is addressed. Emphasis has been given to the hydrodynamics of the coastal regions off Yantai using a numerical model implementing modules of flow, waves, sediment transport and morphodynamics.

METHODS

A description of the numerical model for prediction of the tide and wave induced currents, the resulting sediment transport and morphological changes(morphodynamics) along the coastal areas of the Shandong Peninsula is provided initially. This is followed by the selection of the storm events and a description of the simulated periods.

Model description

The model employed in this study was set-up on the basis of the Delft3D suite, developed by Deltares. A coupled modeling system implementing modules of flow, waves, sediment transport and morphodynamics was set-up. The flow module (FLOW) solves the continuity, motion and transport equations under the shallow water assumption. In relatively shallow areas, wave action becomes important because of whitecapping, wave breaking on the surface and energy dissipation near the bottom. In the surf zone long-shore and cross-shore currents are generated, and the bed shear stress is enhanced. The wave module solves the discrete spectral action balance equation accounting for such processes. In this study wave simulations are performed using the 3rd generation spectral wave model SWAN (SWAN is an acronym for Simulating WAves Nearshore.) (Holthuijsen *et al.*, 1993). The interaction between the flow and wave modules takes into account the effect of waves on current (via forcing, enhanced turbulence and bed shear stress) and the effect of currents on waves (via set-up, current refraction and enhanced bottom friction). The sediment transport module (SED) uses the tide and wave induced currents to determine bed suspended loads and transports, while a morphodynamic module (MOR) continuously updates the model bathymetry based on the calculated sediment fluxes.

A nesting sequence was set-up to study the effects of storms at the northern coast of the Shandong Peninsula (Figure 2). The model bathymetry is based on nautical charts and data from the General Bathymetric Chart of the Oceans (GEBCO) (OC, IHO, and BODC, 2003) with a resolution of about 1.5 km. The larger scale rectangular model (BYS) covers the Bohai/Yellow Seas above the parallel of 34.5° with ca. 5 km grid resolution. The smaller circular model (SNS), centered close to the Yantai urban area, covers the northern part of the Shandong Peninsula, with horizontal grid resolution varying from 0.2 km onshore up to 1 km offshore (Figure 2). The BYS model has an open boundary in the south. Tidal forcing based on the Global Model of Ocean Tides (TPXO) (Egbert and Bennet, 1994) is imposed at this boundary. Water levels and wave spectrum information from the larger scale model are imposed at the open sea boundaries of the small-scale model as time series. The model is forced with tides at the open sea

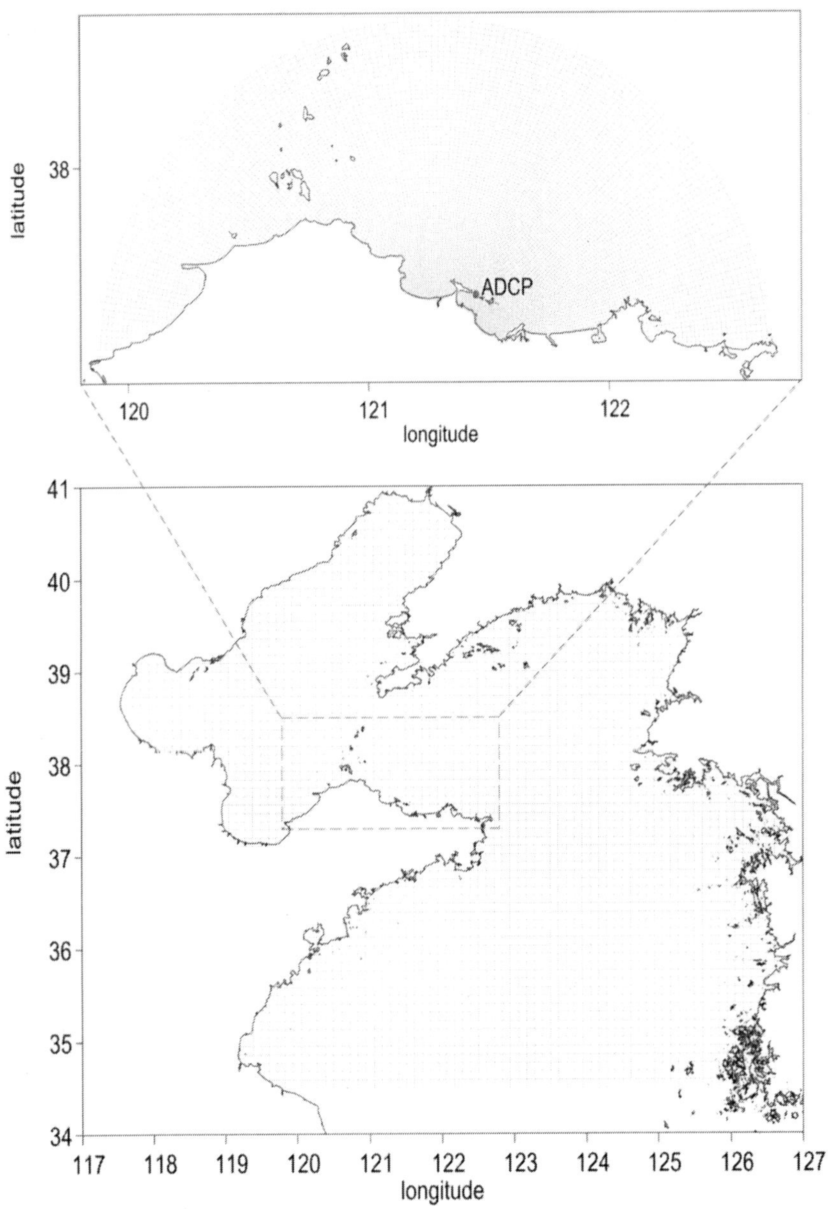

Figure 2. Model grids: BYS Model (down) and SNS Model (up)

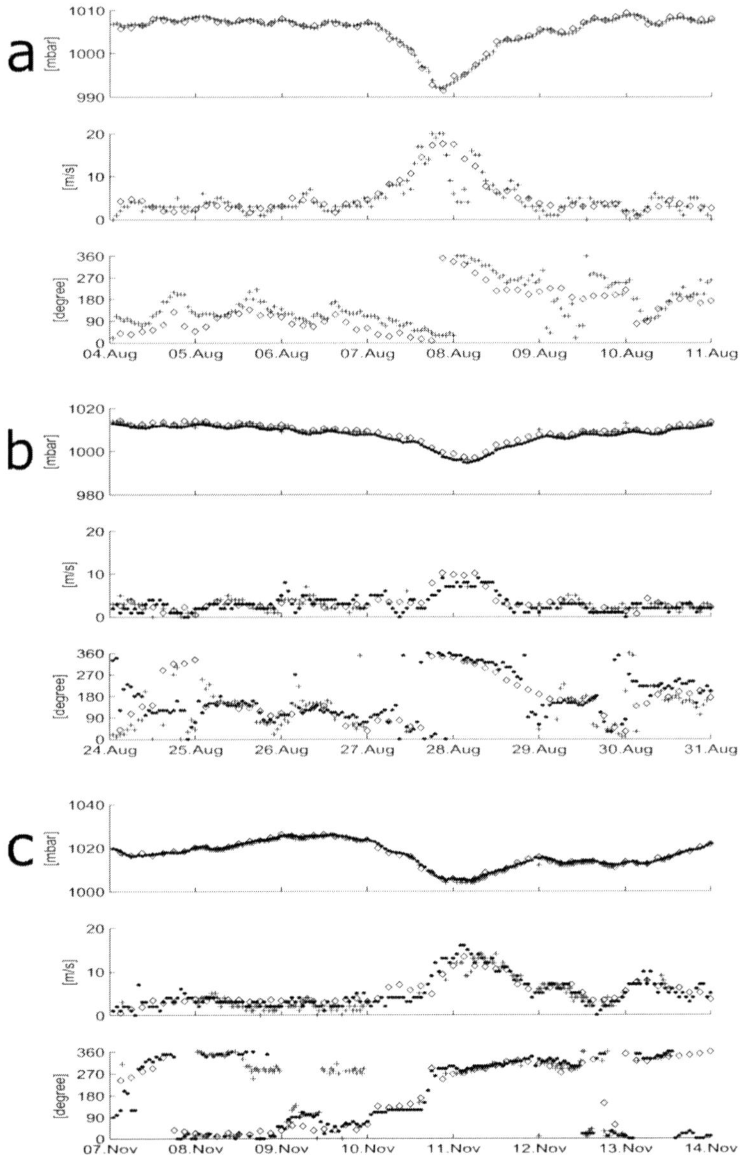

Figure 3. Comparison of measured and predicted air pressure, wind magnitude and direction at the (+) airport meteorological station, (•) ODIN-WESTPAC and (◇) DWD model data. Values are shown at the stations: (a) Yantai for storm 15, (b) Qingdao for storm 17, and (c) Dalian for storm 18

Table 1. Most severe storms in Yantai from *2000* to *2013*

Storm id	Start date	End date	Mean storm wind speed (m/s)	Mean storm wind direction (°N)	Max. water level at ADCP station(m)	Max. sig. wave height at ADCP station(m)
1	02.04.2000	12.04.2000	14.3	322°	1.38	2.66
2	07.10.2003	17.10.2003	13.2	14°	1.15	2.44
3	21.11.2004	01.12.2004	14.2	304°	0.93	2.25
4	04.08.2005	14.08.2005	15.0	76°	1.34	2.80
5	16.10.2005	26.10.2005	14.9	9°	1.12	3.38
6	18.10.2006	28.10.2006	14.9	348°	0.99	2.88
7	01.03.2007	11.03.2007	16.2	344°	1.26	4.08
8	18.12.2008	28.12.2008	13.6	342°	0.77	2.20
9	19.01.2009	29.01.2009	13.4	303°	0.86	2.45
10	09.03.2009	19.03.2009	12.4	329°	1.08	2.01
11	08.04.2009	18.04.2009	14.6	23°	1.17	2.42
12	09.04.2010	19.04.2010	13.8	319°	1.20	2.01
13	21.10.2010	31.10.2010	13.5	7°	0.95	2.33
14	07.11.2010	17.11.2010	13.6	280°	1.06	1.76
15	04.08.2011	14.08.2011	16.2	350°	1.33	3.90
16	27.03.2012	06.04.2012	13.5	343°	0.96	2.05
17	24.08.2012	03.09.2012	14.6	345°	1.24	2.61
18	07.11.2012	17.11.2012	16.1	283°	1.09	1.82
19	02.03.2013	12.03.2013	15.1	342°	1.07	3.29
20	22.05.2013	01.06.2013	14.0	89°	1.66	2.44

boundary and wind and pressure fields from the global model of the German Weather Service (DWD)(Majewski, Frank, and Liermann, 2014) with spatial and temporal resolution of ca. 30 km and 3 hours respectively. Figure 3 shows comparisons of the measured and modeled air pressure, wind magnitude and direction. Measurements at the airport meteorological stations (UW) (University of Wyoming, Department of Atmospheric Science) and the Ocean Data and Information Network for the Western Pacific Region (ODIN-WESTPAC) (Ocean Data and Information Network for the Western Pacific Region) database are compared to results of the DWD model at the locations: Dalian, Qingdao and Yantai. The location of the stations is indicated in Figure 1. Comparions are shown for three of the storms (storms 15, 17 and 18) considered in this study. Details of the storms are given in Table 1. It can be seen that the meteorological model is able to predict wind velocities and directions in good agreement with observations.

Hydrodynamic models

Results of sensitivity studies showed that a two-dimensional depth-integrated approximation (2DH) is capable of describing flow and transport processes in the region reasonably well. Only in summer, during periods in which density currents are stronger, the monthly averaged currents cannot be well approximated by the 2DH model. The large-scale flow model was calibrated on the basis of tidal components at the stations Dalian and Lianyungang. The location of the stations is shown in Figure 1. Figure

4 shows comparisons of the tidal components obtained on the basis of measured and modeled results. There is good agreement at the two stations. At Dalian, in the north, the correlation coefficient between measured and modeled values is 0.99 whereas at Lianyungang, close to the open boundary, the correlation is 0.93. The lower correlation is attributed to the fact that in the global tidal model the amphidromic point near Lianyungang is not in the exact same location as given by the Chinese co-tidal and co-range charts, although the maps generated from model results agrees reasonably well with the ones based on observations. Figure 5 shows comparisons of measured and simulated water levels, wave heights and current velocities. Measured values from a stationary ADCP deployed in the Yantai Bay in September 2013 at a water depth of about 17 m were used (see Figure 2). There is good agreement between measured and computed values. The wave heights are slightly underestimated due to the low temporal resolution of the meteorological forcing, which is unable to capture the peaks of wind velocities values adequately.

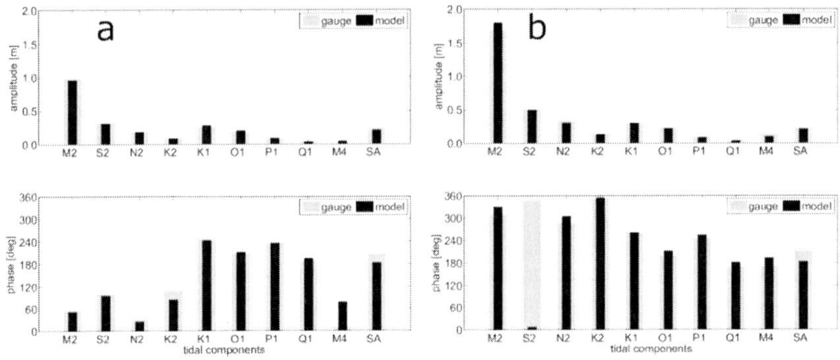

Figure 4. Comparison of tidal harmonics obtained on the basis of measured and modeled values at stations Dalian (a) and Lianyungang (b)

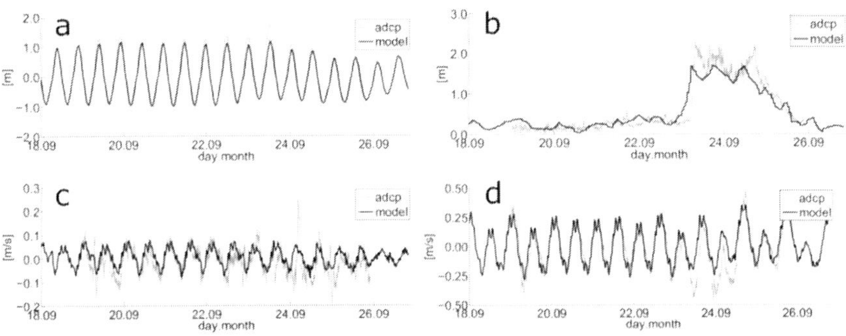

Figure 5. Measured versus computed water levels (a), wave heights (b), east-west currents (c) and north-south currents (d) in the Yantai Bay-September 2013

Sediment transport and morphodynamic models

The transport of the cohesive and non-cohesive sediment fractions is modeled following Partheniades (1965) and van Rijn *et al.* (2000), respectively. The main rivers discharging into the Bohai Sea are modeled as river mouth discharge points. Monthly averaged river discharges and suspended sediment concentrations are imposed. For the Yellow River values based on the Yellow River Sediment Bulletin (Yellow River Conservancy Commission of the Ministry of Water Resources) were used. The bed stratigraphy is represented in the sediment transport model through several sub-layers. The 5m-bed layer under the seabed was subdivided into 0.2 m thick sub-layers. Below this layer another 5m – sand layer composed of a mixture of fine and medium sand is considered. The spatial distribution of the sediment fractions is based on Liu *et al.* (2010). They presented an update of the original distribution of surface sediments in the Yellow Sea and its surrounding seas after Li *et al.* (2005) (see Figure 6). It can be seen that the surface seabed sediment is mainly composed of cohesive material comprising of different mixtures of clay and silt. However in some regions there are large deposits of coarser material (from fine to coarse sand). Based on this distribution, three sediment fractions were defined in the model: the first fraction consists of cohesive material (D_{50} <63 μm); the second of fine sand (D_{50} = 200 μm); and the third of medium sand (D_{50} = 500 μm).

The sediment transport model was warmed up for a year period considering a simplified spatial distribution of the three sediment fractions. To speed up the mixing of the bed composition, the sediment transport was multiplied by a factor of 100, without enabling morphological update as proposed by van der Wegen *et al.* (2010). The evolution of each sediment fraction during the warming-up simulation is shown in Figure 7 for six different locations (see Figure 8j). It can be seen that at the locations (a) and (b) the changes of the volume fractions are very small. At locations (c) and (f) the volume of mud in the sediment decreases as the fine sand volume increases. At location (e) a significant transformation is visible between the sand fractions. Later, only small variations around a mean value settle. Although at the location (d) a "stable" composition has not been fully reached within the warming-up period, it can be observed that towards the end of the simulation period both mud and fine sand volume fractions tend back to their initial values. The final sediment distribution obtained from the warming-up simulation is presented in Figure 8. This

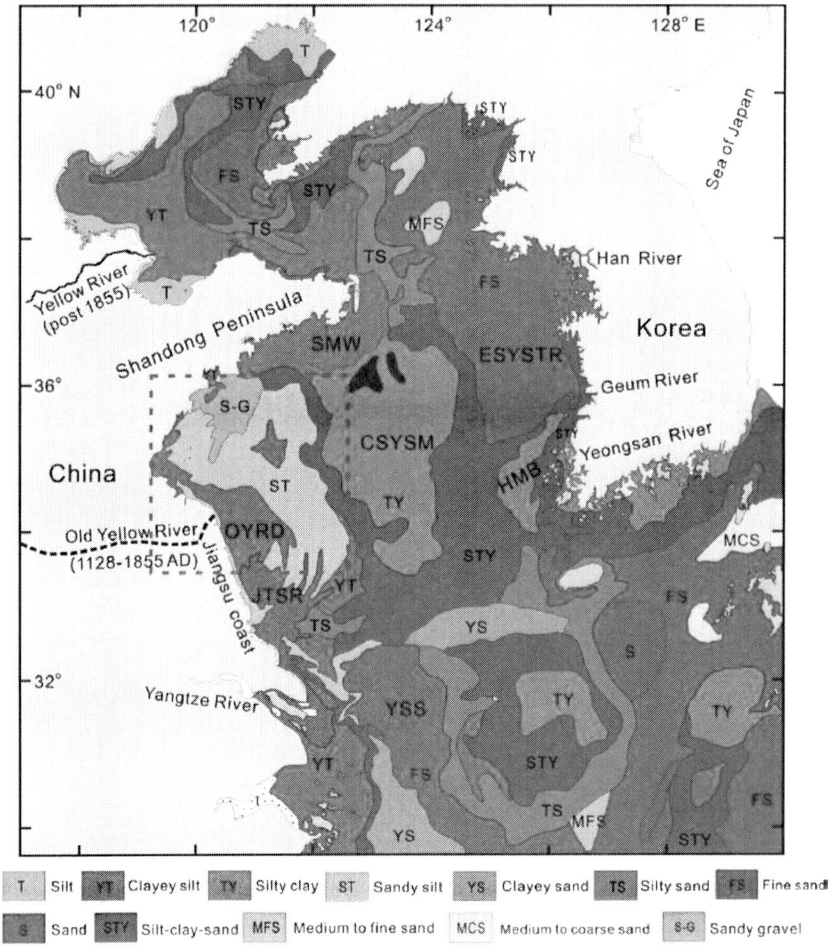

Figure 6. Distribution of surface sediments in the Yellow Sea and adjacent seas (Liu *et al.*, 2010)

result was used as initial sediment distribution for the storm simulations and also for a "long-term" simulation covering a five years period, from 2009 to 2013. Finally the morphodynamic model was used to update the model bathymetry at each computational time-step. The changes in the mass of bed material due to the sediment sink and source terms and transport gradients is calculated and then translated into bed level changes based on the dry bed densities of the various sediment fractions. The total bed elevation change due to the sediment transport at each grid cell is calculated as follows:

$$\Delta S = \Delta S_{bed} + \Delta S_{sus} \tag{1}$$

$$\Delta S_{bed}^{(m,n)} = \left(S_{b,u}^{(m-1,n)} \Delta y^{(m-1,n)} - S_{b,u}^{(m,n)} \Delta y^{(m,n)} + S_{b,v}^{(m,n-1)} \Delta x^{(m,n-1)} - S_{b,v}^{(m,n)} \Delta x^{(m,n)} \right) * \Delta t / A^{(m,n)} \tag{2}$$

$$\Delta S_{sus}^{(m,n)} = (Sink - Source) * \Delta t \tag{3}$$

where ΔS_{bed} is the bed elevation change due to bedload, ΔS_{sus} is the bed elevation change due to suspended load, $S_{b,u}$and $S_{b,v}$ are the bedload transport vector components, Δt is the computational time-step, A is the grid cell area, and *Sink*and *Source* are the suspended-sediment sink and source terms.

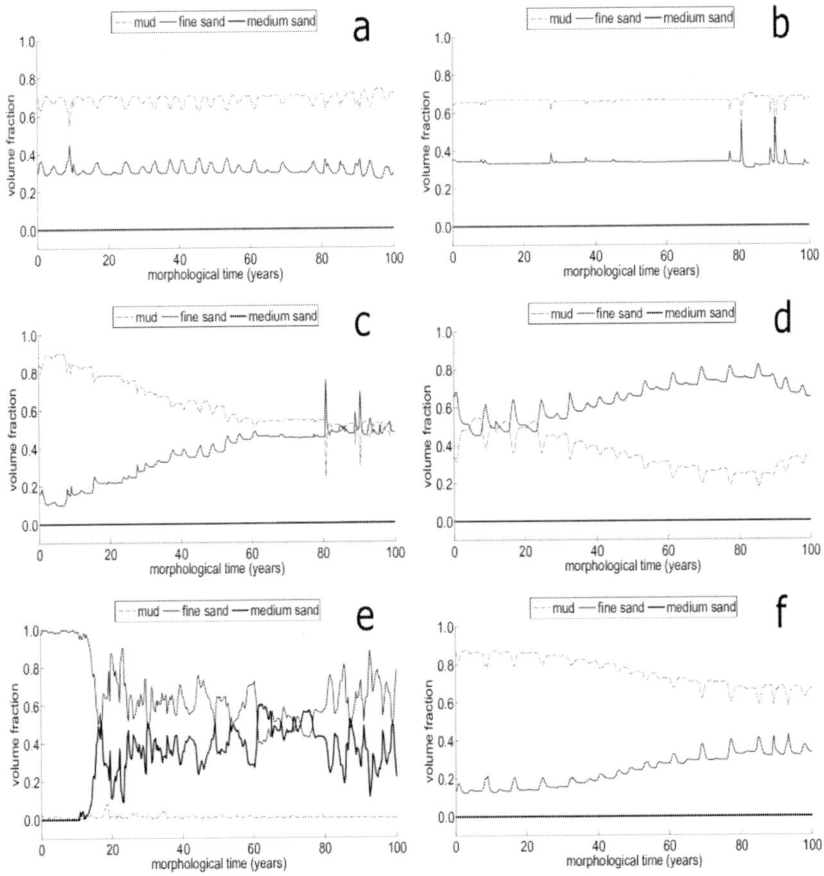

Figure 7. Evolution of the sediment fraction during the warming-up simulation

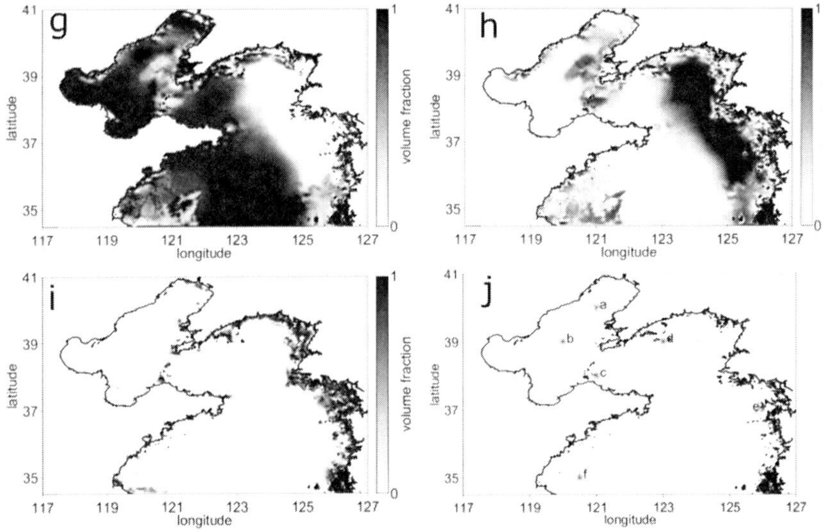

Figure 8. Final sediment distribution for cohesive material (g), fine sand (h) and medium sand (i)

Storm events

The territory of China covers zones of temperate, subtropical and tropical climate. The coastline is affected mainly by tropical or mid-latitude storm surges. While in the East and South China Seas most of the storm surges are of tropical nature, both types can occur in the Yellow and Bohai Seas. However, the frequency of mid-latitude storm surges is greater. In the Bohai/Yellow Sea region the annual climate is distinguished mainly in two seasons: summer and winter. During the summer (April to September) mid-latitude storms coming mainly from south strike the region, while in the winter, storm surges produced by cold-air outbreak coming from north hit the region.

The selection of the most severe storms was done according to the wind magnitude near Yantai and the annual Chinese reports from SOA (China State Ocean Administration Staff, 2000–2013). Air pressure and wind data were taken from the results of the simulations with the German Weather Service's meteorological model from 2000 to 2013. Periods with maximum wind speed exceeding 12 m/s (Beaufort 6) where identified, as this value represents the order of magnitude for storms in the region (Zhao and Jiang, 2011). Table 1 lists the mean wind speeds and direction of the selected storms in chronological order. Storms with significant wave heights up to about 4 m were identified. Figure 9shows wind roses for normal and storm weather conditions. As it can be seen in Figure 9a, the threshold of 12 m/s seems to be a reasonable limit for storms in the area. The predominant wind direction during storms oscillates between NNE and NNW (see Figure 9b).

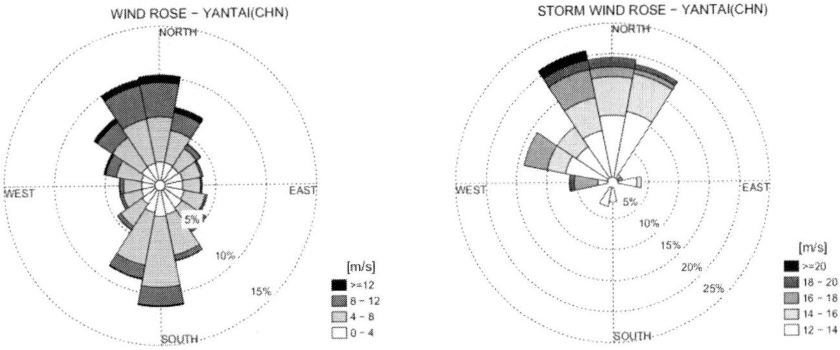

Figure 9. Wind roses for normal conditions (left) and storm conditions (right) in Yantai

RESULTS

The results of the model simulations are analyzed hereafter. First an evaluation of the energy levels along the coast of the Shandong Peninsula is presented. After that the resulting sediment transport patterns are analyzed.

Energy levels on the coast

The total energy from currents and waves is estimated at each grid cell as follows:

$$E = \frac{1}{2}\rho d v^2 + \frac{1}{8}\rho g H_s^2 \qquad (4)$$

where E is the energy given in J/m^2, ρ is the water density in kg/m^3, d is the water depth in m, v is the current velocity in m/s, g is the gravity acceleration in m/s^2 and H_s is the significant wave height in m. The first term on the right-hand side of † refers to currents and the second one to the waves.

The resulting maximum levels of energy in the area were estimated by averaging all the maximum values obtained in the simulations with the 20 storms. Maps of mean energy due to currents only, waves only and currents and waves are shown in Figure 10a, 10b and 10c, respectively. It should be

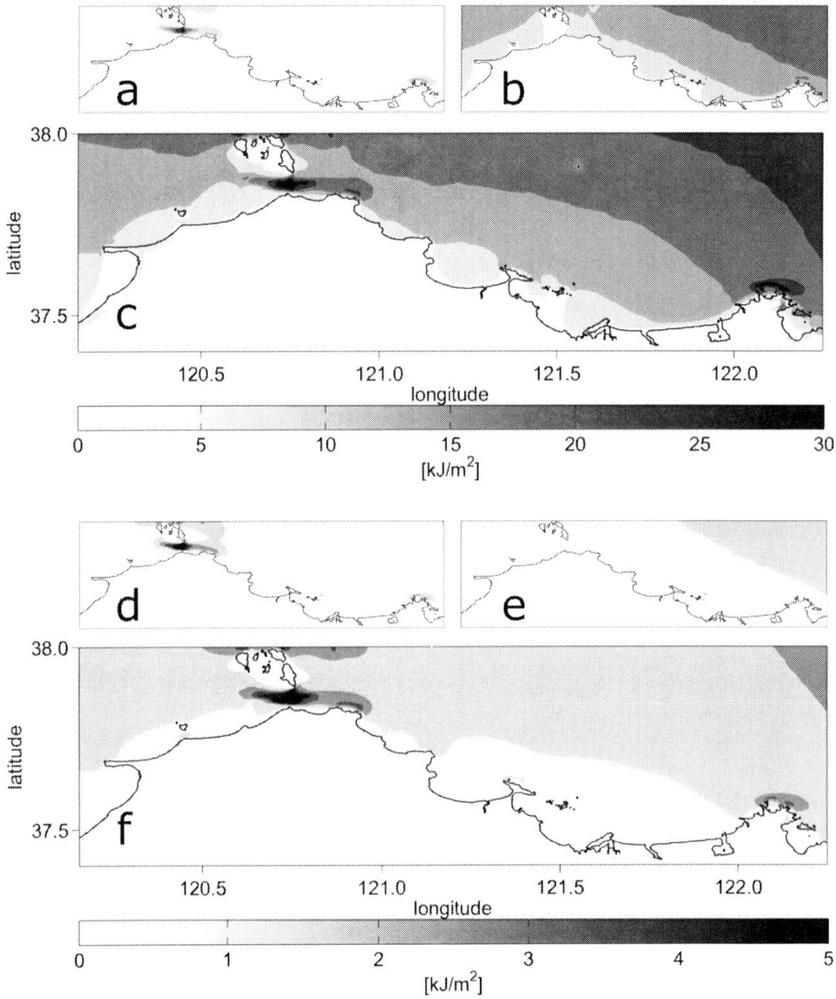

Figure 10. Mean available energy: currents (a), waves (b) and total (c) from storms, and currents (d), waves (e) and total (f) from long term

pointed out that there is a "natural" energy level along the coast. In order to estimate the long-term energy level, simulations covering a five-year period (2009–2013) were carried out. Figure 10d, 10e and 10f show mean values of the whole period, again from currents only, waves only and total energy, respectively. The energy magnitudes both figures are scaled to each case, to 30 and 5 kJ/m^2 respectively. It was found that along most stretches of the coast the maximum energy during storms is mainly due to waves (Figure 10b). Along the coastline of the Shandong Peninsula the contribution due to waves is up to three times higher than the energy due to currents (Figure 10a). Apart from the capes and near channels/straits, where tidal currents predominates, the long-term energy due to currents (Figure 10d) and waves (Figure 10e) is quite low. During storms the hydrodynamics due to currents and waves increases significantly. At the Shandong Peninsula the wind-generated waves play a more important role.

Near coastal area sediment transport

To identify the regions in the area where sediment transport might take place, maps of the resulting energy levels along the coast were constructed (see Figure 10). To cross-check this information with the actual modeled sediment transport, the mean sediment transport vectors from all the simulated storms are indicated in Figure 11 (left). In general an eastward sediment transport is observed. However, close to land, an opposite pattern is observed and the local circulation and wave effects become more important. Similarly to the patterns of sediment transport observed in the offshore regions, the strongest transport takes place at capes, channels and straits and it is oriented towards the east. However by analyzing the east-west component of the transport vectors (Figure 11 (center)), a weak but opposing transport westwards can be observed in some areas of the coast of the Shandong Peninsula. In Figure 11 (right) the normalized vectors of flow and wave currents averaged during the storm periods are presented. By comparing the center and right columns, it can be observed that the westward transport along the coast is usually found where wave currents match its direction (center and east coast).

Having identified the main directions and magnitudes of sediment transport, the resulting morphological changes due to the storms are investigated. Figure 12 shows (a) maximum bathymetrical changes and (b) final bathymetrical changes averaged from all simulated storm events. It can be seen that in most parts of the model area the bed level changes reach a maximum/minimum state during the storm events and afterwards they retreat towards the initial conditions. This is not the case for the near coastal areas. Close to the shore, several zones of deposition can be identified (yellow). Besides, along the whole Shandong Peninsula, there are very narrow erosion zones between the areas of deposition and the coastline (light blue).

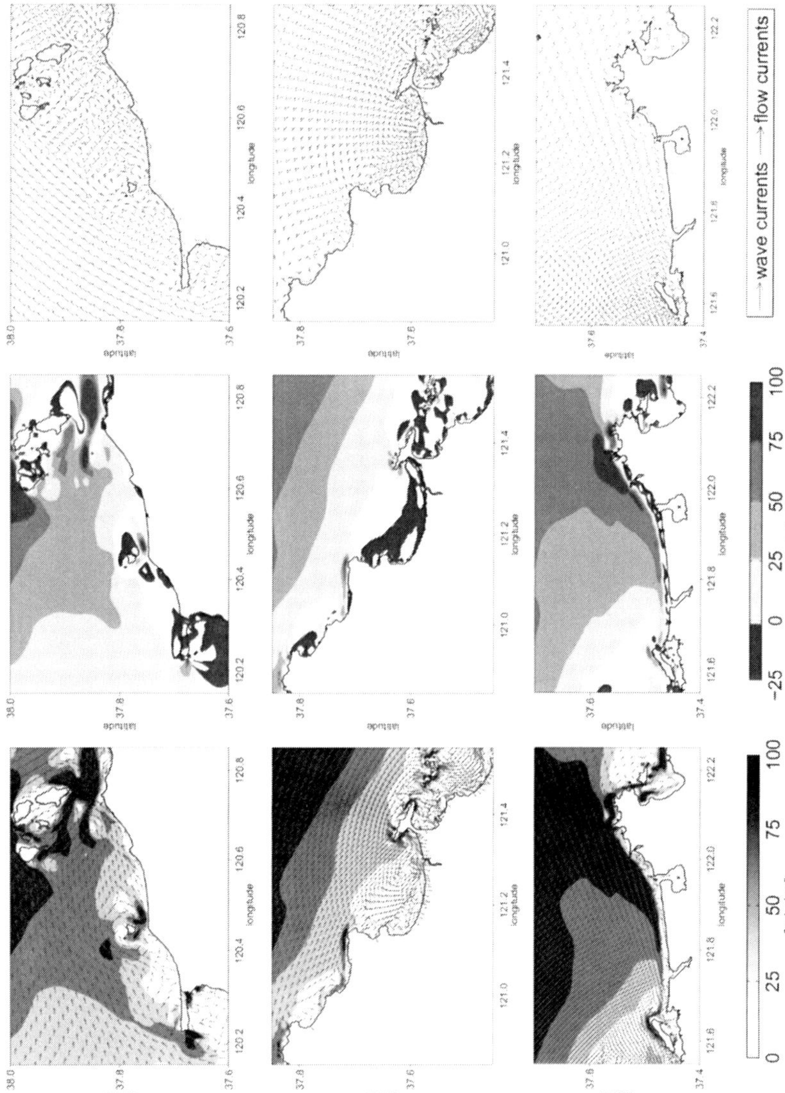

Figure 11. Mean sediment transport (left), mean east-west transport component (center) and mean flow/wave current vectors (right) during storms

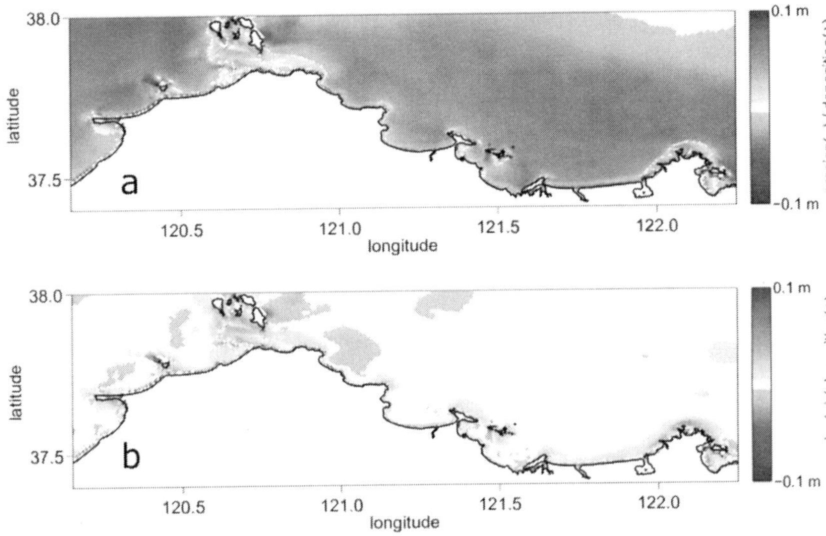

Figure 12. Maximum (a) and final (b) bathymetric changes due to storm events

DISCUSSION

The quantification of the effect of storms on coastal areas is rather difficult. Depending on the characteristics of the coastal area different aspects, such as the regional climate, main driving forces, sediment characteristics and geological composition should be accounted for. Even though several additional studies of storm surges and typhoons were carried out in the Bohai Sea, most of them consider air pressure and wind fields generated empirically. Li *et al.* (2010)summarized results of investigations of the 2003 storm surge in the Bohai Sea. They concluded that storm generated waves strongly affect the tidal system and that the resulting flow field influences the transport of sediments. Ding and Ding (2014) showed that during the same extreme event the storm surge current velocities were greater than the tidal currents. They pointed out that the current patterns get complicated under the influence of wind direction, which might affect the sediment movement close to the coast.

The results obtained in this study have shown that the investigation of the effects of storms on the coast of the Shandong Peninsula requires a distinguished interpretation. Therefore results of a global meteorological model were used instead. Despite of the limited spatial and temporal resolutions of the meteorological model used here, it shall provide a better description of the conditions in terms of currents and waves during normal and storm conditions.

It was found that as a result of the northern wind direction during the storms, the water masses are pushed towards the southern coast of the Bohai Sea causing it to flow out through the northern coast of the Shandong Peninsula. This explains the strong, in general eastward transport of sediments along the northern deeper waters (see Figure 11). When approaching the coast, two different situations can be observed: (1) in most areas the east-west transport gets weaker and very often changes direction; (2) at straits, capes and along the deeper channels the strong eastward transport pattern predominates. The first one might be due to the local circulation and the direction of wave breaking during northerly storms, whereas the second one seems to be related to the local water depth.

Based on the resulting bathymetric changes during the simulated storms, it can be inferred that the movement of sediment close to the shore is initiated by wave action and then transported by tidal currents. Waves can transport coarser and heavier sediment particles in the breaking zone, but afterwards tidal currents redistribute the finer and lighter fractions of the eroded material, in suspension.

Care should be taken in the interpretation of the results. As already mentioned, a 2DH model approximation was adopted. Although the model results show a tendency of recovery of the seabed after storms (after about five days), it is well known that the density gradients could be quite significant in summer and thus play an important role in the long-term sediment transport. That means the actual energy contribution from currents could become important and therefore change the dynamics of suspended sediment and the annual sediment budget. Besides, the main rivers discharging into the Bohai Sea have been modeled in a simplified way. In this study the monthly river discharges and sediment transport averaged over recent years were considered. In other words the annual variation in terms of fresh water and sediment supply was not taken into consideration. In this study the actual variations in river discharges throughout the past decade, particularly the decreasing river discharges and sediment, were not accounted for.

CONCLUSIONS

Results of an extensive assessment of the effect of storms on the morphodynamics of the coastal areas of Yantai in the Shandong Peninsula were analyzed. The most severe storms observed from 2000 to 2013 were identified and reconstructed with the help of a high-resolution morphodynamic model forced with wind and pressure fields from a global meteorological model. A nesting sequence covering the entire Bohai Sea with increasing resolution towards the coast of Yantai was adopted. Altogether 20 severe storms were simulated. The resulting water levels, current velocities, wave heights, sediment transport and bed elevation changes were analyzed.

During the storms water levels reached up to 1.7 m and maximum wave heights of more than 4 m resulted near Yantai. The results showed that the main contribution to the available energy during storm events comes from waves. Tidal currents are more relevant only at capes and channels/straits. Besides, storms tend to produce a stronger eastward transport of sediment on offshore areas of the northern Shandong Peninsula and a weaker westward transport of sediment on shallower areas closer to the shore. The westward transport due to storms is mainly the result of wave-induced currents. Although an overall loss of seabed material during the peak of the storm events is observed, there is a clear tendency for the seabed to recover. In other words, after a given storm event the sediment eroded in the deeper waters will probably be replaced by the dynamics of normal conditions. However nearer to the shore, usually a wide accretion area right in front of a narrower erosion zone is observed. This fact emphasizes the importance of natural sediment deposits along the coast (coastal sand dunes), which can contribute to the coastal stability. To advance the understanding in this field, future research on long-term sediment transport processes should focus on the investigation of the relevance of the density currents using a 3D model approximation. Attention should also be paid to the influence of the Yellow River discharge on the dynamics of suspended sediment in the Shandong Peninsula coast.

ACKNOWLEDGMENTS

The authors would like to thank the German Ministry of Education and Research (BMBF) for the funding of the Project SPLASH (03F0632A) and the German Weather Service (DWD) for providing data access.

LITERATURE CITED

1. China State Ocean Administration Staff, 2000–2013. *Bulletin of marine disasters of China*. China State Ocean Administration.
2. Chin, T. and Feng, S., 1975. A preliminary study on the mechanism of shallow water storm surges. *Scientia Sinica*, 18(2), 242–261.

3. Ding, Y. and Ding, L., 2014. A Numerical Simulation of Extratropical Storm Surge and Hydrodynamic Response in the Bohai Sea. *Discrete Dynamics in Nature and Society*, 1–8.
4. Egbert, G. D. and Bennet, A. F., 1994. TOPEX/POSEIDON tides estimated using a global inverse model. *Journal of Geophysical Research*, 99, C12, 24,821–24,852.
5. Holthuijsen, L. H.; Booij, N., and Ris, R. C., 1993. A spectral wave model for the coastal zone. In: *Second International Symposium on Ocean Wave Measurement and Analysis*, pp. 630–641.
6. Huatong, W. and Wang, J. D., 1985. Modelling the hydrodynamics of the Bohai Sea in China. *Chinese Journal of Oceanology and Limnology*, 3(2), 185.
7. IOC; IHO, and BODC, 2003. *Centenary Edition of the GEBCO Digital Atlas*. Liverpool, U. K. : British Oceanographic Data Centre.
8. Jiang, W.; Pohlmann, T.; Sun, J., and Starke, A., 2004. SPM transport in the Bohai Sea: field experiments and numerical modelling. *Journal of Marine Systems*, 44, 3, 175–188.
9. Li, G.; Yang, Z., and Liu, Y., 2005. *Sediment distribution map of the East China Seas*. China: Science Press.
10. Li, P. L.; Zhao, W.; Zuo, J. C.; Li, L., and Tian, J. W., 2003. A Numerical Study of the Wave, Current and Tide Interactions in the Bohai Sea During a Winter Storm. *Proceedings of the thirteenth International Offshore and Polar Engineering Conference* (Honolulu, USA), pp. 270–276.
11. Li, X.; Zhang, W., and Zhang, J., 2010. Study on Bohai Sea tidal dynamics under extratropical storm surge. *Proceedings of the ninth International Conference on Hydrodynamics* (Shanghai, China), pp. 611–616.
12. Liu, J.; Saito, Y.; Kong, X.; Wang, H.; Wen, C.; Yang, Z., and Nakashima, R., 2010. Delta development and channel incision during marine isotope stages 3 and 2 in the western South Yellow Sea. *Marine Geology*, 278, 54–76.
13. Majewski, D.; Frank, H., and Liermann, D., 2014. *GME User's Guide*. Deutscher Wetterdienst.
14. Ocean Data and Information Network for the Western Pacific Region. http://www.odinwestpac.org.
15. Partheniades, E., 1965. Erosion and Deposition of Cohesive Soils. *Journal of the Hydraulics Division*, 91, 105–139.
16. Shin, F.; Sun, W. and Wei, G., 1997. A WDM method on a generalized curvilinear grid for calculation of storm surge flooding. *Applied Ocean Research*, 19, 275–282.
17. University of Wyoming, Department of Atmospheric Science. http://www.weather.uwyo.edu.
18. van der Wegen, M.; Dastgheib, A.; Jaffe, B. E., and Roelvink, D., 2010. Bed composition generation for morphodynamic modeling: case study of San Pablo Bay in California, USA. *Ocean Dynamics*, 61, 173–186.
19. van Rijn, L. C.; Roelvink, J. A., and Horst, W. T., 2000. *Approximation formulae for sand transport by currents and waves and implementation in DELFT-MOR*. Z3054.40, WL - Delft Hydraulics.
20. Yellow River Conservancy Commission of the Ministry of Water Resources Staff, 2006–2011. *Yellow River Sediment Bulletin*. Yellow River Conservancy Commission of the Ministry of Water Resources.
21. Zhao, P. and Jiang, W., 2011. A numerical study of storm surges caused by cold-air outbreaks in the Bohai Sea. *Nat Hazards*, 59, 1–15.

CHAPTER 14

Towards the Development of an Operational Forecast System for the Florida Coast

Vladimir A. Paramygin, Y. Peter Sheng * and Justin R. Davis

Coastal and Oceanographic Engineering Program, University of Florida, Gainesville, FL 32611-6580, USA

ABSTRACT

A nowcasting and forecasting system for storm surge, inundation, waves, and baroclinic flow for the Florida coast has been developed. The system is based on dynamically coupled CH3D and SWAN models and can use a variety of modules to provide different input forcing, boundary and initial conditions. The system is completely automated and operates unattended at pre-scheduled intervals as well as in event-triggered mode in response to Atlantic-basin tropical cyclone advisories issued by the National Hurricane Center. The system provides up to 72-h forecasts forward depending on the input dataset duration. Spatially, the system spans the entire Florida coastline by employing four high-resolution domains with resolutions as fine as 10–30 m in the near-shore and overland to allow the system to resolve fine estuarine details such as in the Intracoastal Waterway and minor tributaries. The system has been validated in both hindcast and nowcast/forecast modes using water level and salinity data from a variety of sources and has been found to run robustly during the test periods. Low level products (e.g., raw output datasets) are disseminated using THREDDS while a custom defined web-based graphical user interface (GUI) was developed for high level access.

Keywords: forecasting; storm surge; baroclinic; Florida

1. INTRODUCTION

Coastal zones in the U.S. and throughout the world are subject to increasing hazards including storms and storm surge, sea level rise, and harmful algal bloom. Tropical cyclones and associated surge and inundation along the southeastern US coastline area major concern for coastal communities and their economies. Coastal waters in the southeastern US support ecologically and economically significant ecosystems, providing tourism, boating, fishing, and other recreational opportunities with an annual economic benefit of $675+ billion. With 73.5% of the population living in the coastal zone and 77.1% of GDP coming from shore-adjacent counties, this concern about tropical cyclones is particularly important to the State of Florida as it ranks in the top five of US states in the total ocean economy for its reliance on coastal tourism, recreation, and transportation sectors for employment [1,2]. Florida's battle with tropical cyclones is notorious as it has been affected by more hurricanes than any other state. For example, between 1900 and 2010, Monroe County, located along the southwest Florida coast was affected by 32 hurricanes, which is more than any other county in the United States [3].

Management of the Floridian coastal environment is a challenging task for several state and local agencies including Florida Department of Environment Protection, Florida Division of Emergency Management, Water Management Districts, and coastal counties as well as local governments. The work of these agencies is heavily dependent on information made available by such federal agencies as National Oceanic and Atmospheric Administration (NOAA), Federal Emergency Management Agency (FEMA), U. S. Geological Survey (USGS), etc. Within NOAA (the primary agency associated with surge and inundation hazards), the National Weather Service (NWS), National Ocean Service (NOS), as well as the U.S. Integrated Ocean Observing System (IOOS) provide a multitude of data on the national scale. However, data at physical scales relevant to regional and sub-regional management can be scarce, which makes coastal zone management difficult.

There are several operational and/or quasi-operational forecasting systems for the Florida coast. The National Hurricane Center (NHC) provides the official tropical cyclone surge forecasts, based on the SLOSH [4] and ADCIRC [5] model forecasts. These forecasts provide information for emergency operations and evacuation along the U.S. Atlantic and Gulf coasts. However, both SLOSH and ADCIRC forecasts are based on two-dimensional barotropic models. SLOSH is further constrained in that it uses a coarse grid resolution (on the order of a kilometer) and lacks such important processes as tides, waves, and nonlinear inertia. The Center for Ocean Atmospheric Prediction Studies (COAPS) of Florida State University operates a HYCOM-based 3D forecasting system [6] which possesses robust physics (such as atmospheric-ocean interaction) and 3D baroclinic ocean processes. However, the HYCOM forecasting system uses a relatively coarse grid (>500 m) which is inadequate to resolve the complex coastal and estuarine processes. Similar issues can be found in other Florida forecasting systems based on implementations of the ROMS model [7] and the NCOM model [8] which use relatively course grid resolutions along the Florida coast.

To address the need for a high-resolution forecasting system which can simulate water levels, waves, salinity, and baroclinic circulation along the Florida coast the Advanced Coastal Modeling System (ACMS) was developed. This system can provide forecast information which could be used by the state and local agencies to enhance management of coastal ecosystems and coastal communities in the state of Florida.

Example applications of ACMS forecast information include improved protection of coastal communities from coastal inundation; improved coastal and marine planning and decision-making; improved public health advisories; improved storm surge and rip current warnings; safer and more efficient marine operations and emergency response; advanced decision-making regarding commercial, recreational fisheries and shoreline erosion; improved planning to enhance climate resiliency; improved operational management of water control structures and utility infrastructure by Water Management Districts and utility companies, respectively; and improved emergency operations and management during tropical cyclones via information provided to Weather Forecasting Offices (WFOs) and National Estuarine Research Reserves.

The ACMS is based on the dynamically-coupled CH3D [9,10,11,12,13] and SWAN [14] models which account for wave effects (such as wave-induced wind stresses, wave-current interaction which includes radiation stresses throughout the water column and wave-current bottom stresses) and can incorporate a variety of input forcing functions and boundary and initial conditions for driving these models. The system is automated and can be run at pre-scheduled intervals or be triggered by such events as tropical storm advisories by the National Hurricane Center.

Another significant issue to be considered for operational modeling is computational efficiency. Multiple (often over ten) forecasting model runs with high grid resolutions every day require significant computing resources. A forecasting system must be able to produce timely forecasts, since the value of forecast products declines quickly with the time it takes to produce them.

In the remainder of this paper, ACMS is first described, followed by a description of the model setup, and example model validations.

2. MATERIALS AND METHODS

2.1. The ACMS Modeling System

2.1.1. CH3D

CH3D (Curvilinear Hydrodynamics in 3D) is a hydrodynamic model originally developed by Sheng [9,10]. The model can simulate 2-D and 3-D barotropic and baroclinic circulation driven by tide, wind, density gradients, and waves. CH3D uses a boundary-fitted non-orthogonal curvilinear grid in the horizontal direction and a terrain-following sigma grid in the vertical direction. As such, the model can accurately represent complex shoreline and geometries in coastal regions. It uses a robust turbulence closure model to represent vertical turbulent mixing [15] and a Smagorinsky type model for horizontal turbulent mixing. The model uses bathymetry and topography which are referenced to the NAVD88 vertical datum for all domains to accurately simulate the coastal inundation. CH3D has been applied to such water bodies as Charlotte Harbor, Biscayne Bay, Apalachicola Bay, Florida Bay, Indian River Lagoon, Lake Okeechobee, Lake Apopka, Sarasota Bay, St. Johns River, Tampa Bay, Naples Bay, and Rookery Bay in Florida, as well as Chesapeake Bay, New York Bight, Long Island Sound, and the Gulf of Mexico.

2.1.2. CH3D-IMS

CH3D has been coupled to models of wave, sediment transport, water quality, light attenuation, and sea grass dynamics to produce CH3D-IMS [16], an Integrated Modeling System for simulating the response of estuarine and coastal ecosystems to anthropogenic (e.g., increased nutrient loading) and natural (e.g., sea level rise) changes.

2.1.3. CH3D-SSMS

CH3D-SSMS (Storm Surge Modeling System) is a modeling suite that features coupled CH3D and SWAN models for coastal dynamics and large scale surge-wave models that are used to extract boundary conditions for the coastal model [11,17], which has been used extensively to simulate storm surge and inundation due to various tropical cyclones including Hurricanes Charley (2004), Dennis (2005), Isabel (2003), Frances (2004), Ivan (2004), Jeanne (2004), Katrina (2005), Wilma (2005), Katrina (2005), Ike (2009), Sandy (2012), Matthew (2016), and others [11,12,13,17,18,19]. Details of the CH3D model, including equations of motion and boundary and initial conditions, are described in [11]. CH3D-SSMS contains a robust flooding and drying scheme which is an extension of that developed by Davis and Sheng [20].

In a regional storm surge and coastal inundation model Testbed [21], CH3D-SSMS was compared with ADCIRC [5], CMEPS [22], FVCOM [23], and SLOSH [4]. Detailed comparisons of models were made in terms of simulated storm surges during historic storms as well as coastal inundation maps including the surge atlas and the 1% annual chance coastal inundation maps which is also known as the Base Flood Elevation (BFE) according to the Federal Emergency Management Agency (FEMA) of the US [24]. The results of CH3D-SSMS were found to compare well with observed water level data and was as accurate as other models. The computational efficiency of CH3D-SSMS is only inferior to the extremely efficient SLOSH model which uses a very coarse grid (~1 km) and has simpler physics. The model Testbed results demonstrated that, to obtain accurate model results efficiently, it is feasible to use a highly efficient coastal surge-wave model, e.g., CH3D-SWAN, with high resolution in the coastal region, and couple it to large-scale surge-wave models with coarser resolution in the offshore region. Alternatively, one can use an unstructured grid surge-wave model for the coastal and offshore domains with a single grid, but a high-resolution grid in the coastal region often results in stringent computational time step limitation and requires dramatically more computational resources.

CH3D suites continue to be improved as new research enables incorporation of more dynamic features, e.g., vegetation effects on storm surge, into the models. For example, Lapetina and Sheng [25] recently used the vegetation-resolving ACMS to simulate the effects of vegetation, three-dimensionality, and onshore sediment transport on complex storm surge dynamics during Hurricane Ike which inflicted major damage to the Texas coast in 2009. Results of the 3D model are found to be more accurate than 2D model results.

2.2. ACMS

The cornerstone of the ACMS modeling system includes the CH3D (shallow water hydrodynamics) and SWAN (wave) models, running on four domains (Figure 1) that span the entire Florida coast. The two models are dynamically coupled, which enables representation of complex physics such as vertically varying wave-current interaction. ACMS is essentially the integration of CH3D-SSMS and CH3D-IMS which was described in the previous section.

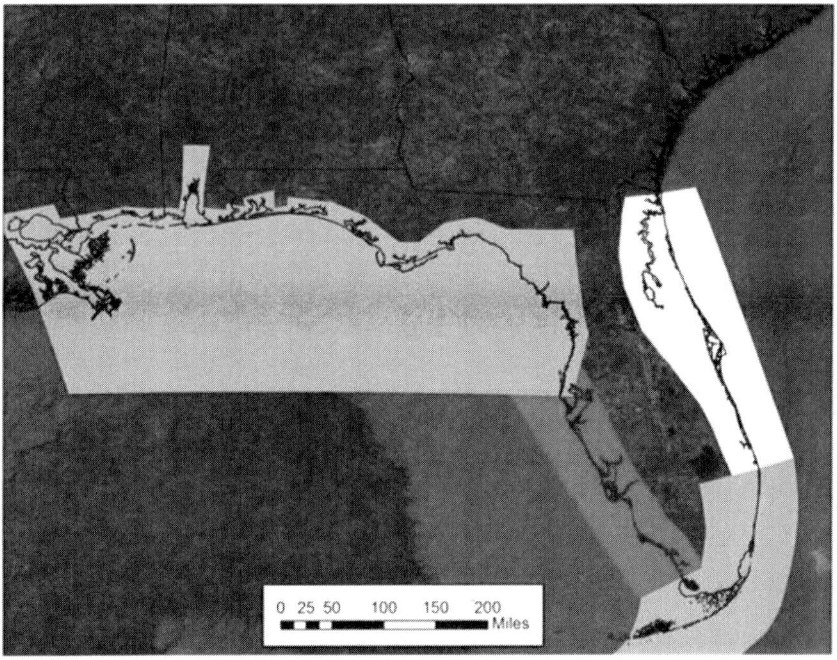

Figure 1. Advanced Coastal Modeling System (ACMS) Florida coast domains: Northern Gulf (NG) of Mexico (pink), Southwest (SW) (blue), Southeast (SE) (green), and East Coast (EC) (yellow).

SWAN is a third-generation phase-averaged wave model that can be applied to nearshore wave modeling. The model can use a variety of computational grid arrangements including non-orthogonal regular, curvilinear, and unstructured triangular grids. SWAN accounts for wave propagation in time and space, shoaling, refraction due to currents and depth, frequency shifting due to currents and dynamic depth, wave generation by wind, energy dissipation by bottom friction, depth-induced breaking and transmission through and reflection from obstacles (full or partial reflection can be considered). SWAN represents waves using a two-dimensional wave action density energy spectrum and the evolution of the spectrum is described by the spectral action balance equation in which a local rate of change of action density in time is related to the propagation of action in geographical space, shifting of relative frequency due to currents and depths, depth-induced and current-induced refraction balanced by the source term in terms of energy density representing the effects of energy generation, energy dissipation and nonlinear wave-wave interactions.

ACMS can use a variety of wind fields such as

- Hurricane Research Division's H*Wind [26];
- Navy's NOGAPS [27];
- GFDL [28];
- NAM (North American Mesoscale) that uses WRF (Weather Research and Forecasting model [29]) and is run by the National Centers for Environmental Prediction [30]; and
- Several synthetic parametric wind models driven by storm parameters that are derived from National Hurricane Center (NHC) predictions.

2.3. ACMS Modules and Workflow

ACMS consists of four main modules (Figure 2): (1) data acquisition and pre-processing module; (2) simulation setup (staging/running/etc.) and job management module; (3) post-processing module; and (4) visualization module. These modules provide automation of such processes as input data acquisition, archiving and cataloging of data and model results, data pre-processing, setting up model simulations, running and monitoring jobs, post-processing of model results, and visualization. Some of the most important properties of the ACMS are full automation, compliance with existing standards for ocean data, and efficient use of available computational resources. Previous implementation of the system showed that it can perform both 3D baroclinic and storm surge simulations simultaneously during tropical cyclones [18], as well as ensemble forecasting of storm surges based on an ensemble of storm tracks generated from the probability distribution of previous track forecasting errors [31].

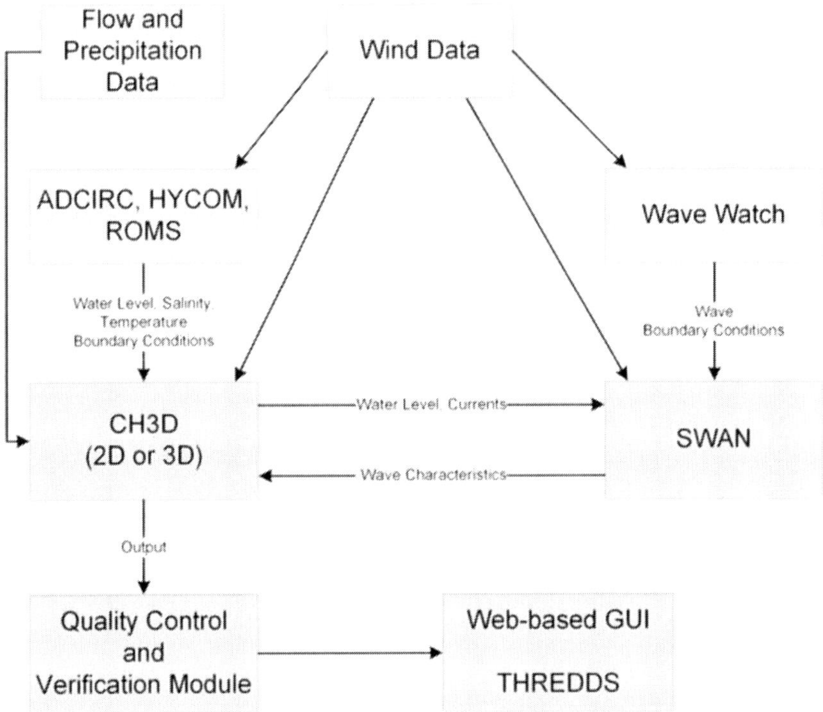

Figure 2. ACMS workflow diagram.

The data acquisition module is responsible for data acquisition and consists of monitors that poll the data providers for new data and acquires the data as it becomes available. Monitors for a variety of datasets are available: NOAA NHC advisories, the U.S. Navy's Automated Tropical Cyclone Forecasting System (ATCF [32]) forecast products, atmospheric inputs (NAM, NOGAPS, GFDL, etc.), boundary and initial conditions for circulation from such models as HYCOM [33] and ROMS, and boundary and initial conditions for waves from wave models such as WaveWatch III (WWIII) [34]. River flow measurements and predictions, salinity measurements, etc. are also collected where available from USGS, National Estuarine Research Reserves, and the National Weather Service River Forecast Center. All the data are obtained as they become available, processed (with QA/QC, subsetting, and necessary format conversions), archived, and cataloged (using a MySQL database).

ACMS supports two modes of operation:

-Event triggered, where a model instance is created as a response to an event (such as an NHC-issued tropical cyclone advisory) or

-Preset cycles, where a model is initiated at fixed times, which usually follow the standard 4 -cycles per day scheme (model initialized at 00:00, 06:00, 12:00 and 18:00 UTC).

The job management module initiates the simulation and polls the data acquisition module. Once all the data necessary for model input is collected, the module generates the necessary input files, sends the job to the computing cluster via HTCondor job management system [35] and monitors the job status for potential computing resource failures in which case the jobs are resubmitted to alternate resources.

The post-processing module extracts output from completed forecast runs, generates aggregate products, calculates statistics, and places products and outputs in the archive. ACMS currently uses NetCDF with CF-1.5 conventions as a data format of choice and Unidata's THREDDS Data Server [36] as the main platform for data distribution. Visualization module (mostly client-based) is written in JavaScript and uses THREDDS server (via WMS feeds and NetCDF subsetting) as a data provider to display data in a user-friendly manner (Figure 3).

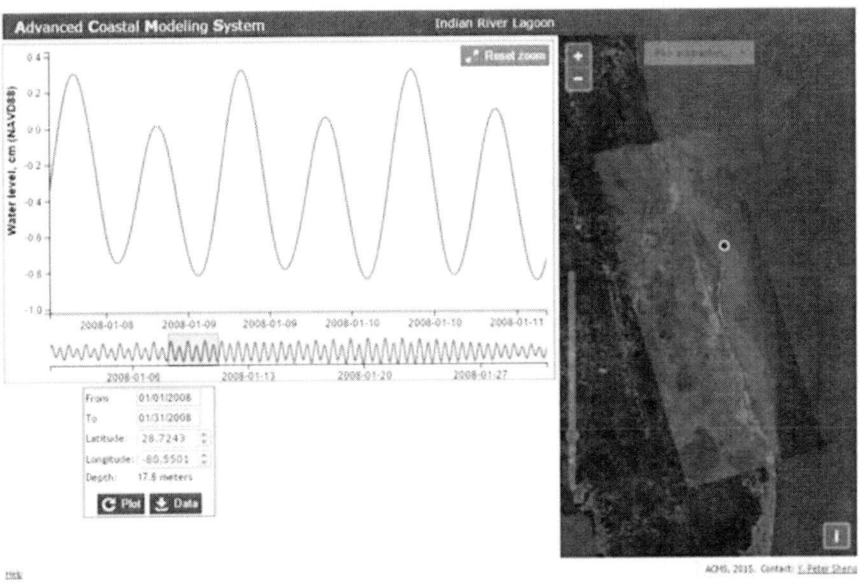

Figure 3. Sample ACMS web-based interface for data preview and download developed for St. Johns River Water Management District with focus on the Indian River Lagoon on the east coast of Florida.

Data availability is the limiting factor for initiating a new forecast cycle. A complete data set such as wind, the waves at the open boundary, the surge at the open boundary, and the flow rates at rivers should be available from the archive for the forecast cycle to be initiated. Data is pulled from the archive by the Data Processing Module and all necessary input files are generated for all the simulations that are scheduled to run within that cycle. Completion of this process triggers the start of the cycle at the Core Module which is responsible for setting up the boundary conditions for all the models involved in the cycle, scheduling, and submitting the simulation to one of the available computational resources. There are mechanisms that enable forecasts even when some of the data is missing. Certain data such as missing atmospheric snapshots or relatively short gaps in time-series data can be reconstructed, interpolated or extrapolated based on available data.

2.4. ACMS Model Setup

Two implementations of ACMS are discussed in this paper:

1. 2D implementation—The 2D barotropic CH3D model is coupled with SWAN and CH3D receives open boundary conditions from a large-scale ADCIRC model running on a coarse grid ~2–5 km, while SWAN receives open boundary condition from WWIII. Despite the coarseness of the grid, ADCIRC produces satisfactory results along the offshore CH3D boundary and runs very quickly to allow syncing with the CH3D model. This implementation is event-triggered by NHC tropical cyclone advisories. ACMS downloads NHC advisories and, whenever it contains forecasted tropical cyclone track coming within 100 miles of a CH3D domain, creates a model instance for that advisory. This implementation is used to quickly forecast storm surge and inundation during tropical storms. Not only do 2D model simulations complete quicker, but all the inputs required for the model are also contained in the advisory, because the surge and wave models are driven by a synthetic parametric model for atmospheric wind and pressure. In most cases, the amount of time between the advisory time stamp and prediction is less than two hours.

2. 3D Implementation—The 3D baroclinic CH3D model is coupled with SWAN, and CH3D receives boundary conditions from a large-scale HYCOM or ROMS model, while SWAN receives open boundary condition from WWIII. The model runs four times a day at 6-h intervals starting at 00:00 UTC. This implementation is intended to provide more comprehensive forecasts including water levels, waves, baroclinic circulation, and salinity. However, this increased fidelity comes at a cost: not only the model runtime increases to 4–6 h (depending on domain and conditions such as networking and transfer speeds, etc.), but the time

required to fetch all the inputs (including open boundary conditions from HYCOM or ROMS, atmospheric predictions from NAM, and river flows, etc.) can be twice as long. The model has the capability to simulate temperature, however, due to limited data available for boundary conditions and verification, current forecasting implementation does not include temperature simulation.

2.4.1. Atmospheric Forcing

Atmospheric forcing in ACMS includes atmospheric pressure, wind, and precipitation. The system can use a variety of wind fields as well as several synthetic parametric models. The 2D implementation uses the parametric model of Xie et al. [37]. Model parameters such as location of the storm, maximum wind, and radii to 34 kt, 50 kt, and 64 kt winds are based on the NHC predicted storm parameters. The 3D implementation is driven by the atmospheric forcing predicted by the NAM model.

2.4.2. Surge-Wave Coupling

In ACMS, the CH3D model is dynamically coupled to the SWAN wave model [14]: wave results obtained by SWAN are passed to CH3D and water depths and currents obtained by CH3D are passed onto SWAN. This accounts for wave setup and wave-current interaction within the CH3D model, which features several formulations for calculating wave stresses [38], including vertically varying formulations (e.g., Mellor [39]) as well as the vertically uniform formulation of Longuet-Higgins and Stewart [40,41]. The 3D implementation of the model uses the formulation by Mellor [39] as it was found to produce more accurate results [38].

The time step used for CH3D simulation is 60 s and time step used for SWAN simulation is 5 min, which is when the two models exchange information.

2.4.3. Coastal-Offshore Coupling

Both coastal CH3D and SWAN models use the same non-orthogonal curvilinear model grid and are dependent upon open boundary conditions provided by larger scale ocean models. ACMS interfaces enable it to receive boundary conditions from a variety of large-scale models such as HYCOM, ROMS, CH3D, and ADCIRC for CH3D, and larger-scale SWAN or WWIII for SWAN.

In the 2D implementation, CH3D obtains open boundary conditions (elevation at the open boundary) from a coarse-grid (~2–5 km) ADCIRC model which is run simultaneously with the CH3D model. Large-scale SWAN model produces boundary conditions for the coastal SWAN model.

In the 3D implementation, CH3D obtains open boundary conditions (vertically varying currents and salinity) from a HYCOM, while SWAN derives its boundary conditions either from a WWIII model or a large-scale SWAN model.

2.4.4. ACMS Model Domains and Forecast Cycles

The four domains used by ACMS span the entire Florida coastline (Figure 1) and extend 50–100 km offshore (Table 1). The forecasts range from 48 h up to 72 h, depending on the configuration and available input. Each cycle is initialized from the previous cycle, and a 6-h nowcasting is performed to fill the 6-h gap between cycles, followed by a forecast.

Table 1. Characteristics of Advanced Coastal Modeling System (ACMS) model domains.

Domain	Minimum Resolution (m)	Approximate Grid Cell Count	Average Offshore Extent (km)
East Coast (EC)	32	339,000	55
Southeast (SE)	21	607,000	60
Southwest (SW)	29	366,000	65
Northern Gulf of Mexico (NG)	47	404,000	75

ACMS model domains use NAVD88 as a vertical datum of choice, which makes computing surge, inundation, and flooding a simple and transparent process since all the topography data is generally referenced to NAVD88. The 3D implementation of the model typically uses six equally spaced sigma layers in vertical—this number was determined by comparing simulations obtained with 4, 6, 8, 16 layers, which found that 6 layers were sufficient to resolve the pycnocline and that going from 6 to 8+ layers provides negligible differences in simulation results.

2.4.5. Boundary Conditions

The water level at the open boundary of CH3D domain is prescribed by combining the water level predicted by a regional ocean circulation model and spatially varying tidal constituents which include M2, S2, N2, K2, K1, O1, P1, Q1, SA, and SSA. These tidal constituents were determined to be important for the Florida coast based on the NOAA tidal gauge data, while other constituents generally are estimated to have an amplitude of less than 1 cm. The constituents at the open boundary are developed via an iterative process in which phases and amplitudes at the open boundary are adjusted during tide-only simulations to provide the best possible fit with observed tides at the coastal stations. Salinity at the open ocean boundary is interpolated from a 1/12 degree HYCOM or ROMS (provided at 6-h intervals).

Open boundary conditions for SWAN are wave height and period obtained from the results of a 0.25 degree WWIII model.

River flow measurements/forecasts and salinity measurements are gathered from a variety of sources such as NOAA Advanced Hydrologic Prediction Service, Florida DEP, Florida Water Management Districts, and National Estuarine Research Reserves. These data serve as boundary conditions for flow and salinity upstream of rivers and creeks and are crucial to accurate predictions of currents and salinity in estuaries, inlets, and near-shore zone. River flow and salinity data used for boundary conditions in estuaries are extrapolated in time based on the trend identified in previous data for stations where flow forecasts are unavailable or the forecasted period is shorter than the length of ACMS forecast. The extrapolation is based on identifying a linear trend during the last 7 days and using it to extrapolate the flow (subject to a maximum increase/decrease of 2 ppt in a 7-day period). This was found to produce slightly better results compared to using the last available value for the entire forecast period.

3. RESULTS

3.1. Model Validation

Using the model setup described in the previous section, a series of validation tests were performed by hindcasting non-storm events (lighter wind speeds, tidally-dominated flow) as well as several tropical cyclones including Andrew (1992), Jeanne (2004), Wilma (2005), and Fay (2008). The storms are selected to represent different hurricanes ranging from slow moving Tropical Storm Fay, which became almost stationary for about a day just off the east coast of Florida near Jacksonville, to fast-paced Hurricane Wilma that went across the entire state in less than six hours with significant variation in intensity and size. Some forecast results are also presented since the ACMS system has been running in the quasi-operational mode since 2012. Only a select few stations (Figure 4) per domain are shown here as it would be impossible to show model-data comparisons for all of them in the scope of this paper.

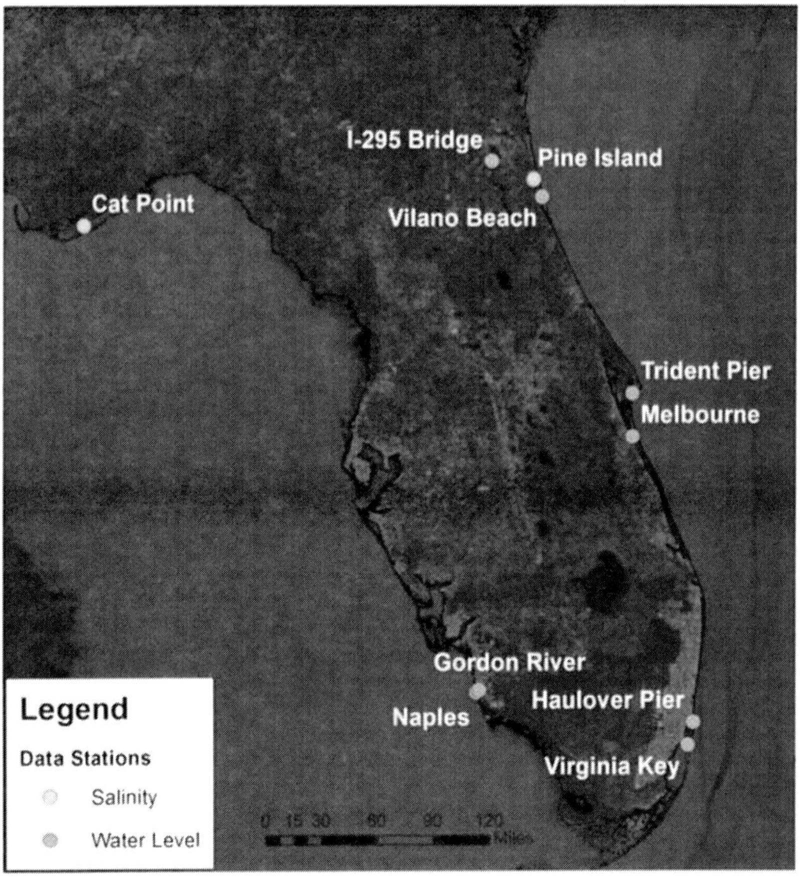

Figure 4. Location of select data stations used for ACMS validation.

3.1.1. Non-Storm Conditions

Simulation of tides is a key feature of a forecasting system because water levels and currents are very important for safe navigation and recreation as well as the increasing coastal inundation during high tides. Under calmer weather conditions, tides usually dominate and determine the coastal circulation. Hence, the ability to accurately predict tidal water level and flows is crucial. Data during 2008–2014 were used for validation purposes and the criterium for tidal validation is to achieve an average RMS error of 7 cm per domain. The total number of stations used for validation of tides is: nine for the EC domain, five for the SE domain, five for the SW and six for the NG domains. Vilano Beach (Figure 5) and Melbourne (Figure 6) stations are examples of tidal simulations on the

EC domain. Data from a NOAA station at Naples, FL (Figure 7) on the west coast was used to validate the tides for the SW domain.

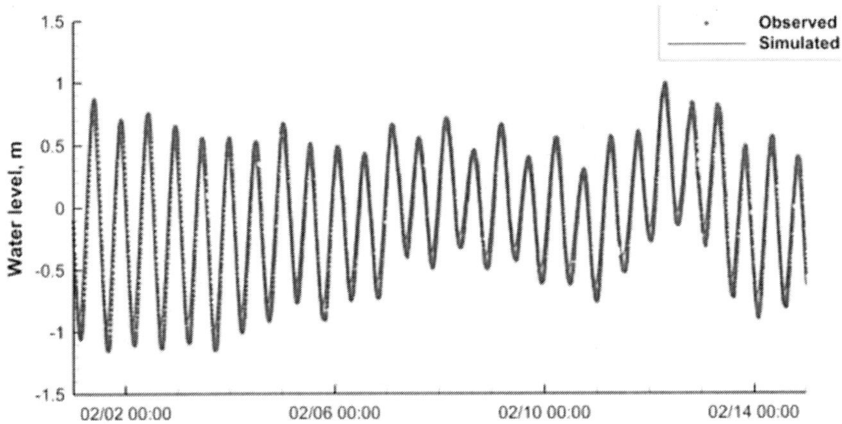

Figure 5. Comparison of simulated (hindcast) and observed water levels at Vilano Beach station for tidally-dominated flow.

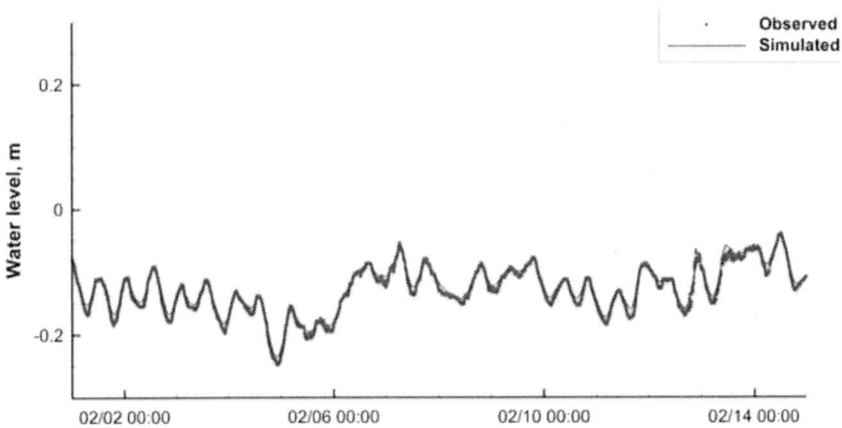

Figure 6. Comparison of simulated (hindcast) and observed water levels at Melbourne station for tidally-dominated flow.

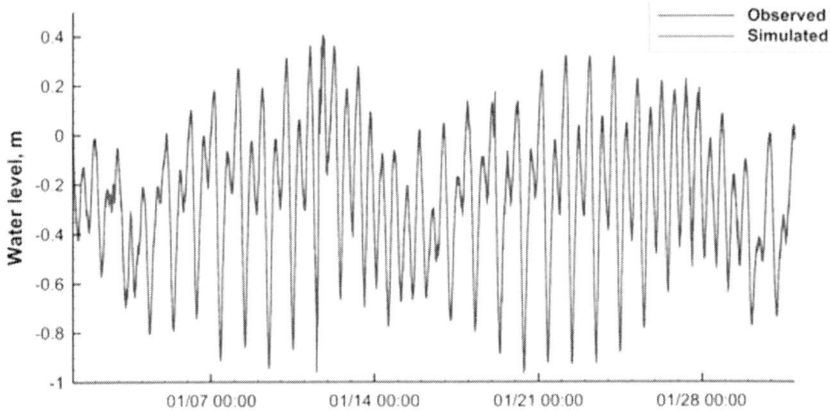

Figure 7. Comparison of simulated (hindcast) and observed water levels at Naples station for tidally-dominated flow.

Tidal Simulation

Overall, ACMS predicted amplitudes and phases of select tidal constituents (M2, S2, N2, K2, K1, O1, P1, Q1, SA, and SSA) are very close to the observed values at all stations with the average RMS error being under 5%.

Nuisance Flooding

The ACMS was used to forecast "king" tides (astronomically high tides) near Miami Beach (Figure 8), and the results were provided to the city of Miami Beach. During a "king" tide, numerous South Florida communities (Miami Beach, Fort Lauderdale, Key West, and Naples, etc.) experience nuisance flooding with streets inundated of 30 cm or more. Nuisance flooding is occurring more frequently as the sea level continues to rise. By 2050 some communities (e.g., Key West) are expected to have nuisance flooding during more than 100 days per year. The City of Miami Beach installed 20 pumps in 2015 to mitigate nuisance flooding during king tides.

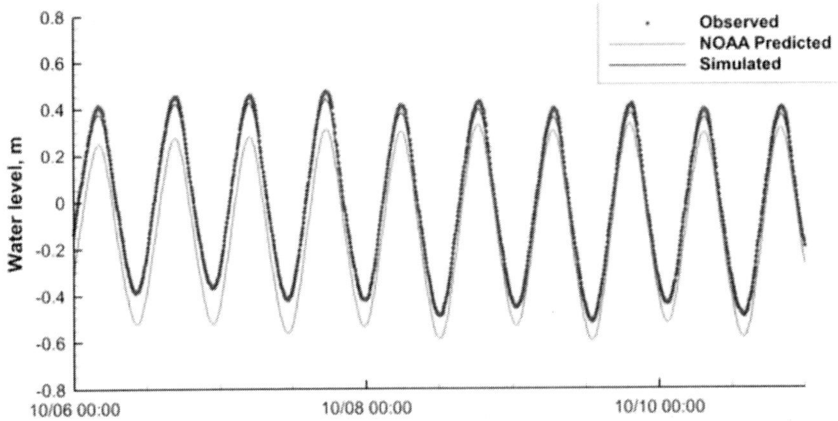

Figure 8. Forecast of "king" tides and comparison with National Oceanic and Atmospheric Administration (NOAA) observed and predicted water levels at Virginia Key station in October 2014 (initialization time: 6 October, 2014 00:00Z).

Salinity Simulation

Limited salinity data was available for validation with just two stations for the EC and SE domains, five stations clustered inside the Naples/Rookery Bay region on the west coast of Florida and a few in the Apalachicola Bay maintained by the Apalachicola National Estuarine Research Reserve (ANERR). The quality of salinity predictions can vary drastically depending on availability and accuracy of river flow predictions. However, most stations show satisfactory agreement. The RMS error in the SW domain during September and October 2014 varied between 2 ppt and 7 ppt with a correlation coefficient (r^2) between 0.52 and 0.9. The NG domain comparisons (Figure 9) show slightly better agreement with the RMS error between 2 ppt and 5 ppt. The RMS errors for stations in the EC domain vary between 3 ppt and 9 ppt, most likely due to the inaccuracy and limited availability of river flow data that provides fresh water inflow into the model domains.

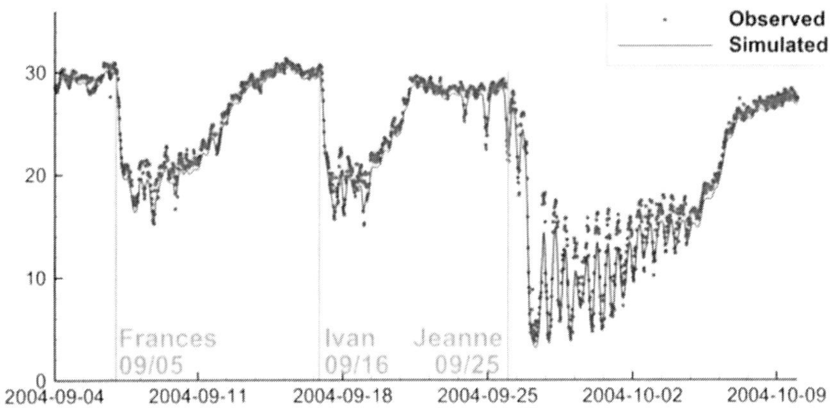

Figure 9. Comparison of simulated and observed salinity Cat Point station (NG domain), orange vertical lines indicate land fall times of the three storms: Frances, Ivan, and Jeanne.

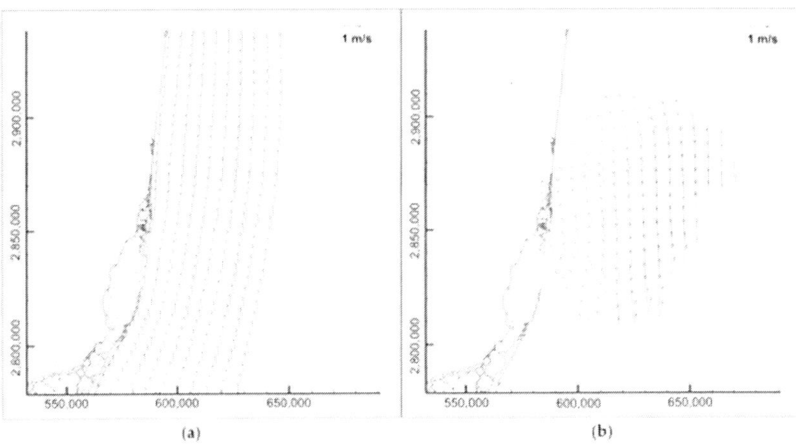

Figure 10. Comparison of estimated surface currents (**a**) vs. surface currents measured by the HF Radar (**b**) east of the Biscayne Bay (SE domain, 20 March 2014 12:00Z). Coordinate space is UTM zone 17N (m).

Surface Current Simulation

There exist limited datasets of observed surface currents collected by the high frequency radar (HF Radar, [42,43]) along the southeast coast of Florida near Biscayne Bay. The ACMS simulated currents in the central part of the coastal domain appear to be comparable (Figure 10) with the observed data. However, it should be noted that these currents data have not been fully analyzed to remove errors associated with interference of the radar signals.

3.1.2. Simulation of Surge, Wave, and Inundation during Tropical Cyclones

Model validations for tropical cyclone conditions were carried out in a hindcast mode for the following four storms: Hurricane Andrew (1992), Hurricane Jeanne (2004), Hurricane Wilma (2005), and Tropical Storm Fay (2008).

Hurricane Andrew

Andrew was a small but ferocious hurricane that brought unprecedented economic devastation to the southern Florida peninsula. Overall damage in the U.S. is estimated at ~$26.5 billion (1993 USD) making it one of the five costliest storms in U.S. history. The tropical cyclone struck southern Dade County, Florida (Figure 11) especially hard, with violent winds and storm surges characteristic of a category 5 hurricane [44] on the Saffir/Simpson Hurricane Scale, and with a central pressure of 922 mb. Unfortunately, the amount of data available for Andrew is limited to a single station at Haulover Pier, but the model predicted water level at this station compared very well with the observed data (Figure 12), especially when the wave model is coupled to the surge model in the simulation.

Hurricane Jeanne

Hurricane Jeanne (Figure 13), while known as a very deadly storm claiming more than 3000 lives in Haiti alone, weakened significantly before making its landfall on the east coast of Florida near Stuart [45]. After which, it further weakened to a tropical cyclone making its way across Florida peninsula towards Tampa. Because of its smaller size (about 50 miles at landfall) the area affected by it was relatively small. As such, a very limited amount of data is available for comparison. The water level response predicted by ACMS at Trident Pier (Figure 14) matches the observed data well, both in terms of the peak water level and phase.

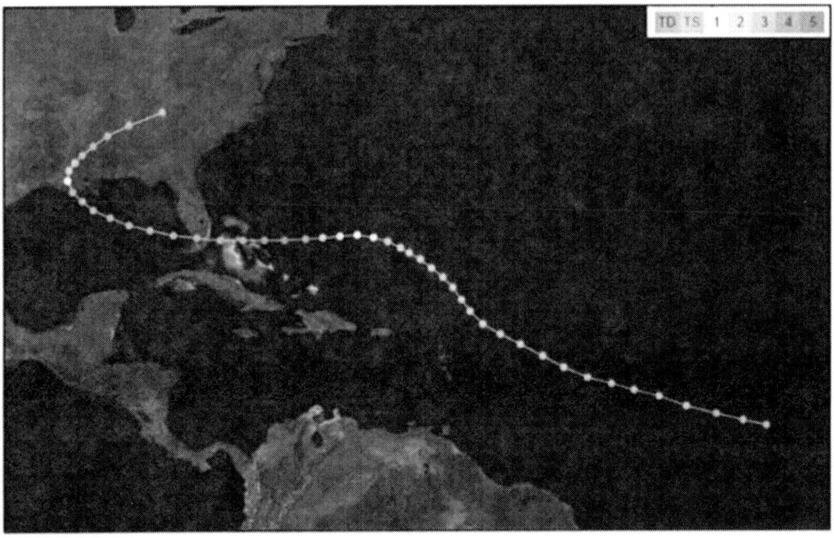

Figure 11. Track of Hurricane Andrew (1992).

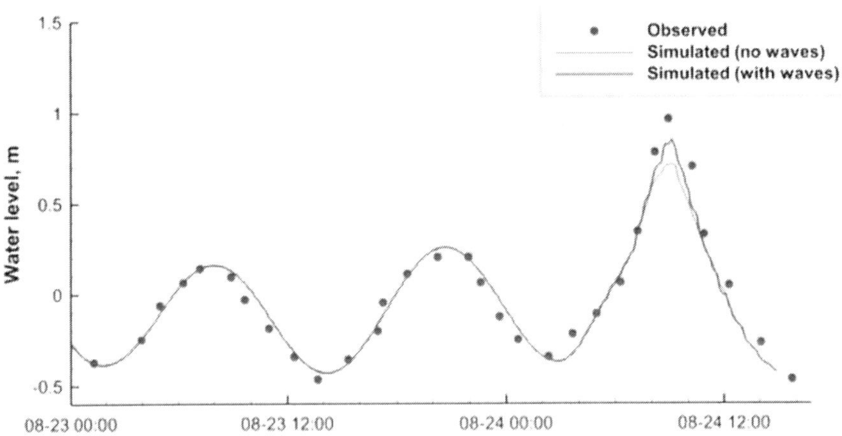

Figure 12. Comparison of observed and simulated (hindcast) water levels at the Haulover Pier station during Hurricane Andrew (1992). "No waves" curve shows results based on CH3D model that does not include wave effects and "with waves" shows results based on the coupled CH3D-SWAN model.

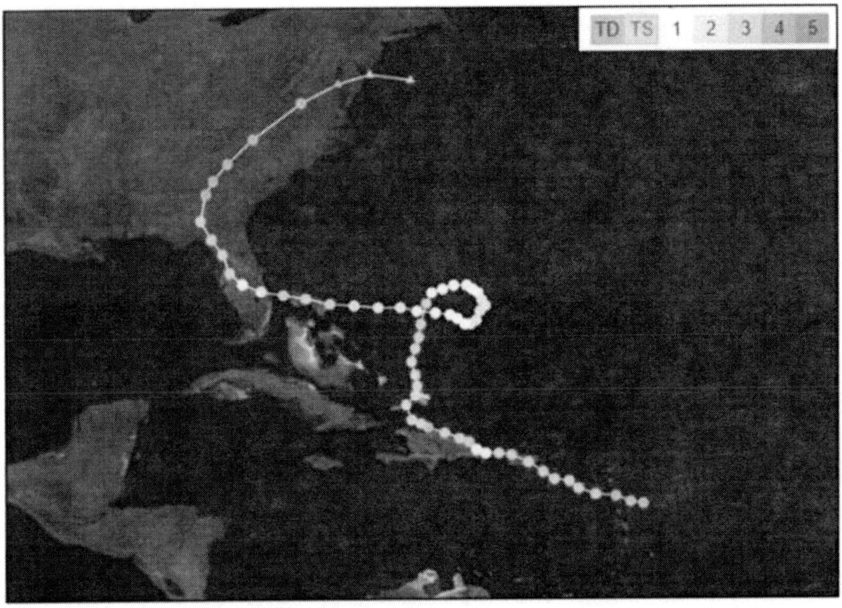

Figure 13. Track of Hurricane Jeanne (2004).

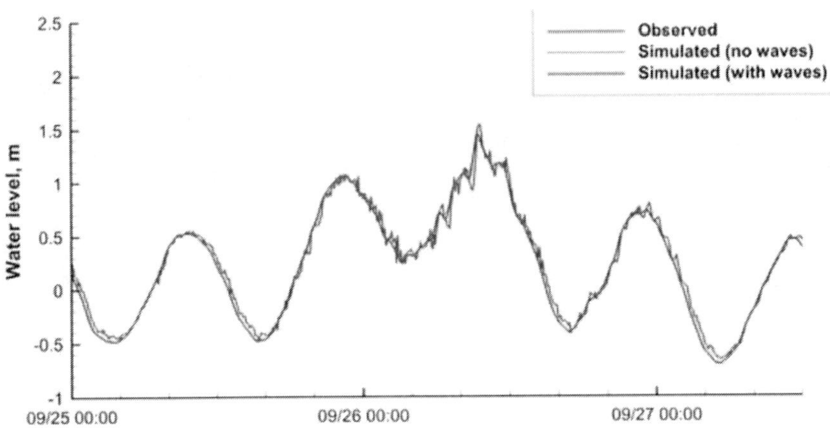

Figure 14. A comparison between simulated (hindcast) and measure waver levels at the Trident Pier Station during Hurricane Jeanne (2004).

Hurricane Wilma (2005)

Hurricane Wilma (Figure 15) was the most intense tropical cyclone ever recorded in the Atlantic basin. In the U.S. it made landfall near Cape Romano, Florida with winds of 120 mph [46] and quickly crossed Florida emerging on the east coast just 5 h later. There is significant amount of data available for Wilma, including over 20 storm gauges that were installed along the west coast of Florida by the USGS. Below is a comparison of peak surge heights at these gauges (Figure 16) and comparison of simulated and observed data at Trident Pier station on the east coast that was affected by the storm after Wilma crossed the Florida peninsula (Figure 17). Maximum storm surge during Wilma was also compared to a number of high water marks and the correlation coefficient between recorded and predicted water marks was 0.78.

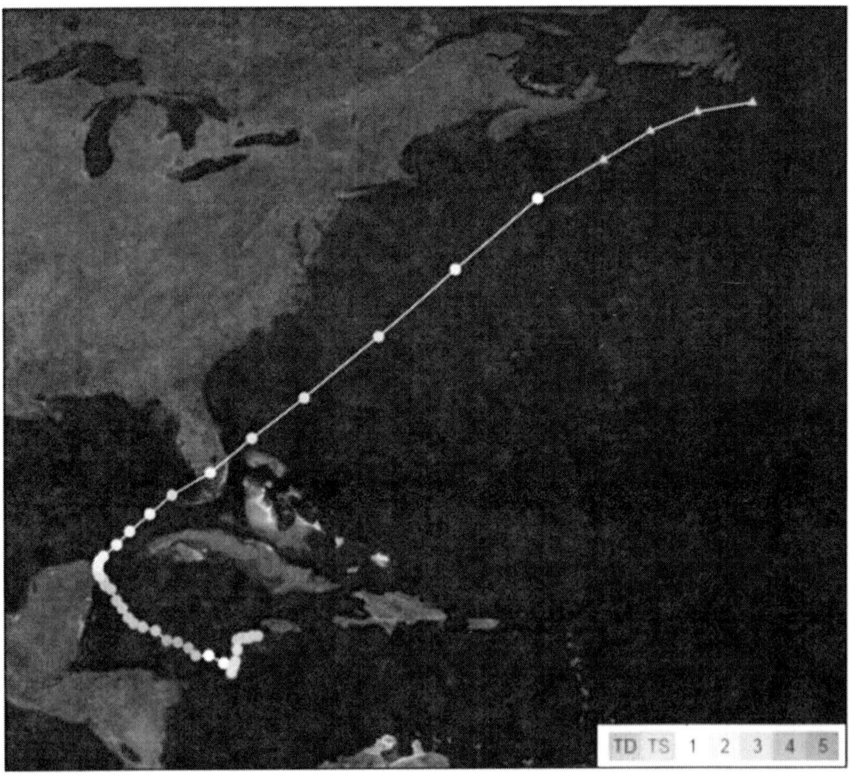

Figure 15. Track of Hurricane Wilma (2005).

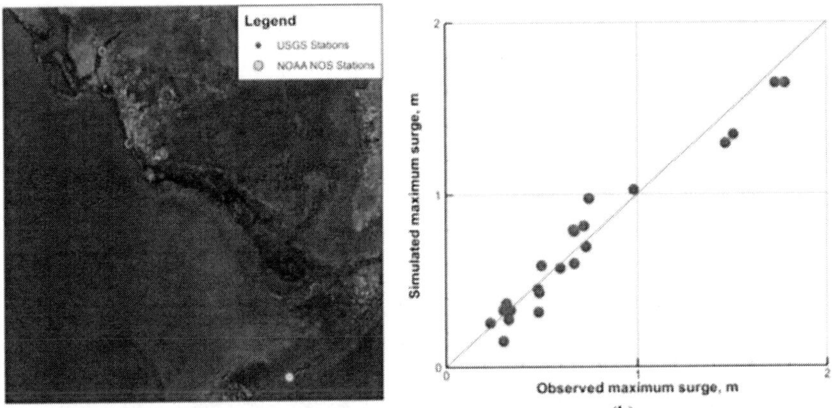

Figure 16. Map of data stations with observed data (**a**) and comparison between observed and simulated (hindcast) and peak surges during Hurricane Wilma (**b**).

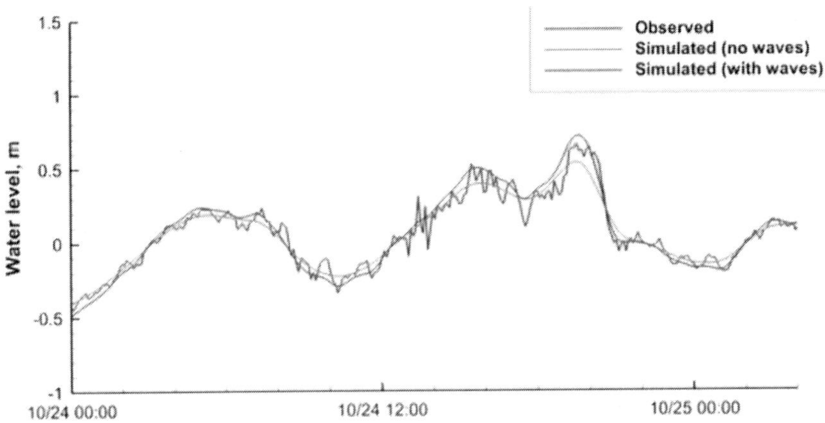

Figure 17. A comparison between simulated (hindcast) and observed water levels at the Trident Pier Station during Hurricane Wilma (2005).

Tropical Storm Fay

The last storm presented for validation purposes is Tropical Storm Fay [47], (Figure 18). It is the weakest storm of the four presented, however, one notable feature of this storm is that it was slowly moving just off the east coast of Florida zigzagging across the coastline over a period of about 24 h making a total of four landfalls in Florida. Fay produced torrential rainfall dropping as much as 27 inches of rain near Melbourne, Florida. The rainfall significantly affected the river flows and salinity making it an interesting case study. Water level comparison at I-295 bridge station (Figure 19) near Jacksonville and salinity comparison at Pine Island station (Figure 20) show that predicted values compare quite well with observed data. It is worth noting that, even though the semidiurnal variations in salinity are only partially captured, the observed significant drop in salinity over the 3-day period was well captured by the model simulation.

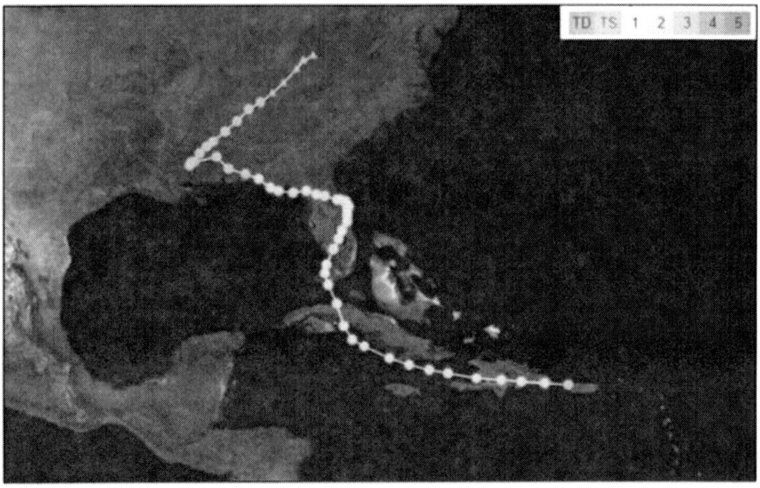

Figure 18. Track of Tropical Cyclone Fay (2008).

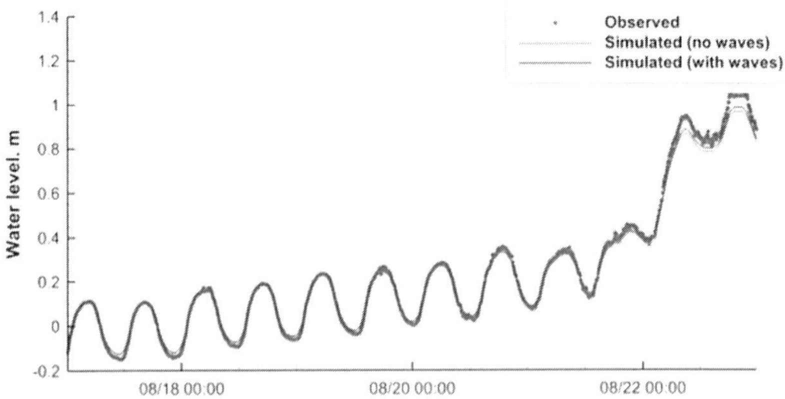

Figure 19. A comparison between the simulated (hindcast) and observed water levels at the I-295 Bridge Station (EC domain) during Tropical Storm Fay (2008).

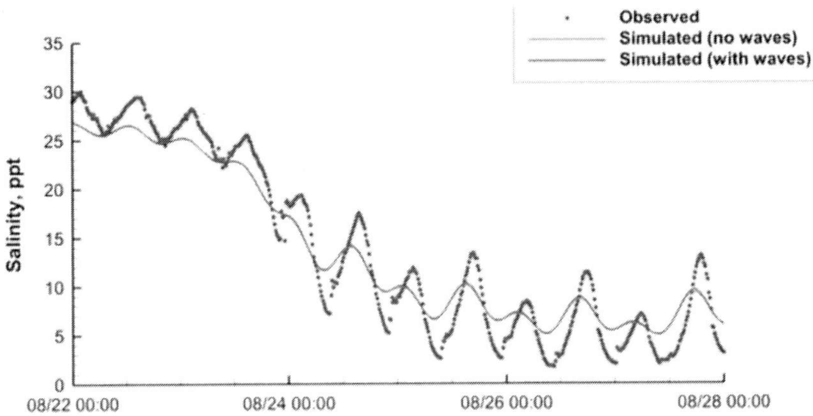

Figure 20. A comparison between simulated (hindcast) and observed salinity at the Pine Island Station (EC domain) during Tropical Cyclone Fay (2008).

3.2. 2015 Hurricane Season Forecasting

ACMS was tested in a quasi-operational mode during the 2015 hurricane season spanning from 1 June 2015 to 30 November 2015. Unfortunately, arrangements for hardware location, networking, etc. are such that it is currently impossible to guarantee a 24/7 uptime for the system as it is located in a research rather than an operational environment and is subject to power and network outages, hardware failures, etc. The system was functioning about 85% of the time. Several statistics were calculated (Table 2) based on these forecasts: root mean square error (RMSE), central frequency (CF), and positive/negative outlier frequency (POF/NOF). These are some of the criteria that are used by NOS for model skill assessment [48]. The error calculations are for the 0–24 h forecast window. Acceptable error limits used for calculation of CF are 15 cm for water level and 3 ppt for salinity and acceptable error limits used for calculation of POF/NOF are 30 cm for water level and 6 ppt for salinity [48].

3.3. Computational Efficiency and Timing

One notable feature of ACMS is its computational efficiency. Currently, model codes run in parallel (implemented via OpenMP) on Intel-based machines running Red Hat Linux (RHEL6/7) with quad-core CPUs (Intel i5-4690 CPU @ 3.50 GHz). Computers are not shared by simulations of different domains or with any other resource-intensive tasks (i.e., each simulation is assigned to a dedicated quad-core machine). The wall times for the full ACMS setup described earlier in this paper for each domain are shown in Table 3.

Table 2. Errors statistics calculated based on ACMS forecasts during 1 June 2015–30 November 2015. Units for root mean square error (RMSE) are cm for water level data and ppt for salinity. Green indicates that the errors are within acceptable limits (>90% for CF and <1% POF/NOF), red indicates otherwise.

Data Station	ACMS Domain	Data Type	Source	RMSE cm/ppt	CF %	POF/NOF %
8728690 Apalachicola	NG	WL	NOAA	7	92.4	0.1/0.3
8727520 Cedar Key	NG	WL	NOAA	6	93.2	0.2/0.3
8726724 Clearwater Beach	SW	WL	NOAA	6	92.0	0.1/0.1
8720219 Dames Point	EC	WL	NOAA	6	93.8	0.2/0.2
8720030 Fernandina Beach	EC	WL	NOAA	5	94.6	0.2/0.1
8725520 Fort Myers	SW	WL	NOAA	4	97.2	0.1/0.0
8720357 I-295 Bridge	EC	WL	NOAA	4	95.6	0.1/0.2
Key West	SE	WL	NOAA	5	95.6	0.1/0.1
Lake Worth Pier	SE	WL	NOAA	6	94.1	0.2/0.2
Mayport	EC	WL	NOAA	4	97.8	0.1/0.0
Mckay Bay	SW	WL	NOAA	5	96.2	0.1/0.1
Naples	SW	WL	NOAA	4	98.2	0.0/0.0
Old Port Tampa	SW	WL	NOAA	5	96.7	0.4/0.2
Panama City	NG	WL	NOAA	5	95.7	0.2/0.7
Pensacola	NG	WL	NOAA	7	93.3	0.4/0.9
Port Manatee	SW	WL	NOAA	6	94.8	0.3/0.2
Racy Point	EC	WL	NOAA	12	86.8	1.4/0.8
Red Bay Point	EC	WL	NOAA	10	91.2	1.1/0.8
S. Riverwalk	EC	WL	NOAA	7	94.0	0.2/0.6
St Petersburg	SW	WL	NOAA	6	94.4	0.3/0.2
Trident Pier	EC	WL	NOAA	5	93.4	0.3/0.3
Vaca Key	SE	WL	NOAA	5	94.8	0.2/0.3
Virginia Key	SE	WL	NOAA	5	96.2	0.2/0.1
Bing's Landing	EC	WL	FLDEP	6	93.2	0.2/0.3
Binney Dock	EC	WL	FLDEP	6	94.7	0.2/0.3
Dry Bar	NG	WL	FLDEP	5	95.3	0.2/0.2
East Bay	NG	WL	FLDEP	6	92.2	0.3/0.6
Gordon River Inlet	SW	WL	FLDEP	5	94.4	0.3/0.1
Melbourne	EC	WL	FLDEP	4	96.6	0.2/0.1
Naples Bay	SW	WL	FLDEP	4	96.3	0.1/0.1
Pilot Cove	NG	WL	FLDEP	7	92.0	0.3/0.5
Ponce de Leon	EC	WL	FLDEP	6	94.5	0.3/0.2
St. Lucie Inlet	EC	WL	FLDEP	5	95.8	0.2/0.2
Tolomato River	EC	WL	FLDEP	7	93.3	0.1/0.3
Vilano Beach	EC	WL	FLDEP	7	94.0	0.1/0.2
Bing's Landing	EC	S	FLDEP	2.1	94.1	0.6/0.4
Dry Bar	NG	S	FLDEP	6.2	77.5	1.7/3.3
Melbourne	EC	S	FLDEP	2.7	88.2	0.7/0.9
Tolomato River	EC	S	FLDEP	1.8	92.9	0.7/0.4
Cat Point	NG	S	ANERR	2.8	87.0	0.8/0.6
Henderson Creek	SW	S	RNERR	1.7	93.7	0.3/0.3
Fakahatchee Bay	SW	S	RNERR	2.3	90.2	0.5/0.4
Faka Union Bay	SW	S	RNERR	2.6	88.2	0.6/0.4
Pellicer Creek	EC	S	GNERR	4.1	81.5	1.5/0.9
San Sebastian	EC	S	GNERR	3.1	86.7	0.8/0.6
Pine Island	EC	S	GNERR	3.3	81.1	1.3/0.8

NOAA—NOAA Tides and Currents [49]; FLDEP—Florida Department of Environmental Protection [50]; ANERR—Apalachicola Bay NERR [51]; RNERR—Rookery Bay NERR [51]; GNERR—Guana-Tolomato-Matanzas NERR [51]; Data types: WL—Water level, S—Salinity.

Table 3. Average wall-clock time for ACMS simulations (a simulation consists of a 6-h hindcast/nowcast and a 72-h forecast) per domain. Times can vary from cycle to cycle, depending on conditions and number of wetted grid cells.

Domain	Wall Clock Time (Hours:Minutes)			
	4 CPU Cores		32 CPU Cores	
	2D	3D	2D	3D
East Coast	0:58	3:40	0:22	00:28
Southeast	1:45	5:42	0:37	00:43
Southwest	1:10	4:11	0:29	00:32
Northern Gulf of Mexico	1:29	5:04	0:34	00:37

As noted previously, given the computational efficiency of ACMS provides good balance between the provided accuracy and required resources. In order to run in real time, ACMS needs only 1–4 processing cores, compared to other systems which often require a high-performance computing system with hundreds or thousands of processors. Current ACMS setup uses a single quad-core system (as described above) per model domain. The codes within the system are effectively parallelized and the wall time can be reduced by using more CPU cores. For example, using 32 CPU cores yields approximately 30–40 min wall times (Table 3).

As such, the 2D and 3D implementations of ACMS differ significantly in wall time required to perform the model simulations, however, this difference only slightly affects the total forecast cycle time. The amount of data required and time needed to fetch these data is significantly different for the 2D and 3D implementations. It can take up to 12 h to obtain all the required boundary conditions for the 3D implementation. The only external data required for the 2D model is a hurricane track as it uses a synthetic parametric wind model to generate wind and pressure fields, which generally takes only minutes to obtain. Pre/post-processing steps take 2–5 min and hence the entire forecast cycle using a 2D model can be completed in less than two hours using the current computational setup.

4. DISCUSSION AND FUTURE WORK

This paper details how the ACMS was setup, tested, and validated in tidal and hindcast scenarios. The system is shown to be robust and results match well with data for hindcasting. Simulated water levels and salinity for hindcasted periods match well with data and 24-h forecasts that were performed during the 2015 hurricane season are within acceptable limits for most stations. RMS errors for water levels are found to vary between 4 cm and 12 cm. Additional efforts will be undertaken to analyze the sources of error for stations that do not satisfy the criteria (e.g., two stations located upstream the St. Johns River: Racy Point and Red Bay Point) and attempt will be made to reduce the errors. The computational performance of the system is also discussed. Model efficiency allows the production of robust forecasts in a limited-resource environment and choice of NetCDF-based data standards gives flexibility in distributing the data using a THREDDS server and derived products using in-house web-based user interface. The system will be tested in operational settings (completely unattended) and operational performance, uptime, etc. will be a subject of a future publication.

Data availability and its accuracy significantly affect the results of simulations. It is especially important for salinity, as missing river flow or salinity boundary conditions can drastically change the quality of predictions inside an estuary. Additional quality control and data extrapolation methods will be explored along with an option to include a watershed model into ACMS suite for more robust predictions of upstream river flows and subsequently salinity inside estuaries. Temperature is another important variable. While ACMS is capable of simulating temperature, such simulations are currently in the preliminary phase. Feasibility of accurate temperature predictions constrained by limited data is being explored. The key to this is data availability with very little data available for use as boundary conditions and for validation. Efforts will be made to obtain such data, and setup and validate the temperature model.

Forecasting results from the 2016 hurricane season will be analyzed using more statistical categories and NOS criteria, including maximum duration of positive and negative outliers and worst case outlier frequency, as well as timing of maximums and minimums.

Sheng et al. [52,53] developed a vegetation model which incorporates the effects of vegetation on mean flow and turbulence in the water column. The vegetation resolving model showed that, during hurricanes, total inundation volume can be reduced by up to 40% due to the presence of vegetation. The reduction of storm surge and coastal inundation depends on the characteristics (type, distribution, height, and density, etc.) of vegetation as well as hurricane characteristics (intensity and forward speed, etc.). The vegetation module will be added to the ACMS, however, additional efforts are needed to optimize the algorithms and ensure that this addition does not hinder the efficiency of ACMS.

ACKNOWLEDGMENTS

This work was sponsored by the Southeast Coastal Ocean Observing Regional Association (NOAA/NOPP, IOOS.11(033)UF.PS.MOD.1), St. Johns River Water Management District, South Florida Water Management District, and U.S. Fish and Wildlife Service.

AUTHOR CONTRIBUTIONS

V.A.P., J.R.D. and Y.P.S. conceived and designed model simulation, developed various parts of the model. V.A.P. setup, performed the model simulation, and wrote the paper with Y.P.S.

CONFLICTS OF INTEREST

The authors declare no conflict of interest. The founding sponsors had no role in the design of the study; in the collection, analyses, or interpretation of data; in the writing of the manuscript, and in the decision to publish the results.

REFERENCES

1. NOEP (National Ocean Economics Program). Coastal and Ocean Economic Summaries of the Coastal States. 2014. Available online: http://www. oceaneconomics. org/download (accessed on 17 July 2015).
2. NOEP (National Ocean Economics Program). State of the U.S. Ocean and Coastal Economies Coastal States Summaries—2016 Update. 2014. Available online: http://www. oceaneconomics.org/download (accessed on 17 July 2015).
3. NOAA Hurricane Research Division. Detailed List of Continental United States Hurricane Impacts/Landfalls, 1851–1960, 1983–2015. Available online: http://www.aoml.noaa.gov/hrd/hurdat/UShurrs_detailed.html (accessed on 25 August 2016).
4. Jelesnianski, C.P.; Chen, J.; Shaffer, W.A. SLOSH: Sea, Lake, and Overland Surges from Hurricanes; US Department of Commerce, National Oceanic and Atmospheric Administration, National Weather Service: Washington, D.C., USA, 1992.
5. Luettich, R.; Westerink, J.J.; Scheffner, N.W. ADCIRC: An Advanced Three-Dimensional Circulation Model for Shelves Coasts and Estuaries, Report 1: Theory and Methodology of ADCIRC-2DDI and ADCIRC-3DL. DRP-92–6; U.S. Army Engineers Waterways Experiment Station: Vicksburg, MS, USA, 1992; p. 137.
6. Bleck, R.; Benjamin, S. Regional weather prediction with a model combining terrain following and isentropic coordinates. Part I: Model description. Mon. Weather Rev. 1993, 121, 1770–1785.

7. Shchepetkin, A.F.; McWilliams, J.C. The regional oceanic modeling system (ROMS): A split-explicit, free-surface, topography-following-coordinate oceanic model. Ocean Model. 2005, 9, 347–404.
8. Barron, C.N.; Kara, A.B.; Hurlburt, H.E.; Rowley, C.; Smedstad, L.F. Sea surface height predictions from the Global Navy Coastal Ocean Model (NCOM) during 1998–2001. J. Atmos. Ocean. Technol. 2004, 21, 1876–1894.
9. Sheng, Y.P. A Three-Dimensional Numerical Model of Coastal and Estuarine Circulation and Transport in Generalized Curvilinear Grids; Technical Report No. 587; Aeronautical Research Associates Of Princeton: Princeton, NJ, USA, 1986.
10. Sheng, Y.P. Evolution of a Three-Dimensional Curvilinear-Grid Hydrodynamic Model for Estuaries, Lakes and Coastal Waters: Estuarine and Coastal Modeling I; ASCE: Reston, VA, USA, 1990; pp. 40–49.
11. Sheng, Y.P.; Paramygin, V.A.; Alymov, V.; Davis, J.R. A Real-Time Forecasting System for Hurricane Induced Storm Surge and Coastal Flooding. In Proceedings of the Ninth International Conference on Estuarine and Coastal Modeling, Charleston, SC, USA, 31 October–2 November 2006; pp. 585–602.
12. Sheng, Y.P.; Alymov, V.; Paramygin, V.A. Simulation of storm surge, waves, and inundation in the Outer Banks and Chesapeake Bay during Hurricane Isabel in 2003: The importance of waves. J. Geophys. Res. 2010, 115, C4.
13. Sheng, Y.P.; Zhang, Y.; Paramygin, V.A. Simulation of Storm Surge, Wave, and Coastal Inundation during Hurricane Ivan in 2004. Ocean Model. 2010, 35, 314–331.
14. Ris, R.C.; Holthuijsen, L.H.; Booij, N. A third-generation wave model for coastal regions: 2. Verification. J. Geophys. Res. 1999, 104, 7667–7681.
15. Sheng, Y.P.; Villaret, C. Modeling the effect of suspended sediment stratification on bottom exchange process. J. Geophys. Res. 1989, 94, 14429–14444.
16. Sheng, Y.P.; Kim, T. Skill assessment of an integrated modeling system for shallow coastal and estuarine ecosystems. J. Mar. Syst. 2009, 76, 212–243.
17. Sheng, Y.P.; Paramygin, V.A. Forecasting storm surge, inundation, and 3D circulation along the Florida Coast, Estuarine and Coastal Modeling. In Proceedings of the 11th International Conference ASCE, Fort Lauderdale, FL, USA, 20–22 May 2010.
18. Sheng, Y.P.; Paramygin, V.A.; Zhang, Y.; Davis, J.R. Recent enhancements and application of an integrated storm surge modeling system: CH3D-SSMS. Estuarine and Coastal Modeling. In Proceedings of the Tenth International Conference ASCE, New Orleans, LA, USA, 9–12 March 2008; pp. 879–892.
19. Sheng, Y.P.; Tutak, B.; Davis, J.R.; Paramygin, V. Circulation and flushing in the lagoonal system of the Guana Tolomato Matanzas National Estuarine Research Reserve (GTMNERR), Florida. J. Coast. Res. 2008, 55, 9–25.
20. Davis, J.R.; Sheng, Y.P. Development of a parallel storm surge model. Int. J. Numer. Meth. Fluids 2003, 42, 549–580.
21. Sheng, Y.P.; Davis, J.R.; Figueiredo, R.; Liu, B.; Liu, H.; Luettich, R., Jr.; Paramygin, V.A.; Weaver, R.; Weisberg, R.; Xie, L.; et al. A Regional Testbed for Storm Surge and Inundation Models. In Proceedings of the 12th International Conference on Estuarine and Coastal Modeling, St. Augustine, FL, USA, 7–9 November 2011.
22. Peng, M.; Xie, L.; Pietrafesa, L.J. A numerical study of storm surge and inundation in the Croatan–Albemarle–Pamlico Estuary System. Estuar. Coast. Shelf Sci. 2004, 59, 121–137.
23. Chen, C.; Cowles, G.; Beardsley, R.C. An Unstructured Grid, Finite Volume Coastal Ocean Model: FVCOM User Manual, 2nd ed.; SMAST/UMASSD TR-06-0602; MIT: Cambridge, MA, USA, 2006; p. 315.
24. National Academy of Sciences. Mapping the Zone—Improving Flood Map Accuracy; Committee on FEMA Flood Maps, National Research Council: Washington, DC, USA, 2009.
25. Lapetina, A.; Sheng, Y.P. Simulating complex storm surge dynamics: Three-dimensionality, vegetation effect, and onshore sediment transport. J. Geophys. Res. Oceans 2015, 120, 7363–7380.

26. Powell, M.D.; Houston, S.H.; Amat, L.R.; Morisseau-Leroy, N. The HRD real-time hurricane wind analysis system. J. Wind Eng. Ind. Aerodyn. 1998, 77, 53–64.
27. NOGAPS. Available online: http://coaps.fsu.edu/woce/html/models/fnoc.htm (accessed on 8 January 2017).
28. About the GFDL Hurricane Model Ensemble. Available online: http://data1.gfdl.noaa.gov/hurricane/gfdl_ensemble/about.html (accessed on 8 January 2017).
29. Michalakes, J.; Dudhia, J.; Gill, D.; Henderson, T.; Klemp, J.; Skamarock, W.; Wang, W. The Weather Reseach and Forecast Model: Software Architecture and Performance. In Proceedings of the 11th ECMWF Workshop on the Use of High Performance Computing In Meteorology, Reading, UK, 25–29 October 2004.
30. NAM Products. Available online: http://www.nco.ncep.noaa.gov/pmb/products/nam (accessed on 8 January 2017).
31. Davis, J.R.; Para mygin, V.A.; Forrest, D.; Sheng, Y.P. Towards the probabilistic simulation of storm surge and inundation in a limited resource environment. Mon. Weather Rev. 2010, 138, 2953–2974.
32. Miller, R.J.; Schrader, A.J.; Sampson, C.R.; Tsui, T.L. The Automated Tropical Cyclone Forecasting System (ATCF). Weather Forecast. 1990, 5, 653–660.
33. Halliwell, G. Evaluation of vertical coordinate and vertical mixing algorithms in the HYbrid Coordinate Ocean Model (HYCOM). Ocean Modelling. 2004, 7, 285–322.
34. Tolman, H.L. User Manual and System Documentation of WAVEWATCH III Version 3.14. NOAA/NWS/NCEP/MMAB Technical Note 276; National Weather Service: Camp Springs, MD, USA, 2009; p. 194.
35. HTCondor—Home. Available online: https://research.cs.wisc.edu/htcondor (accessed on 1 July 2016).
36. THREDDS Data Server. Available online: http://www.unidata.ucar.edu/software/thredds/current/tds (accessed on 15 June 2016).
37. Xie, L.; Bao, S.; Pietrafesa, L.J.; Foley, K.; Fuentes, M. A Real-Time Hurricane Surface Wind Forecasting Model: Formulation and Verification. Mon. Weather Rev. 2006, 134, 1355–1370.
38. Sheng, Y.P.; Liu, T. Three-dimensional simulation of wave-induced circulation: Comparison of three radiation stress formulations. J. Geophys. Res. 2011, 116, C05021.
39. Mellor, G.L. The depth dependent current and wave interaction equations: A revision. J. Phys. Oceangr. 2008, 38, 2587–2596.
40. Longuet-Higgins, M.S.; Stewart, R.W. Radiation stress and mass transport in gravity waves with application to 'surf beats'. J. Fluid Mech. 1962, 13, 481–504.
41. Longuet-Higgins, M.S.; Stewart, R.W. Radiation stresses in water waves: A physical discussion with applications. Deep-Sea Res. 1964, 11, 529–562.
42. Parks, A.B.; Shay, L.K.; Johns, W.E.; Martinez-Pedraja, J.; Gurgel, K.-W. HF radar observations of small-scale surface current variability in the Straits of Florida. J. Geophys. Res. 2009, 114, C08002.
43. Shay, L.K.; Cook, T.M.; Peters, H.; Mariano, A.J.; Weisberg, R.; An, P.E.; Soloviev, A.; Luther, M. Very High Frequency Radar Mapping of Surface Currents. IEEE J. Coast. Eng. 2002, 27, 155–169.
44. Landsea, C.W.; Franklin, J.L.; McAdie, C.J.; Beven, J.L., II; Gross, J.M.; Jarvinen, B.R.; Pasch, R.J.; Rappaport, E.N.; Dunion, J.P.; Dodge, P.P. A reanalysis of hurricane Andrew's intensity. Am. Meteorol. Soc. 2004, 85, 1699.
45. Lawrence, M.B.; Cobb, H.D. Tropical Cyclone Report Hurricane Jeanne 13–28 September 2004; National Hurricane Center: Miami, FL, USA, 2005.
46. Pasch, R.J.; Blake, E.S.; Cobb, H.D., III; Roberts, D.P. Tropical Cyclone Report Hurricane Wilma. 15–25 October 2005; National Hurricane Center: Miami, FL, USA, 2006.
47. Stewart, S.R.; Beven, J.L., II. Tropical Cyclone Report Tropical Cyclone Fay (AL062008) 15–26 August 2008; National Hurricane Center: Miami, FL, USA, 2009.

48. Kelley, J.G.W.; Zhang, A.; Chu, P.; Lang, G.A. Skill Assessment of NOS Lake Huron Operational Forecast System (LHOFS); NOAA Technical Memorandum NOS CS 23: Silver Spring, MD, USA, 2010.
49. NOAA Tides and Currents. Available online: https://tidesandcurrents.noaa.gov (accessed on 8 January 2017).
50. Florida Department of Environmental Protection/Division of State Lands/Bureau of Survey and Mapping. Available online: http://www.fldep-stevens.com (accessed on 8 January 2017).
51. Centralized Data Management Office. Available online: http://cdmo. baruch. sc.edu/get/landing.cfm (accessed on 8 January 2017).
52. Sheng, Y.P.; Lapetina, A.; Ma, G. The reduction of storm surge by vegetation canopies: Three-dimensional simulations. Geophys. Res. Lett. 2012, 39, L20601.
53. Lapetina, A.; Sheng, Y.P. Three-dimensional modeling of storm surge and inundation including the effects of coastal vegetation. Estuaries Coasts 2014.

CHAPTER 15

Application of State of the Art Modeling Techniques to Predict Flooding and Waves for a Coastal Area within a Protected Bay

Malcolm L. Spaulding [1,*], Annette Grilli [1], Chris Damon [2], Grover Fugate [3], Tatsu Isaji [4] and Lauren Schambach

[1]*Ocean Engineering, University of Rhode Island, Narragansett, RI 02882, USA*

[2]*Environmental Data Center, University of Rhode Island, Kingston, RI 02881, USA*

[3]*Rhode Island Coastal Resources Management Council, Wakefield, RI 02879, USA*

[4]*RPS/ASA, South Kingstown, RI 02879, USA*

ABSTRACT

Flood Insurance Rate Maps (FIRMs) are developed by the Federal Emergency Management Agency (FEMA) to provide guidance in establishing the risk to structures and infrastructure from storm surge sand associated waves in the coastal zone. The maps are used by state agencies and municipalities to help guide coastal planning and establish the minimum elevation and construction standards for new or substantially improved structures. A summary of the methods used and a comparison with the results of 2013 FIRM mapping are presented for Warwick, Rhode Island (RI), a coastal community located within Narragansett Bay. Because of its location, Warwick is protected from significant coastal erosion and wave attacks, but is subject to surge amplification. Concerns surrounding the FEMA methods used in the 2013 FIRM analysis are put in context with the National Research Council's (NRC) 2009 review of the FEMA coastal mapping program. New mapping is then performed using state of the art, fully coupled surge and wave modeling, and data analysis methods, to address the NRC concerns. The new maps and methodologies are in compliance with FEMA regulations and guidelines. This new approach makes extensive use of the numerical modeling results from the recent US Army Corp of Engineers, North Atlantic Coast Comprehensive Study (NACCS, 2015). Revised flooding maps are presented and compared to the 2013 FIRM maps, to provide insight into the differences. The new maps highlight the importance of developing better estimates of surge dynamics and the advancement in nearshore mapping of waves in flood inundated areas by the use of state of the art, two-dimensional, wave transformation models.

Keywords: coastal flooding; inundation; and waves; Flood Insurance Rate Maps (FIRMs); coupled wave and surge modeling; base flood elevation

1. INTRODUCTION

The Federal Emergency Management Agency (FEMA) leads the national effort for the development of Flood Insurance Rate Maps (FIRMS), for the purposes of establishing the risk that structures and infrastructure face from coastal inundation and wave attacks, and setting the associated insurance rates. Coastal municipalities often use these maps to guide coastal development and to establish the minimum elevation for construction standards. Additionally, the maps are regulatory, universally available, and developed within strict FEMA guidelines. This work is typically performed by contractors and organized by individual FEMA regions. The regions are given substantial latitude in establishing the methods used in the analysis, but all must follow the FEMA guidelines; for example, the Atlantic and Gulf of Mexico guidelines are provided in [1].

A Flood Insurance Study (FIS) was completed in 2012 and implemented in 2013 for Kent County, RI [2] (Figure 1). The storm surge levels were estimated by the linear interpolation of 100-year return period water levels from adjacent National Ocean and Atmospheric Administration (NOAA)-National Ocean Survey (NOS) water level stations [3]. For this study, data from Newport, RI, and Providence, RI, stations were used. The

10-, 2-, 1-, and 0.2-percent-annual-chance flood water levels were determined by the application of extreme value analysis (Generalized Extreme Value (GEV) using L Moments to estimate the distribution's parameters) to the historical water level data at the two stations. Water levels corresponding to selected return periods were provided for each station. The water levels at intermediate locations, relative to these two reference stations, were determined by linear interpolation. No estimates of values for other confidence levels (e.g., 95%) were provided. The contractor used FEMA's Coastal Hazard Analysis and Modeling Program (CHAMP)/Wave Height Analysis for Flood Insurance Studies (WHAFIS) [4] to estimate the waves in the flood inundated areas. This program computes the wave crest elevations along representative transects, using a 1-D wave action equation. Transects are selected by considering major topographic, vegetative, and cultural features, with variable spacing recommended to be tens to hundreds of meters, depending on the density of the development and nearshore topography. WHAFIS is forced by an appropriate depth-limited wave height at the seaward end of each transect. The 100-year Still Water Elevation Level data (SWEL), including, if applicable, the static wave setup calculated using Goda's semi-empirical formulation [4], and the wave height estimates, are then assembled to provide predictions of the Base Flood Elevation (BFE) (Figure 2). The wave heights are then used to establish the flooding zones (VE, AE, and X, see Figure 2 for definitions of the zones). Data from each transect are laterally interpolated using engineering judgment, according to the FEMA guidelines [1], to develop a 2-D map of the flooding zones and associated BFEs.

Figure 1. FEMA (2012) [2] Kent County, RI, study area. Also, shown are the locations of WHAFIS transects, numbered from 1 to 34.

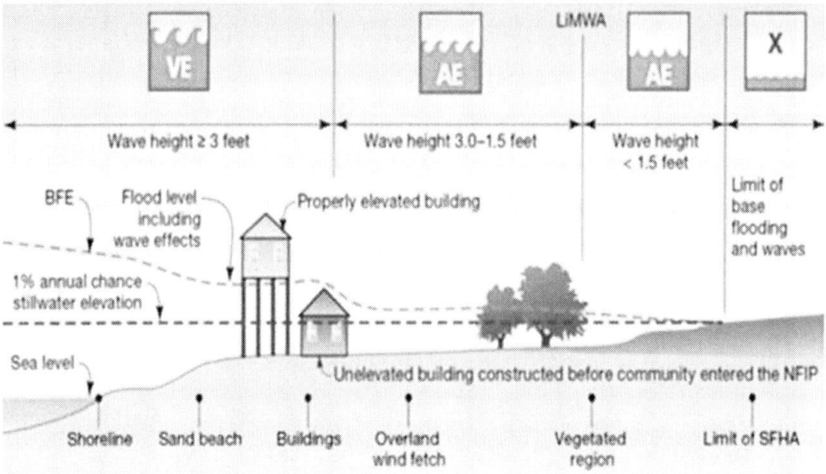

Figure 2. FEMA definition schematic for flooding and structures located on grade and elevated territories. The still water elevation (SWEL) for 1% annual chance (100-year) is shown by the dashed red line and the Base Flood Elevation (BFE) by the blue dashed line. FEMA zones (VE, AE seaward, and landward of the Limit of Moderate Wave Action (LiMWA)), based on wave heights, are also noted at the top of the figure [6].

The National Research Council (NRC) performed an in-depth review of the FEMA flood mapping program in 2009 [5]. Recommendations from their report on coastal flooding (Section 5 of their report) include the following: FEMA should use coupled 2-D surge and wave models to reduce uncertainties associated with the use of a 2-D surge model and the 1-D WHAFIS model. Wave crests calculated by CHAMP/WHAFIS have not been sufficiently validated, creating potentially significant uncertainties in BFE estimates. Problems identified in the NRC review, relevant to the Warwick, RI, study area, include: (1) wave transformation is a 2-D process that cannot be represented in a 1-D model; (2) surges and waves are completely decoupled, which may lead to over or underestimates of the BFE; (3) one-dimensional transects do not reflect 2-D terrain; and (4) the manual interpolation of 1-D results to two dimensions is subjective.

It is clear from analyzing the methods used in the FEMA 2013 flooding maps, performed for Kent County, RI, summarized above [2], that the NRC's recommendations were not followed.

The goal of the present study is to develop new FIRMs for Kent County, using state of the art methods, following the NRC recommendations. The

City of Warwick, RI (portion of Kent County on the coast), has been selected for this application. It is located inside Narragansett Bay and represents a protected coastal area. A companion paper, Spaulding et al. (2017) [7], provides an application of the same methods used here, but for a fully exposed coastal shoreline in Charlestown, RI. It should be noted that there are a variety of other models available in the international community that could potentially be used to address coastal flooding and the associated wave climate, including, but not limited to, FVCOM, FUNWAVE, TELEMAC, CH3D, Delft3D, SWASH, XBeach, and CMS Flow and Wave. These models might result in a more accurate prediction of coastal flooding, but unfortunately, they are not approved for use by FEMA for flood insurance studies in the US. In keeping with the FEMA guidelines, only models approved by FEMA for use in coastal flood mapping studies have been employed in the present effort [8]. FEMA guidelines for coastal flood mapping of the study area have also been followed [1]. While there are many alternatives in assembling a model system to perform the flooding study, we have used two well-known, widely used FEMA approved models in the current approach (ADCIRC and STWAVE). Each model is described in more detail below.

Section 2 provides an overview of the methods used in the analysis for estimating surge and waves. The results and discussion, including a validation of the estimates, are provided in Section 3. The conclusions are included in Section 4, followed by the references. Supplemental material is provided in Appendix A.

2. METHODS

2.1. Surge and Tide Water Levels

The US Army Corp of Engineers (US/ACOE) completed a multi-year numerical modeling effort under the North Atlantic Coast Comprehensive Study in 2015 (NACCS) [9], with a focus on the area impacted by Hurricane Sandy. In the NACCS study, simulations were performed using fully coupled surge and wave models (the ADvanced CIRCulation model (ADCIRC) and STeady state spectral WAVE (STWAVE) forced by the 3rd generation wind-wave model WAM) for 1050 synthetic tropical storms and 100 historical extratropical storms, with a primary focus on the region from Cape Hatteras to Cape Cod. An unstructured grid with a resolution down to 50 m was used to represent the storm surge in the area flooded during a storm. The surge model, ADCIRC, was validated with data from the National Ocean and Atmospheric Administration-

National Ocean Service (NOAA NOS) water observation stations for hurricanes Gloria, Josephine, Irene, and Sandy, and two extratropical storms in 1996 (storms #070 and #073) (see [9] for details on model validation). The wave model, STWAVE, (200 m resolution) was similarly validated with data from offshore wave buoy observations [9]. Given the constraints based on the volume of data generated, the results of model simulations were archived at selected save points, approximately 18,000 in total for the study area and about 1000 in RI. These included the water elevation, mean and peak wave period, zero moment wave heights, mean wave direction, wind speed and direction, and short time series of water level and waves for each storm event. Return period analyses were also available at each save point. The data were made available for Rhode Island via ArcView on line or as GIS layers under the STORMTOOLS® initiative [10,11]. Figure 3 shows the location of the *save points* for the Kent County study area; Warwick, RI, is shown in the red circle. An analysis was performed in the NACCS study to estimate the water level versus return period for the tropical storms for surge only and surge & tides (96 random tidal phases, linear superposition) [9]. Predictions were provided for the mean, and upper and lower 95% confidence interval values, and are available at all save points.

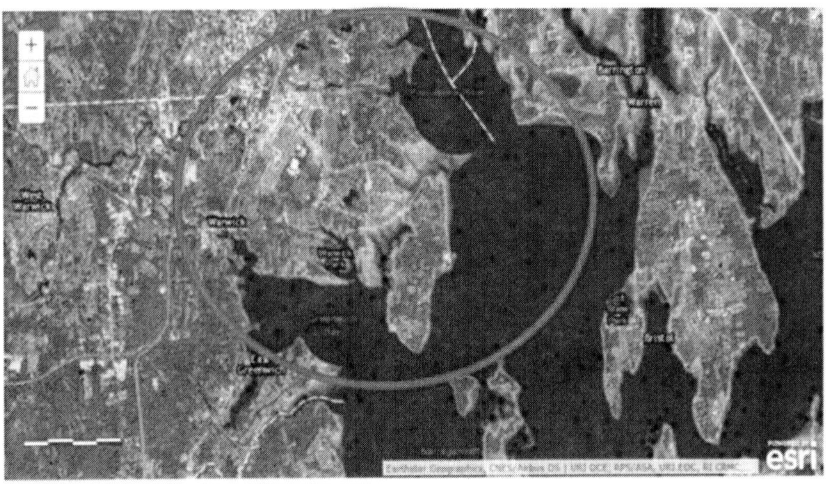

Figure 3. Location of the NACCS save points along the coastline, mid-Narragansett Bay [9]. The location of the Warwick, RI, study area is shown in the red circle. The boundaries of Kent County are shown by the dashed white lines.

Flooding maps were generated from the NACCS simulations for selected return periods and are available at the STORMTOOLS web site [11]. The strategy utilized in this application is to apply extremal analysis at a primary water level station to determine the water levels for the varying return period, and then hydrodynamic model simulations to determine the spatial scaling of the water levels for storms/return periods, referenced to this primary gauging station. As an alternative, water levels can be directly determined from the NACCS *save point*, where return period information is available (e.g., Figure 3). Figure 4 shows the 100-year flooding contour map for the upper 95% confidence interval for the surge & tide case of RI. The water levels (m, referenced to MSL) were spatially interpolated from the *save points* noted, in part, in Figure 3, to generate this figure. The Warwick, RI, study area is shown in the red circle. The surge amplification in the bay is almost linear with the distance up the bay, referenced to the mouth of the bay [10] and approximately constant in the cross bay direction.

The NACCS simulations included surge and wave coupling, but given the low wave heights in the bay, the model predicted very little wave set up in Warwick, RI [4].

2.2. Wave Modeling

Wave estimates for flood inundated areas were made by applying STWAVE [12,13] to the study area on a 10 m high resolution grid, using NACCS results as boundary and initial conditions. The STWAVE model is a phase average, steady state, spectral wave model and simulates depth-induced wave refraction and shoaling, current induced refraction and shoaling, depth and steepness-induced wave breaking, wind-wave growth, and wave-wave interaction and white-capping, that redistributes and dissipates energy in a growing wave field. The model does not include wave run-up, as this is a time dependent process. To be inclusive, the study domain included most of Narragansett Bay (Figure 5), from the bay mouth to the well north of Warwick, RI. The surge level was defined by NACCS surge results for the 100-year storm, as defined above and shown in Figure 4, interpolated on the computational grid. The wave model was forced by predictions of the NACCS wave model for the mean 100-year return period condition (Figure 6). The 100-year wave spectral parameters provided at the NACCS save points were interpolated at the offshore boundary of the high resolution 10 m grid and were used as boundary conditions to the high resolution wave simulations: the wave period at the boundary location (38 m depth) was set at 20 s and the wave height at 9 m. Swell- and wind-generated gravity waves in shallow water are specified by a Texel, Marsen, and Arsloe (TMA) spectrum

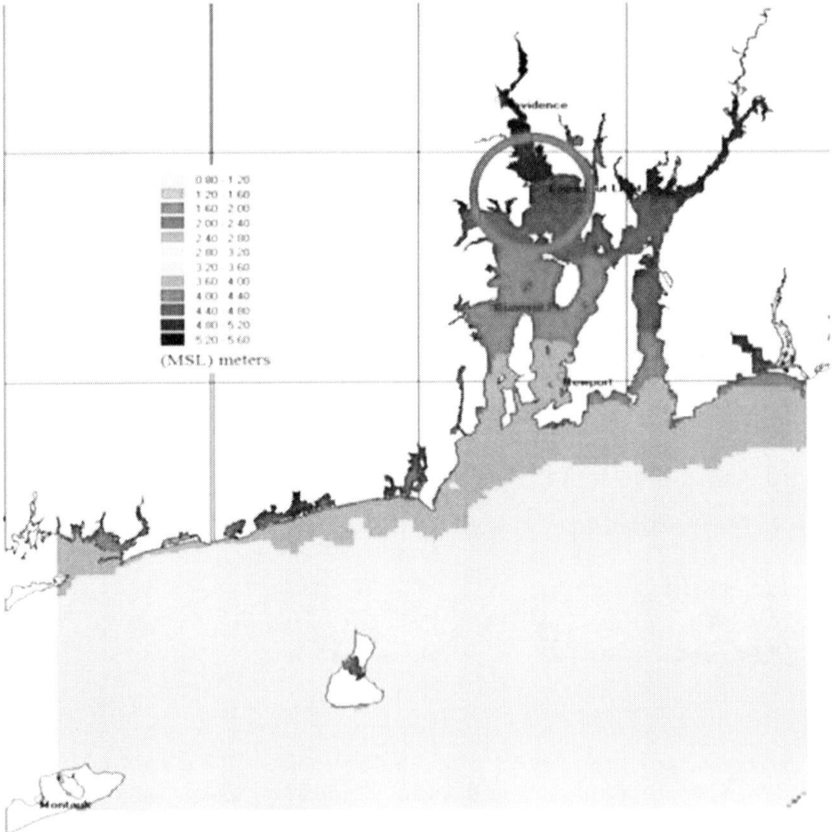

Figure 4. Water level contour map for 100-year, upper 95%, surge plus tide case, based on NACCS data. The Warwick, RI, study area is shown in the red circle. Water levels are in meters relative to the Mean Sea Level [7].

[12,13], with a direction of 180 degrees clockwise relative to the North (waves from the south), a spectral peak enhancement factor of eight, and a directional spreading factor of 20. The TMA spectrum is expanded into a directional spectrum using the standard wrapped normal directional spreading function [14]. Parameters of the directional spectrum are selected to simulate a narrow spectrum around the main direction of propagation to be conservative and, at least before propagation, focusing most of the energy in the targeted direction. A sensitivity study, however, shows insignificant differences in waves at the target site using the standard value of 3.3 and 4 for the enhancement factor and directional spreading factor, respectively. In addition to simulating local wind-wave

regeneration, the model was forced at the surface, assuming a 100-year wind speed of 35 m/s from the south (180 degrees). The 100-year wind speed value was determined by an extremal analysis at the offshore US Army Corp of Engineers, Wave Information Study (WIS) [15] hindcast point #63079 (WIS station 79). This high wind speed from the south was adopted as a conservative estimate. The results of the simulations are referred to as the NAST method in the following, referring to a combination of NACCS and STWAVE.

Frictional losses from overland flows are addressed by assigning Manning roughness coefficients to each grid and depend on the land cover and vegetation type. Manning roughness was specified based on a 30 m resolution map of ground cover, provided by the RI Geographic Information System (RIGIS) [16]. The land use was translated into Manning roughness based on standard hydraulic literature [17], as implemented by Wamsley et al. [18], resulting in a Manning coefficient map with variations from 0.02 in open water to 0.1 over woody wetlands.

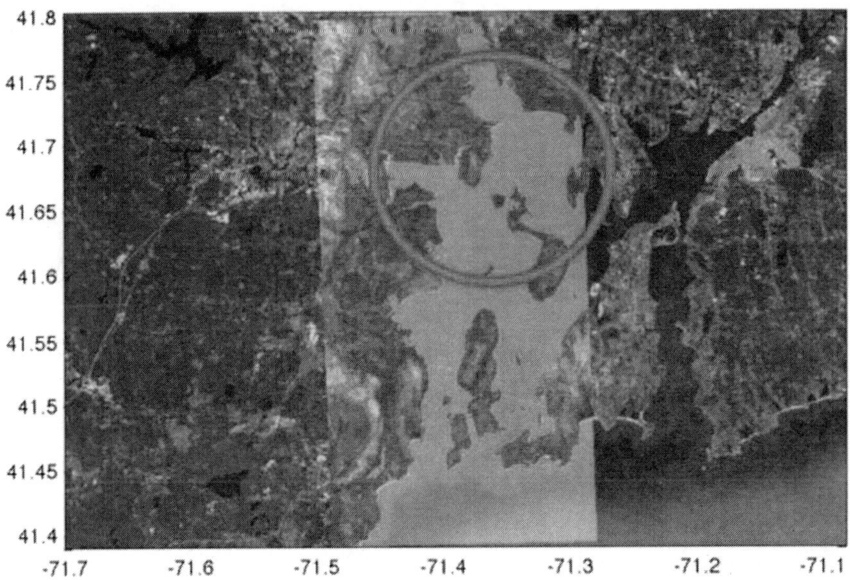

Figure 5. Study domain (highlighted area) for the STWAVE application to Narragansett Bay. Study area shown in red circle.

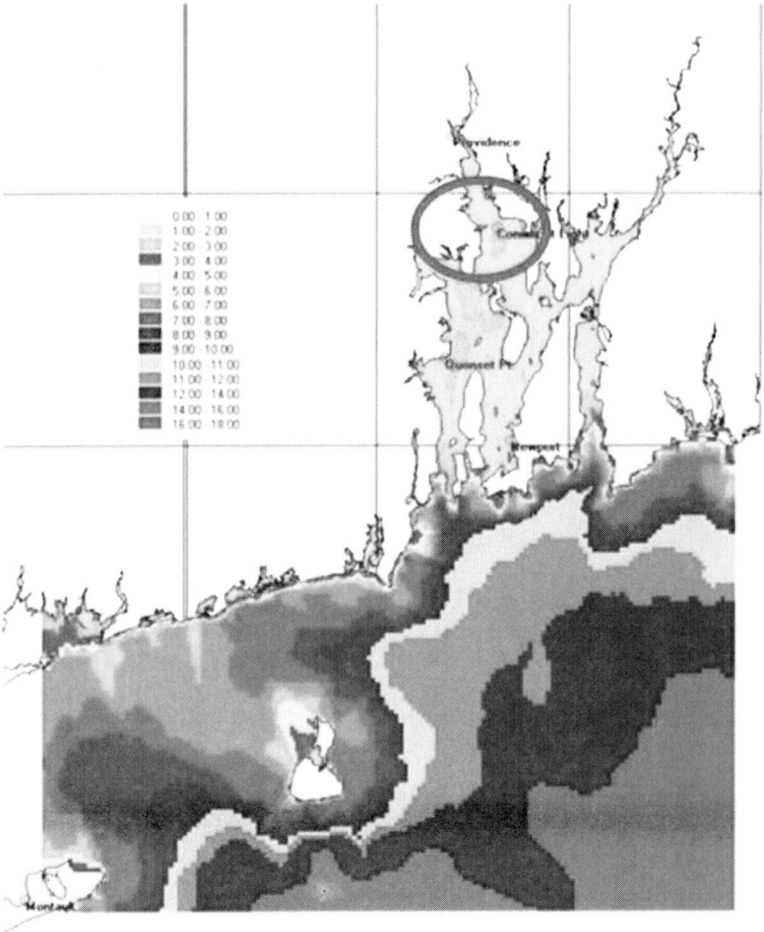

Figure 6. Contour map of the 100-year return period significant wave height (m, MSL referenced) for the mean surge plus tide case for RI coastal waters based on NACCS data (200 m STWAVE grid) [7]. Warwick study area is shown in the red circle.

2.3. Digital Elevation Model (DEM)

To represent the topography and bathymetry in the flooded area, the wave analysis used the 2011 Light Detection and Ranging (LIDAR) data Digital Elevation Model (DEM) for RI [16] and the flooding maps to determine the water inundation depths. The data is available at a 1 m horizontal resolution, NAVD88 benchmarked, with a vertical root mean square error (RMSE) of 15 cm. The DEM was benchmarked against 20

Blind Control Points (BCP) that were distributed throughout the state and performed as part of the 2011 LIDAR survey, with an RMSE of 9.4 cm. A detailed comparison was also performed against FEMA elevation certificates, filed with the city, that were available for selected structures using the value of the Lowest Adjacent Grade (LAG) (Figure 7). Their collective RMSE was 32.6 cm. For a comparison, FEMA BFEs are rounded to the nearest 1 ft (32.8 cm). The LIDAR data were interpolated onto a 10 m grid for the wave simulations.

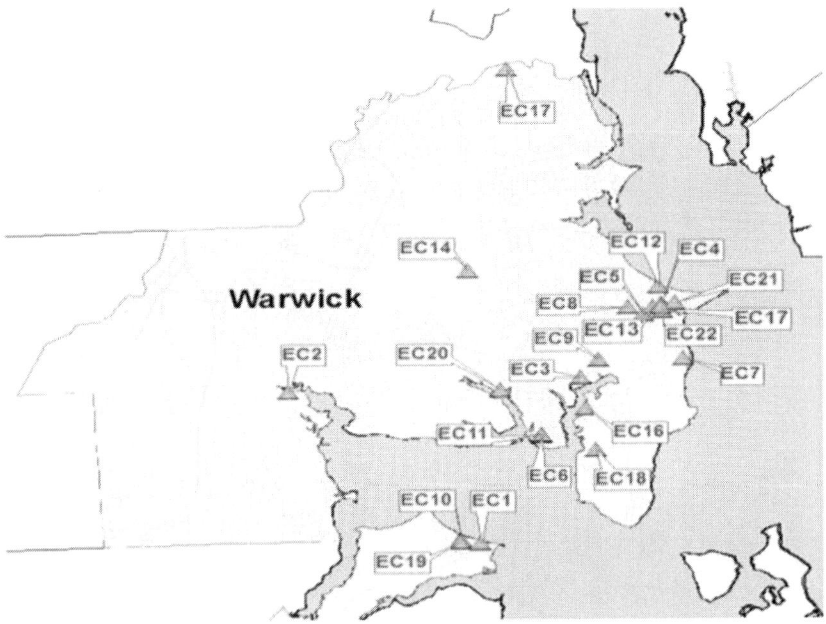

Figure 7. Location of FEMA elevation certificates for Warwick, RI study area provided by the City of Warwick, RI.

3. RESULTS AND DISCUSSION

3.1. Surge and Tide Water Levels

To assess the performance of the NACCS model predictions for the study area, the model predicted water levels versus return period were compared to an analysis performed by the National Ocean and Atmospheric Administration-National Ocean Survey (NOAA NOS), based on the historical data records at Newport (Station ID #8452660) and Providence, RI (Station ID #8454000) [18]. These are the only NOAA NOS primary water level stations in RI and both are inside Narragansett Bay. Figure 8 and Figure 9 show the comparisons for Newport and Providence, RI, respectively. Predictions were provided by NOAA for the mean, and lower and upper 95% confidence intervals, and from NACCS for the mean and upper 95% confidence interval, for the surge only and surge plus tidal cases. Table 1 provides the water levels for NACCS and NOAA for Providence and Newport, RI. The values that FEMA used for these two stations are also provided. The scaling of the value at Providence, relative to Newport, is also provided at the bottom of the table.

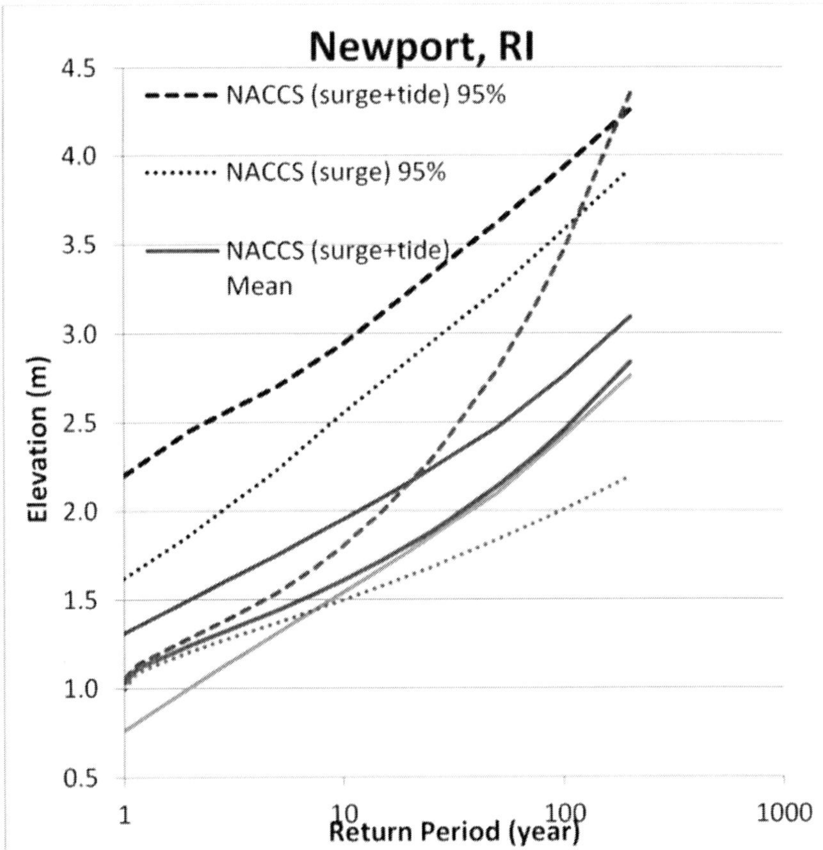

Figure 8. Surge elevation (meters, relative to NAVD88) vs. return period (years) for Newport, RI (Station ID #8452660). Estimates are provided from NOAA for the mean, and lower and upper 95% confidence intervals, and for NACCS for the mean and upper 95% confidence interval, for the surge only and surge plus tidal cases [7].

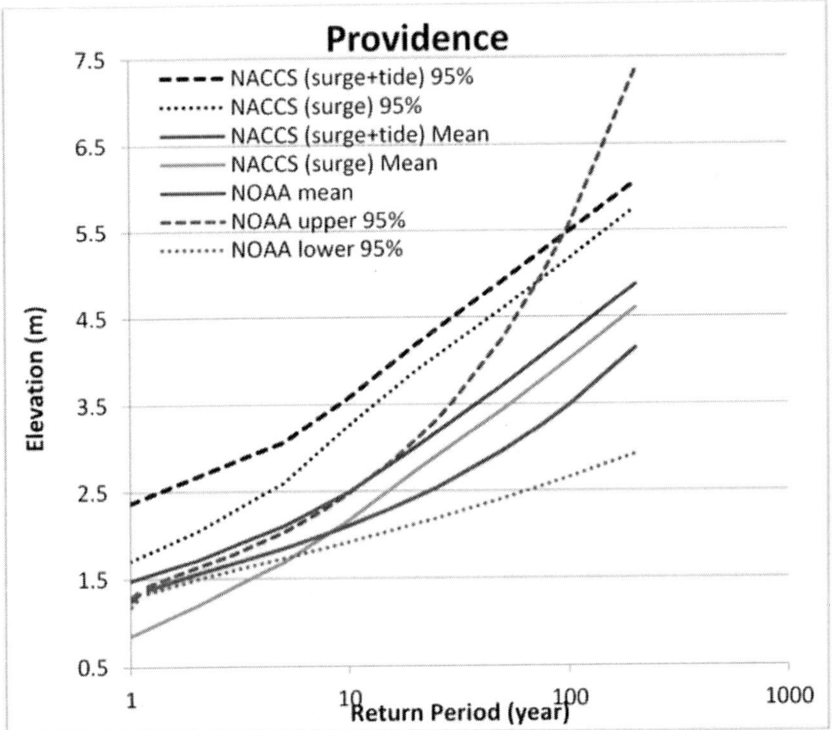

Figure 9. Surge elevation (meters, relative to NAVD88) vs. return period (years) for Providence, RI (Station ID #8454000). Estimates are provided from NOAA for the mean, and lower and upper 95% confidence intervals, and for NACCS for the mean and upper 95% confidence interval, for the surge only and surge plus tidal cases [19].

Table 1. Water levels (m) for the 100-year return period for Providence and Newport, RI, from the NACCS, NOAA, and FEMA [10,18].

Station	Water Levels (m)						
	NACCS Surge + Tide		NACCS Surge		NOAA		FEMA
	Mean	Upper 95%	Mean	Upper 95%	Mean	Upper 95%	FIS
Providence	4.29	5.49	4.01	5.17	3.48	5.56	4.2
Newport	2.76	3.93	2.42	3.58	2.47	3.47	3.21
Scaling Relative to Newport							
Providence	1.55	1.4	1.66	1.44	1.41	1.6	1.31
Newport	1	1	1	1	1	1	1

The water levels versus return period for the mean value for NACCS surge only and the NOAA mean, are in very good agreement for Newport, RI (Figure 8). The NACCS is higher (typically about 30 cm) than the NOAA values for the surge plus tide case. The structure of the upper 95% confidence limit is dramatically different between the NOAA and NACCS data. NOAA uncertainties are strongly asymmetric with the return period, with the upper confidence deviating more from the mean than the lower confidence interval. This is a result of the limited length of the historical record (approximately 70–85 years) and the presence of several large storm events (1938 and 1954) in the record. The uncertainty in the NACCS, as shown by the upper 95% confidence limit, is approximately independent of the return period. This is attributed to the fact that the analysis is based on synthetic tropical storms, rather than historical events. For the most frequently used 100-year return period water level, the mean value for NOAA is 2.47 m, while the corresponding values from NACCS are 2.42 m (surge only) and 2.76 m (surge plus tide). The upper 95% water level from NOAA for this case is 3.47 m, compared to NACCS values of 3.58 m (surge only) and 3.93 m (surge plus tide). The estimates are thus reasonably comparable.

Similar comparisons for the 100-year event at Providence give 3.48 m for the mean from NOAA, and 4.01 m (surge only) and 4.29 m (surge plus tide) from NACCS (Figure 9). For the upper 95% confidence limit, the NOAA value is 5.56 m, compared to the NACCS values of 5.17 m (surge only) and 5.49 m (surge plus tide). The NACCS values are significantly more conservative than NOAA for the mean. The opposite is true for the upper 95% limit. This difference in values is clearly shown in Figure 9. A review of the historical record at Providence shows that the 1938 storm is the most extreme event and dominates the uncertainty at the upper 95% limit. The 1954 storm is absent from the Providence record, since water level measurements were not collected during this event.

It is noted that the above analysis is based on a return period analysis using only the tropical storms from NACCS and for all data from the historical record. Employing all storms (e.g., tropical and extratropical storms) in the NACCS data, the results of the return period analysis are essentially the same as the tropical storm case only. This is attributed to the fact that tropical storms provide the highest water levels, and hence dominate the upper end of the low frequency tail of the distribution. This is consistent with the NOAA NOS historical observations at the site, with hurricanes responsible for six of the top 10 surge levels [19].

Hashemi et al. [20] developed an artificial neural network (ANN) model using the NACCS models predictions, which also serves to validate NACCS model predictions. Model inputs included the tropical storm strength, forward speed, radius to maximum winds, and storm path. The ANN model predicted the peak water levels at Newport and Providence, for each of these events. Seventy percent (70%) of the storms (approximately 1000 tropical storms) were used to train the algorithm, with 15% utilized for validation, and 15% for testing. The model demonstrated a very good performance for all three phases, with correlation coefficients greater than 0.92. The root mean square error (RMSE) was approximately 30 cm. The ANN model was then applied to predict the peak storm elevation for the largest tropical storms that have historically impacted the area (1938, 1944, 1954, 1991, 2011, and 2012), using the same parameters as for the NACCS storms, but derived from the historical data. Predictions were again performed at Newport and Providence, RI. The model's performance gave a RMSE of 32 cm for all storms, except 2012 (Sandy). The ANN model under predicted the peak water levels for this storm, by about 50 cm. This reduced performance is attributed to the fact that storms with characteristics like Sandy are extremely rare in the historical record and were not included in the NACCS synthetic tropical storm data base.

Model predictions, in another evaluation [7,10], were compared to a wide array of point observations collected during the 1938, 1954, and 2012 storms, and reported by FEMA in their Kent County FIS [2]. In this case, the slope of the linear regression line from Newport to the head of the bay in Providence, was based on the storm of interest. Spaulding et al. (2017) [7] show a comparison of model predictions to observations for these events. While the data quality is typically quite variable, the analysis does well in capturing the amplification of the storm water levels with distance up the bay. The amplification is clearly strongly storm-dependent and driven primarily by the storm track, relative to the bay, and storm strength. As an example, amplifications for 1938, 1954, and 2012 were 1.3, 1.56, and 1.1, respectively.

In the interest of developing conservative estimates of flooding and addressing uncertainties inherent in the modeling and extreme analysis methods, the NACCS surge & tide, 95% confidence limit case was selected for this study.

Figure 10 shows the flood inundation levels (m) for the Warwick study area for the 100-year storm event. Note that either the spatial scaling method, noted earlier, or the selection of the data from the save points immediately offshore of the study area, can be used to specify the flooding. In this case, linear scaling from the NACCS save point data at Newport and Providence has been used (Table 1, namely 1.4, Providence relative to Newport). This provides a very good approximation to the pattern shown in Figure 4. The water level increases from approximately 4.6 m at the southern end of the study area, to 5.2 m at the northern end.

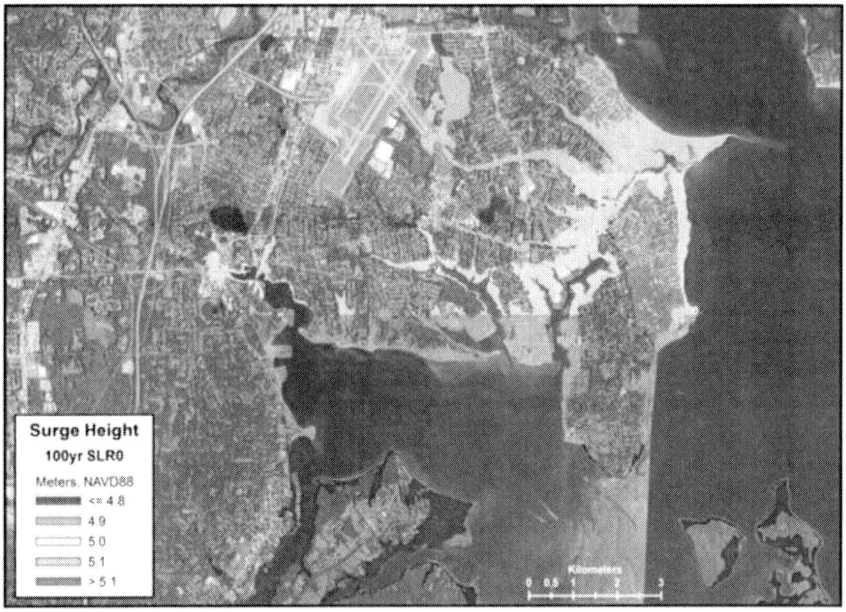

Figure 10. Surge elevation (m) relative to NAVD88 in the inundation zone for Warwick, RI for 100-year event.

3.2. Wave Modeling

To validate the input to the wave model, an analysis was performed comparing extreme values for the 100-year wind and significant wave heights, based on the US Army Corp of Engineers, WIS hindcast point #79 and an analysis of the NACCS peak wave heights for save point ID #6859, for the synthetic tropical storms [9]. This is the *save point* in closest proximity to WIS #79. The WIS analysis is based on a hindcast of winds and waves from 1980 to 2012. In this analysis, a generalized extreme value (GEV) method was applied to both analyses, so the results are internally consistent for the method. The results presented in Spaulding et al. (2017) [7] show that NACCS estimates are consistent with the WIS data for the mean wave height. Based on this comparison, the results from the NACCS study provide reasonable estimates of the 100-year wave height and period. It has been impossible to validate the model predicted wave heights inside Narragansett Bay, since no wave data is available.

3.3. Simulations

Figure 11 shows model predicted wave heights from the mouth of the bay to the lower reach of the Providence River. Results are provided in terms of the controlling wave height, which is the wave height corresponding to the average of 1% of the highest waves. Assuming a Rayleigh distribution, the controlling wave height is estimated from the significant wave height provided as an output from the STWAVE model. It is the wave height used in calculating the BFEs in FEMA's protocol. The upper panel shows the predictions for the entire study area and the lower panel, the area immediately surrounding Warwick, RI. The study domain is shown in Figure 5.

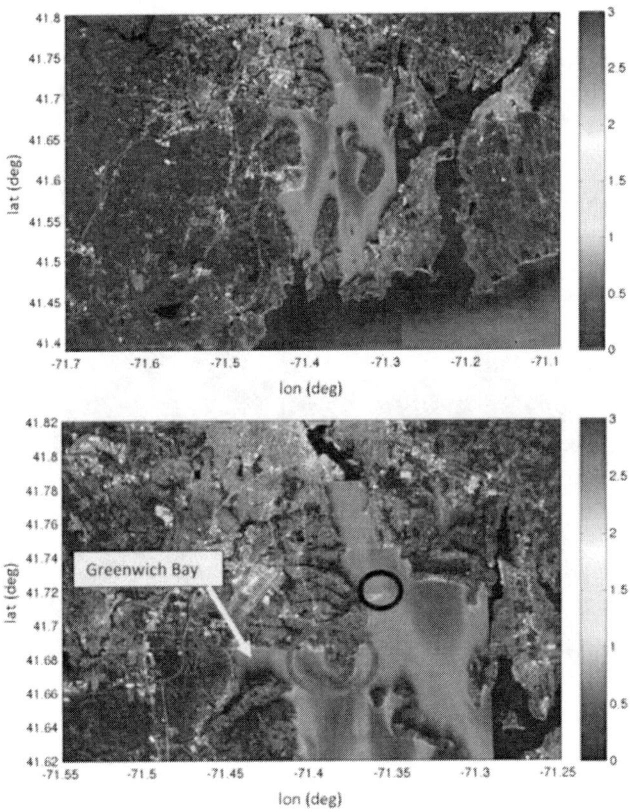

Figure 11. STWAVE predicted controlling wave height (m) for 100-year storm for entire study domain (**upper panel**) and for area in the immediate vicinity of Warwick, RI (**lower panel**). Locations of Greenwich Bay, Oakland Beach and Warwick Neck Point (red circle), and Conimicut Point (black circle) are noted.

Figure 11 clearly shows that the wave heights are largest when the fetch, in the direction of the predominant 100-year wind from the south, is largest. Wave heights in protected embayments (e.g., Greenwich Bay) are substantially lower, since they are shielded by the bay geography.

Figure 12 shows the model predicted controlling wave height for Warwick, RI. The figure shows that the waves in the study area are quite low given the location of Warwick well inside Narragansett Bay (Figure 11). There are several locations where the wave heights are approximately 2 m (circled in red/black). Most of the wave heights, however, are a meter or lower. Correspondingly, the wave heights in areas within Greenwich Bay are limited by the east-west orientation of the bay, with the bay mouth to the east of the south east.

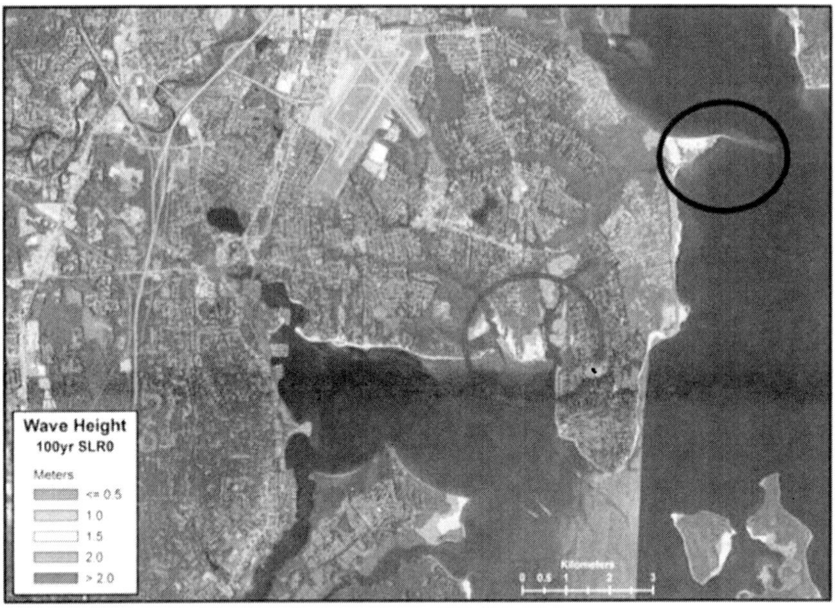

Figure 12. Controlling wave height (m) for Warwick, RI, 100-year event. Areas with substantially larger wave heights are shown; red circle for Oakland Beach and black circle for Conimicut Point.

Figure 13 shows the combination of the surge level (Figure 11) plus the controlling wave height (Figure 12) for Warwick, RI.

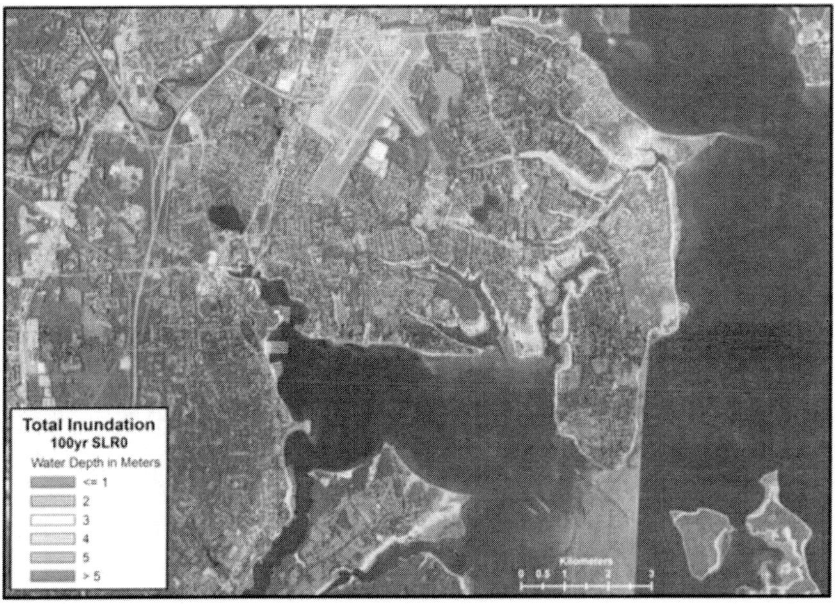

Figure 13. Total water depth (m) (inundation plus wave height relative to ground level) for Warwick, RI, 100-year event.

In the interest of better understanding the impact of improvements in the modeling methodologies applied on the final FIRM maps, Figure 14 shows the FEMA (upper panel) and NAST (lower panel) BFE (relative to NAVD88) flooding zones for the study area. Figure 15 provides a detailed comparison of FEMA and NAST predictions of SWEL, waves, and BFEs by transect. A tabulated version of the comparison is provided in Table A1. A comparison of NAST predictions and FEMA results for the lower West Passage of the bay are provided in Schambach et al. (2017) [21].

Restricting attention to the flooded area, the present study (NAST) results show that the region impacted is slightly larger than FEMA's estimates. This is consistent with the fact that the mean SWEL in the present study is approximately 4.75 m, compared to FEMA's value of 3.66 m; a difference of 0.91 m. This difference is consistent with the use of the NACCS surge & tides, upper 95% confidence interval value for the 100-year return period in the present study, compared to the value that FEMA used in its 2012 FIS (Table 1).

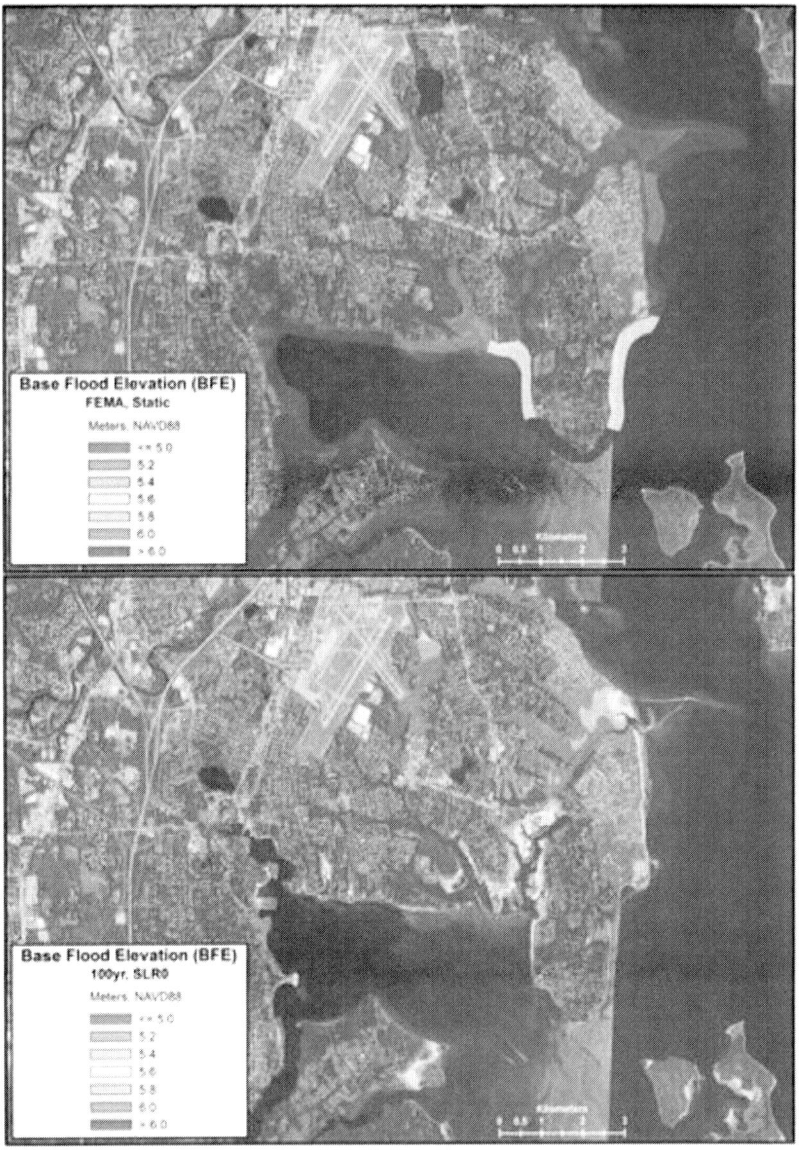

Figure 14. FEMA (**upper panel**) and NAST (**lower panel**) predicted BFEs relative to NAVD88. Note that FEMA mapping includes part of the near shore coastal waters in their BFE zones, while the NAST method does not.

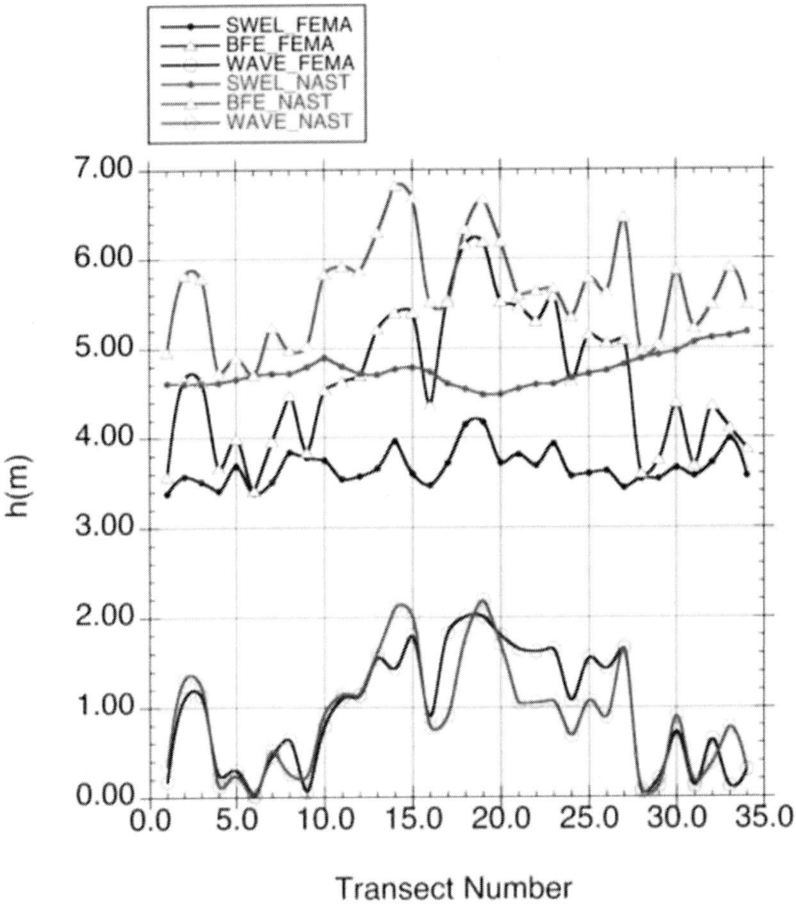

Figure 15. FEMA (black) and NAST (red) SWEL (plain dot), BFE (triangle) (referenced to NAVD88), and controlling wave crest elevations (circle) by transect number (see Figure 5 for locations). All are in meters.

A close inspection of the FEMA maps (upper panel in Figure 14) shows discrete, step changes in along shore variations in the BFEs at selected locations. Not surprisingly, these occur at the locations of the transects shown in Figure 1 and are clearly the result of the subjective interpolation of BFE values from adjacent transects. This highlights one of the key NRC criticisms of the FEMA methods [5] (see Section 1). The NAST approach shows a continuous variation in BFEs, in response to the 2D wave transformation processes. It is noted that the spatial scale of these variations (Figure 14, lower panel) is much smaller than the spacing between FEMA transect lines (Figure 1).

To provide additional insight into the differences between the two maps, Figure 15 shows the FEMA and NAST SWEL, BFE (referenced to NAVD88), and wave crest height, at the seaward end of each transect. The data used in developing this figure is provided in Table A1. Focusing on the FEMA results, the SWEL generally increases with distance up the bay, increasing from 3.4 to 4.0 m (change of 0.6 m), from transect one to 33, with a slight drop to 3.6 m at transect 34. This compares to a continuous increase from 4.6 to 5.2 m (change of 0.6 m) over the full range of transects for the NAST results. Although the increase in NAST SWEL in the northern transects shows a smoother increase than FEMA, the general trends in SWEL in NAST and FEMA's values are very similar. NAST SWEL values are on average 1.1 m higher than FEMA's values. The higher values predicted by NAST are a direct result of using the NACCS surge & tide values at the 95% confidence interval, compared to FEMA's return period analysis, based on historical data. The variations in FEMA and NAST wave crest heights show little variation with distance up the bay, but exhibit substantial variations between transects. The pattern in variations is quite similar between the two. The highest wave crest height of 2 m is located at Warwick Neck Point (transects 18–19), with secondary peaks at Oakland Beach (transects 14–15) and Conimicut Point (transects 25–27). These result from long southerly fetch distances. Crest heights are substantially lower inside Greenwich Bay (transects 1–14) and north of Conimicut Point (transects 28–34); both areas are geographically shielded from the predominant southerly wind direction. It is noted that the FEMA SWEL variations are correlated with the wave crest variations reflecting the wave induced setup. The NAST results show a limited impact of wave set up. Statistical comparisons of FEMA and NAST estimates of SWEL, waves, and BFEs, are provided at the bottom of Table A1. The standard deviations and RMS values are 0.28 and 1.12 m for SWEL, 0.34 and 0.34 m for waves, and 0.48 and 1.1 m for BFEs, respectively. The variances and RMS differences are small for waves, while the variances and RMS values for SWEL and BFE (as a result of the underlying SWEL values) are substantially larger. The latter reflects the use of the upper 95% confidence level for the NAST SWEL and FEMA's use of the mean value for the 100-year return period water level.

4. CONCLUSIONS

Flood inundation maps have been developed for Warwick, RI, using state of the art, fully coupled high resolution surge and wave models (ADCIRC, WAM, and STWAVE), applied to the study area as part of the NACCS. The model predictions were extensively validated with data from tropical and extratropical storms during the study. NACCS simulations were performed for 1050 synthetic tropical storms and 100 historical extratropical storms. Peak values were archived at selected save points and return period analyses were performed for each storm data set. Additional validation, of both the water levels and wave heights versus return period, with historical water level data at Newport and Providence, RI, for the former, and hindcasts (1980–2012) of wave conditions at a selected location off the coast (RI Sound) for the latter, from the USACE Wave Information Study (WIS) [14], were performed as part of this study, to gain additional insight into the performance of the model. It was impossible to validate the wave model performance inside Narragansett Bay, since no wave data were available. For the comparisons that were performed, the results are generally in reasonable to good agreement.

Simulations were then performed using STWAVE on a high resolution grid (10 m) of the flood inundated study area. The model was driven by offshore water levels and wave height, period, and direction, assuming a shallow water, wave spectral distribution model from the NACCS study along the open boundaries. The model was also forced by 100-year return period winds. The model included wave dissipation based on a comprehensive mapping of the roughness from both structures and ground cover. STWAVE predicted the 2-D wave transformation processes and clearly showed the importance of the 2-D nature of the wave field.

4.1. Key Findings Include

- FEMA predicted 100-year water levels (SWEL) range from 3.4 to 3.8 m for the study area. Our study results are larger (4.6 to 5.2 m); a difference of approximately 1.1 m. NAST results are approximately 30% higher than FEMA results. The larger values predicted by our analysis are attributed to the use of the upper 95% confidence limit water levels used in the present study, compared to FEMA's return period analysis, based on the historical data. It is noted FEMA's values are larger than the NOAA mean value, but smaller than the upper 95% for both Newport and Providence; all using the same historical data set. Hence, the difference is due to use of a different extremal value analysis method. Since NOAA has primary authority for collecting and analyzing the water level data, it is unclear why FEMA would choose an alternative method that gives predictions at variance with NOAA's values.

- The comparison of FEMA and NAST wave predictions show comparable results for wave heights (0.5 to 2 m), with the spatial variations controlled by the maximum fetch distance to the south. On average, across all transects, NAST model predictions are approximately 7% lower than FEMA. This small difference is explained by the fact that FEMA's method and the current STWAVE-based approach are both dominated by fetch distances along the central axis of the West Passage of Narragansett Bay, from the south.
- FEMA BFE maps show strong variations between transect lines, reflecting the subjective nature of the interpolation of the 2-D maps from the 1-D transects. NAST results, on the other hand, show the importance of 2-D wave transformation processes and the spatial scales are consistent with the topography and bathymetry of the study area.
- It has proven impossible to recreate the 2-D structure of the wave field in the flood inundated area that FEMA has developed, given the results for the FIS transects. It seems to be primarily controlled by topography. The FIS transect spacing is too large to determine the spatial structure and the 2-D wave processes are not represented by the WHAFIS method used.
- The mean BFE value, averaged over all transects, is approximately 22% larger, based on the NAST analysis, compared to the FEMA results. The difference is attributed to the 30% higher SWEL for the NAST data, compared to the FEMA value and slightly lower (7%) wave crest heights.

Our analysis presented herein has implemented all of the NRC recommendations raised in their review of the FEMA coastal mapping program [5]. Thus, it represents an advancement over the methods used in the FEMA FIS study [2]. Additionally, the NAST maps represent a more conservative approach for assessing the risk to coastal structures and are more protective of public safety.

4.2. Open Coast vs. Protected Bay Area

Comparing the results of the prior paper on developing flood insurance maps for an open coastal area [7] to the present paper, in which exactly the same methods are applied to a protected coastal area, we note that the differences in NAST versus FEMA results are dominated by waves for the open coastal area and by the water level for the protected bay. The former is due to problems in specifying the wave heights in the immediate offshore area and the use of the 1-D wave transformation model to predict the wave field over the flood inundated area in the FEMA approach. The assumption about the dune profile during the 100-year event using FEMA's standard methodology versus an analysis of historical observations (NAST approach), is also a contributing factor. For the protected area, the wave predictions using both the FEMA and NAST methods are consistent, although the small spatial scale variations are an issue. These are principally caused by the fact that the FEMA estimates of wave heights at the end of each WHAFIS transect use different fetch directions, thus ignoring the correlation between extreme winds and direction. The differences in water level are the result of using the upper 95% confidence limit for the water level versus return period in NAST, compared to using the mean value by FEMA.

4.3. Some Lessons Learned

In developing alternate FIRMS for the two study areas using well established, state of the art tools, and comparing the results to recently generated (last five years) FEMA maps, we note the following lessons:

- Given the uncertainty in the various modeling and return period analysis methods employed to determine the water flood level, it seems prudent to develop a quantitative measure of the uncertainty of the results. The use of the upper 95% confidence limit value addresses both issues, is readily available, and simple to implement and understand.
- It is clear that the application of WHAFIS and the subsequent interpolation of its 1-D results to generate a 2-D map is problematic. The boundary conditions for the end of the WHAFIS transects and the assumption about the profile for the eroded dune are both issues. As noted in the NRC review and documented here, wave transformation is a 2-D process that cannot be represented in a 1-D model. WHAFIS should be replaced by a suitable 2-D model (e.g., STWAVE).
- Given the evolution of fully coupled, time dependent flood inundation and wave models on the international scale, it would be appropriate for FEMA to update the list of accepted models to include those providing a more comprehensive treatment of the underlying physics and improved predictive performance.

ACKNOWLEDGMENTS

The application of the models to Charlestown, RI, was supported by Housing and Urban Development (HUD), Grant # B-13-DS-44-0001, Hurricane Sandy CDBG Disaster Recovery—CFDA #14.269 and administered through the State of Rhode Island, Executive Office of Commerce, Office of Housing and Community Development (OHCD). Joseph Dwyer, an MS student in the University of RI, Kingston, RI, Environmental Science and Management program, performed an analysis comparing elevation data from the certificates of elevation to the LIDAR-based digital elevation model.

AUTHOR CONTRIBUTIONS

Malcolm L. Spaulding developed the idea for STORMTOOLS/CERI and its application to generate FIRMs. He led the project effort; Annette Grilli was primarily responsible for wave modeling; Implementation in GIS and providing access to the output was provided by Chris Damon; Grover Fugate advised on the design of the system to meet the needs of coastal planners; Tatsu Isaji performed the water level analysis on NACCS data; and Lauren Schambach performed the transect inter-comparisons.

CONFLICTS OF INTEREST

The authors declare no conflict of interest. The project sponsors had no role in the design of the study; in the collection, analyses, or interpretation of data; in the writing of the manuscript, or in the decision to publish the result.

REFERENCES

1. Federal Emergency Management Agency (FEMA). *Atlantic Ocean and Gulf of Mexico Coastal Guidelines Update*; FEMA: Washington, DC, USA, 2007.
2. Federal Emergency Management Agency. *Flood Insurance Study, Kent County, RI, FEMA Flood Insurance Study Number 44003CV000B*; FEMA: Washington, DC, USA, 2012.
3. Federal Emergency Management Agency. Updated Tidal Profiles for the New England Coastline. March 2012. Available online: https://www.fema.gov/media-library/assets/documents/85240 (accessed on 16 January 2017). []

4. Federal Emergency Management Agency. WHAFIS Documentation Version 4.0. 2007. Available online: https://www.fema.gov/wave-height-analysis-flood-insurance-studies-version-40 (accessed on 16 January 2017). []
5. National Research Council. *Mapping the Zone: Improving Flood Map Accuracy*; Committee on FEMA Flood Maps; Board on Earth Sciences and Resources/Mapping Science Committee; National Research Council: Washington, DC, USA, 2009. []
6. FEMA Definitions. Available online: https://www.fema.gov /sites/default/ files/images/flood_zones_limwa.jpg(accessed on 16 January 2017).
7. Spaulding, M.L.; Grilli, A.; Damon, C.; Oakley, B.; Fugate, G.; Isaji, T.; Schambach, L. Application of state of art modeling techniques to predict flooding and waves for an exposed coastal area. *J. Mar. Sci. Eng.* **2017**, *5*, 10. [] []
8. FEMA Approved Models. Available online: https://www.fema.gov/coastal-numerical-models-meeting-minimum-requirement-national-flood-insurance-program (accessed on 17 January 2017).
9. Cialone, M.A.; Massey, T.C.; Anderson, M.E.; Grzegorzewski, A.S.; Jensen, R.E.; Cialone, A.; Mark, D.J.; Pevey, K.C.; Gunkel, B.L.; McAlpin, T.O. *North Atlantic Coast Comprehensive Study (NACCS) Coastal Storm Model Simulations: Waves and Water Levels*; Report: ERDC/CHL TR-15-44; Coastal and Hydraulics Laboratory U.S. Army Engineer Research and Development Center: Vicksburg, MS, USA, 2015. []
10. Spaulding, M.L.; Isaji, T.; Damon, C.; Fugate, G. Application of STORMTOOLS's simplified flood inundation model, with and without sea level rise, to RI coastal waters. In Proceedings of the ASCE Solutions to Coastal Disasters Conference, Boston, MA, USA, 9–11 September 2015.
11. STORMTOOLS Website. Available online: http://www.beachsamp.org /resources /stormtools/ (accessed on 16 January 2017).
12. Massey, T.C.; Anderson, E.M.; McKee-Smith, J.; Gomez, J.; Rusty, J. *STWAVE: Steady State Spectral Wave Model. User's Manual for STWAVE, Version 6.0*; US Army Corp of Engineers, Environmental Research and Development Center: Vicksburg, MS, USA, 2011. []
13. Smith, J.M.; Sherlock, A.R.; Resio, D.T. *STWAVE: Steady-State Wave Model User's Manual for STWAVE, Version 3.0*; ERDC/CHL SR-01-01; U.S. Army Engineering Research and Development Center: Vicksburg, MS, USA, 2001. []
14. Borgman, L.E.; Panicker, N.N. *Design Study for a Suggested Wave Gage Array off Point Mugu, California*; Technical Report Hel 1-14; Hydraulic Engineering Laboratory University of California at Berkeley: Berkeley, CA, USA, 1970. []
15. USACE Wave Information Study. Available online: http://wis.usace.army .mil/hindcasts.shtml?dmn=atlWIS (accessed on 17 January 2017).
16. RI-GIS Digital Elevation Model. Available online: http:// www.rigis.org /data/topo/2011 (accessed on 16 January 2017).
17. Arcement, G.J.; Sc hneider, V.R. *Guide for Selecting Manning's Roughness Coefficients for Natural Channels and Flood Plains*; U.S. Geological SurveyWater-Supply Paper 2339; U.S. Geological Survey: Denver, CO, USA, 1989.
18. Wamsley, T.V.; Cialone, M.A.; Smith, J.M.; Ebersole, B.A.; Grzegorzewski, A.S. Influence of landscape restoration and degradation on storm surge and waves in southern Louisiana. *Nat. Hazards* **2009**, *51*, 207–224. [] []
19. Zervas, C. *Extreme Water Levels of the United States, 1893–2010*; NOAA Technical Report NOS CO-OPS 067; NOAA: Silver Spring, MD, USA, 2013.
20. Hashemi, R.; Spaulding, M.L.; Shaw, A.; Farhadi, H.; Lewis, M. An efficient artificial intelligence model for prediction of tropical storm surge. *J. Nat. Hazards* **2016**. [] []
21. Schambach, L.; Grilli, A.R.; Spaulding, M.L. Predicting the 100-year storm inundation along the Narragansett Bay shoreline. *Nat. Hazard* **2017**. in preparation. []

Index